Smart Nanomaterials: Synthesis, Properties and Applications

Smart Nanomaterials: Synthesis, Properties and Applications

Editor: Rich Falcon

NY RESEARCH P R E S S

New York

Published by NY Research Press
118-35 Queens Blvd., Suite 400,
Forest Hills, NY 11375, USA
www.nyresearchpress.com

Smart Nanomaterials: Synthesis, Properties and Applications
Edited by Rich Falcon

International Standard Book Number: 978-1-63238-556-7 (Hardback)

Cataloging-in-Publication Data

Smart nanomaterials : synthesis, properties and applications / edited by Rich Falcon.
    p. cm.
Includes bibliographical references and index.
ISBN 978-1-63238-556-7
1. Nanostructured materials. 2. Nanostructured materials--Analysis.
3. Nanocomposites (Materials). I. Falcon, Rich.
TA418.9.N35 S63 2017
620.5--dc23

Printed in the United States of America.

Contents

Preface

Comprehensive insights into the emerging field of smart nanomaterials have been provided in this book. It discusses the synthesis, properties and applications of smart nanomaterials. Smart nanomaterials use nano-scale engineering and superior system integration of existing materials to continuously develop better materials and better products. Defense, automobile industries etc. benefit from the development of these materials. This book unfolds the innovative aspects of developing smart nanomaterials, helping the reader to explore the unexplored. As this field is emerging at a fast pace, this book will help the readers to better understand the concepts of synthesizing smart nanomaterials.

This book is the end result of constructive efforts and intensive research done by experts in this field. The aim of this book is to enlighten the readers with recent information in this area of research. The information provided in this profound book would serve as a valuable reference to students and researchers in this field.

At the end, I would like to thank all the authors for devoting their precious time and providing their valuable contributions to this book. I would also like to express my gratitude to my fellow colleagues who encouraged me throughout the process.

Editor

A facile synthesis of a novel optoelectric material: a nanocomposite of SWCNT/ZnO nanostructures embedded in sulfonated polyaniline

Rajesh K. Agrawalla[a], Rima Paul[a], Pratap K. Sahoo[b], Amit K. Chakraborty[a] and Apurba Krishna Mitra[a]*

[a]Physics, NIT Durgapur, Durgapur 713209, India; [b]Physics, NISER, Bhubaneswar, India

Functionalized single-walled carbon nanotubes (f-SWCNTs) hybridized with freshly prepared zinc oxide (ZnO) nanocrystals have been found to be good luminescent material with tuned emission properties. A three-phase nanocomposite of sulfonated polyaniline embedded with such SWCNT/ZnO nanostructures has been prepared by a simple solution mixing chemical process and characterized by using high-resolution transmission electron microscopy, X-ray diffractometry, Raman spectroscopy, Fourier transform infrared spectroscopy, and thermogravimetric analysis. The study of UV-visible absorption and photoluminescence spectroscopies reveal that the ternary polymer nanocomposite is a luminescent material with enhanced emission intensity. Also an increase in DC conductivity indicates that the nanocomposite is also a good conductive material, satisfying Mott's variable range hopping model for a two-dimensional conduction. Such a three-phase nanocomposite may find extensive application in dye-sensitized solar cells, sensors, and supercapacitors.

Keywords: SPANI; SWCNT/ZnO hybrid; polymer nanocomposite

1. Introduction

Polyaniline (PANI) is the most studied conducting polymer in recent years due to its numerous potential applications. Carbon nanotubes (CNTs) possess remarkable physical properties and have been extensively used as reinforcing fibers in PANI matrix to improve the characteristic properties of the polymer. However, a major drawback of PANI is its insolubility in water and common organic solvents. Water solubility is essential for many electrical applications. Sulfonation is one of the common methods to improve the solubility and processability of PANI in water. The prepared sulfonated PANI (SPANI) becomes water-soluble at all pH values. The various copolymers of PANI nanoparticles with sulfonic acid (–SO$_3$H) groups have been synthesized [1–5]. Such polymeric materials possess many interesting features such as sensing [1] and toxic ion removal [2–4] characteristics. However, the presence of the strong electron-withdrawing sulfonic group significantly reduces the conductivity of SPANI compared to that of pure PANI, limiting its electrical applications. Therefore, an improvement in the electrical conductivity of the water-soluble SPANI is very much necessary. Researchers have combined CNTs with PANI [6–10] and SPANI [11,12] to form composites in which the idea was to exploit

*Corresponding author. Email: akmrecdgp@yahoo.com

excellent electrical properties of CNTs to increase the conductivity of the nanocomposite thus produced. The PANI/CNT composite has improved electrical, optical, and thermal properties. It finds application in biosensors [9], supercapacitors [13], metal–semiconductor devices [14], and actuators [15].

Zinc oxide (ZnO) is one of the hardest materials in the II–VI semiconductor family and it does not suffer from any dislocation degradation [16]. It is a wide band gap compound semiconductor possessing strong luminescence properties and has various applications as a luminescent material [17–19]. Kondawar et al. [20] reinforced ZnO nanoparticles in PANI matrix and measured the electrical conductivity of the composite. They observed higher conductivity for the composite when the PANI–ZnO ratio was 1:2, compared with other ratios. A polaron band was observed in the UV-visible spectra of PANI/ZnO nanocomposite owing to ZnO nanostructures [21]. Dhingra et al. [22] prepared PANI/ZnO composite by direct solid-state mixing of doped PANI powder and ZnO nanoparticles and observed an enhancement in the UV emission for the obtained nanocomposite. ZnO has also been combined with CNTs by different physical as well as chemical methods to obtain CNT/ZnO hybrid nanostructures [23–26]. Paul et al. [25] synthesized the single-walled carbon nanotubes (SWCNT)/ZnO nanohybrid by simple wet chemical method and investigated its photoluminescence (PL) characteristics.

While there are published reports on CNT-based PANI or SPANI composites [6–15], ZnO-based PANI composites [20–22], and CNT/ZnO hybrids [23–25]; to the best of our knowledge, there is hardly any published work till date on ternary polymer composites containing SPANI, SWCNT, and ZnO nanoparticles. We report here a simple chemical synthesis of polymer nanocomposite containing SPANI and SWCNT/ZnO hybrid nanostructures by a solution mixing process. The SPANI was prepared by sulfonation of emeraldine salt of PANI through treatment with chlorosulfonic acid in an inert solvent. The SWCNT/ZnO nanohybrid was prepared by a simple wet chemical process. The synthesized SPANI/SWCNT/ZnO nanocomposite is a promising optical as well as conducting material and may find applications in sensors, optoelectronics, display devices, and supercapacitors.

A major area of application of polymer/CNT composites is in photovoltaic cells, particularly as counter electrodes in dye-sensitized solar cells [27–30]. In organic solar cells, CNTs have been used as electron acceptors in the photoactive layer, and the hybrid nanostructure of SWCNT/ZnO could be a better material for the purpose, in conjunction with SPANI.

2.   Materials and methods

2.1.   Materials

The SWCNTs used in our work were supplied by Chengdu Organic Chemicals Co. Ltd, China with average diameter, length, and purity, as stated by the manufacturer being 1–2 nm, 1–3 µm, and 95 wt.%, respectively. Aniline, 1,2-dichloroethane (DCE) and ammonium persulfate were supplied by Merck Specialties Pvt. Ltd., Mumbai. The chlorosulfonic acid was supplied by LOBA Chemie Pvt. Ltd., Mumbai. Except SWCNTs, all other chemicals were used as received without further purification.

2.2.   Purification and functionalization of SWCNTs

The as-received SWCNTs were purified by heating in a muffle furnace at 350°C in air for 6 h followed by soaking and stirring in 6 M HCl for 12 h. The acid-treated SWCNTs were filtered using vacuum filtration system (Millipore, pore size ~ 0.22 μm) and washed thoroughly with deionized water. The purified SWCNTs were further treated in a mixture of concentrated HNO_3/H_2SO_4 in 1:3 volume proportion for 4 h followed by washing with dilute NaOH aqueous solution and filtration with Millipore filtration apparatus until the pH became neutral, thereby attaching carboxylic acid (-COOH) groups to obtain the functionalized single-walled carbon nanotubes (f-SWCNT).

2.3. Synthesis of SWCNT/ZnO hybrid nanostructures

We mixed 0.779 gm potassium hydroxide (KOH) pellets in 60 ml of deionized water and ultrasonicated using 250 W Piezo-U-Sonic ultrasonic apparatus at 50°C for 30 min. This was followed by mixing of 0.8475 gm of zinc nitrate hexahydrate in 40 ml of deionized water and subsequent stirring. Subsequently, we mixed 0.358 gm of f-SWCNTs in the zinc nitrate solution and the solution was stirred for 2 h at room temperature in magnetic stirrer (REMI 2 MLH). The KOH solution was added dropwise to this solution till the pH became 7. The grey viscous precipitate was filtered using Whatman filter paper followed by drying under IR lamp to obtain SWCNT/ZnO hybrid nanostructures.

2.4. Synthesis of sulfonated polyaniline (SPANI)

We mixed 0.2 M aniline hydrochloride and 0.25 M ammonium persulfate solutions in equal volumes and left overnight for polymerization to take place. The salt precipitate (PANI) was collected by filtration using a Whatman filter paper. It was then mixed with 1,2-DCE and heated to 80°C under stirring using the magnetic stirrer. Chlorosulfonic acid diluted with DCE was added dropwise to the reaction mixture and stirred at 80°C for 1 h. The semi-solid precipitate of SPANI obtained by filtration was then mixed with 400 ml of deionized water and heated to 60°C and stirred for 2 h to promote hydrolysis. The solution was further diluted with deionized water and filtered through a cellulose membrane using a vacuum filtration system (Millipore). The sample of SPANI collected over the filter membrane was dried in air at room temperature.

2.5. Synthesis of SPANI/SWCNT and SPANI/SWCNT/ZnO composites

Aqueous dispersion of SWCNTs was prepared by sonication in the ultrasonicator. The dispersion was mixed with SPANI aqueous solution with stirring at 60°C for 3 h. The resulting solution was cooled and filtered using vacuum filtration system (Millipore). The sample collected by filtration was dried in air at room temperature to get SPANI/SWCNT composite with 6 wt.% of SWCNTs. The whole process of solution mixing was repeated with SWCNT/ZnO nanohybrids in place of SWCNTs, to obtain SPANI/SWCNT/ZnO three-phase nanocomposite. We added 0.045 gm of SWCNT/ZnO nanohybrid into deionized water followed by mixing with SPANI solution to obtain the desired nanocomposite. The SPANI/SWCNT/ZnO nanocomposite contains 15 wt.% of SWCNT/ZnO nanohybrids. The weight percentage of SWCNTs and SWCNT/ZnO nanohybrids in the respective composites were estimated by weighing the composite powder samples obtained after drying.

2.6. *Characterization of the samples*

The micrographs of the sample were obtained using high-resolution transmission electron microscope (HRTEM model JEOL JEM-2010; operating acceleration voltage 200 kV) as well as field-emission scanning electron microscope (FESEM model Zeiss Sigma VP, operating accelerating voltage 5 kV). The X-ray diffractometry (XRD) patterns were obtained using Philips PANalytical X-Pert Pro diffractometer. The molecular structures of the samples were characterized by a Perkin–Elmer Spectrum RX I Fourier transform infrared (FTIR) Spectrometer. The laser source used in the spectrometer was a He–Ne laser (633 nm). Raman spectroscopy was performed using EZ-Raman – M field portable Raman analyzer (Enwave Optronics, Inc.). A diode laser of wavelength 785 nm was used as excitation source. The thermogravimetric analysis (TGA) was carried out with PerkinElmer Pyris 1 TGA thermogravimetric analyzer at the heating rate of 10°C/min in nitrogen atmosphere. The optical absorbance spectra were recorded using a HITACHI U-3010 UV-visible absorption spectrophotometer. PL spectra of the samples were acquired using a PerkinElmer LS 55 Fluorescence spectrophotometer. The electrical conductivity of the samples was measured by a Four-Probe set-up (DFP-02, Scientific Equipment).

3.   Results and discussions

3.1. *HRTEM and FESEM microscopy*

The HRTEM micrograph of SPANI/SWCNT/ZnO nanocomposite is shown in Figure 1 and its FESEM images with energy dispersive X-ray analysis (EDAX) are shown in Figure 2. We observe that SWCNT bundles are decorated with ZnO nanoparticles and are also coated by SPANI to form SPANI/SWCNT/ZnO ternary composite. The average size of the ZnO nanoparticles is found to be 3–5 nm by direct measurement in the HRTEM micrograph. The presence of ZnO in the composite is confirmed by EDAX analysis. The 'carbon' comes from the polymer and SWCNT, while 'sulfur' comes from sulfonation of the polymer. 'Silicon' is present as the sample was put on a silicon substrate.

3.2. *XRD study*

The structural characteristics of SPANI, SWCNT, SPANI/SWCNT binary composite, SWCNT/ZnO hybrid and SPANI/SWCNT/ZnO ternary composite have been analyzed by X-ray diffractograms shown in Figure 3(a) and (b). In the XRD pattern of SPANI, there are three prominent humps at $2\theta = 25^0$, 43.5^0, and 51^0. No crystalline peak of SPANI was observed as SPANI is amorphous. In SWCNT, the peaks are at 26^0, 42.5^0, 44^0, 51^0, and 53.5^0 and these correspond to (002), (100), (101), (102), and (004) reflections of the graphitic planes of the nanotubes, respectively (JCPDS card no. 75-1621). The pattern of SPANI/SWCNT composite is dominated by the features of SWCNTs. The peaks for SWCNT/ZnO hybrid as well as SPANI/SWCNT/ZnO composite are obtained at the positions which correspond to (100), (002), (101), (102), (110), (103), (200), (112), and (201) planes of hexagonal ZnO crystallites (JCPDS card no. 80-0075) as well as (002), (100), (101), (102), and (004) reflections of the graphitic planes of SWCNTs [25]. However, the sharp peaks of ZnO nanocrystallites are suppressed in the diffractogram of the ternary nanocomposite.

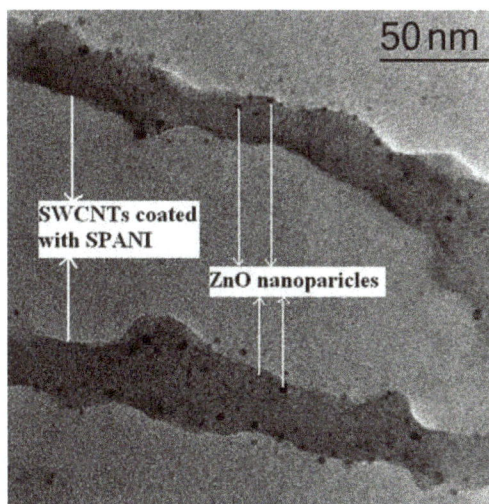

Figure 1.   HRTEM micrograph of SPANI/SWCNT/ZnO nanocomposite.

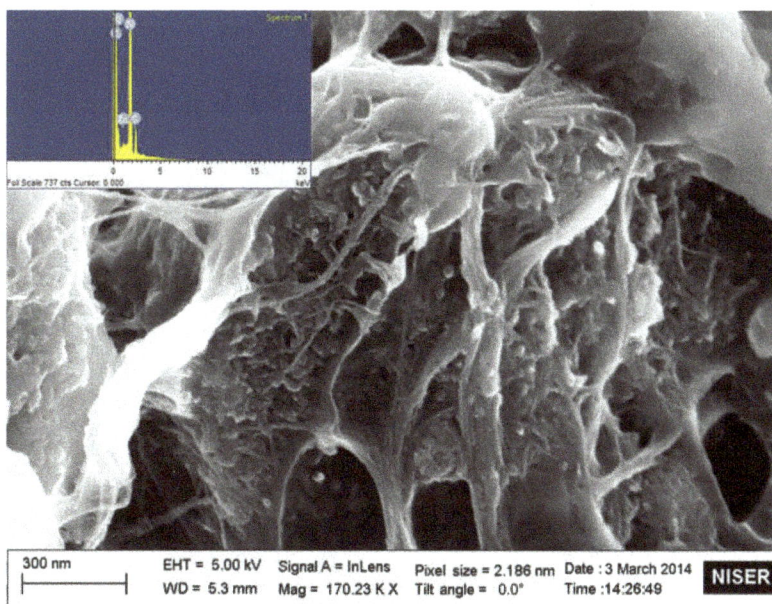

Figure 2.   FESEM image with EDAX spectrum of SPANI/SWCNT/ZnO composite.

3.3.   *FTIR Spectra*

In FTIR spectra, in Figure 4, the stretches at different points indicate the absorption bands. The stretch near 3450 cm^{-1} indicates O–H stretching vibration arising from the absorption of water by KBr used for analysis. The spectrum of pristine SWCNT has stretches near 1600 cm^{-1} due to C = O vibration formed due to acid treatment. In SWCNT/ZnO hybrid, the stretch at 460 cm^{-1} arises due to Zn–O vibration. SPANI shows the characteristic

(a)

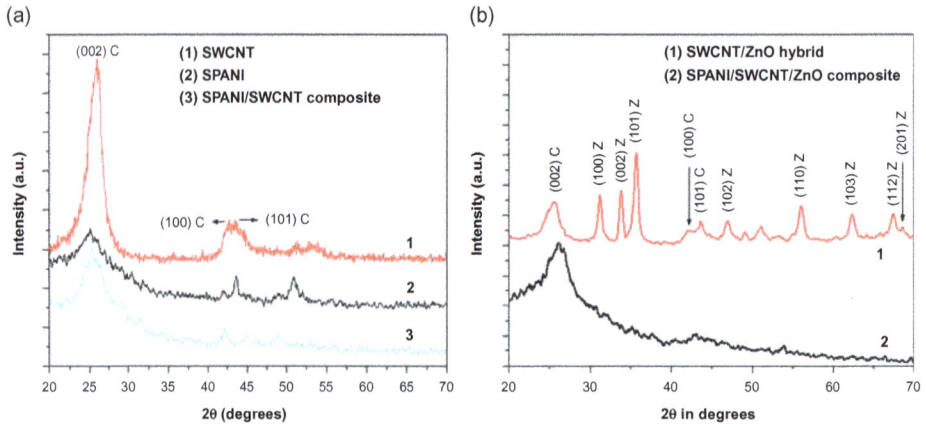

(b)

Figure 3.   (a) XRD patterns of SWCNT, SPANI, and SPANI/SWCNT composite. (b) XRD patterns of SWCNT/ZnO hybrid and SPANI/SWCNT/ZnO composite captioned C (CNT) and Z (ZnO).

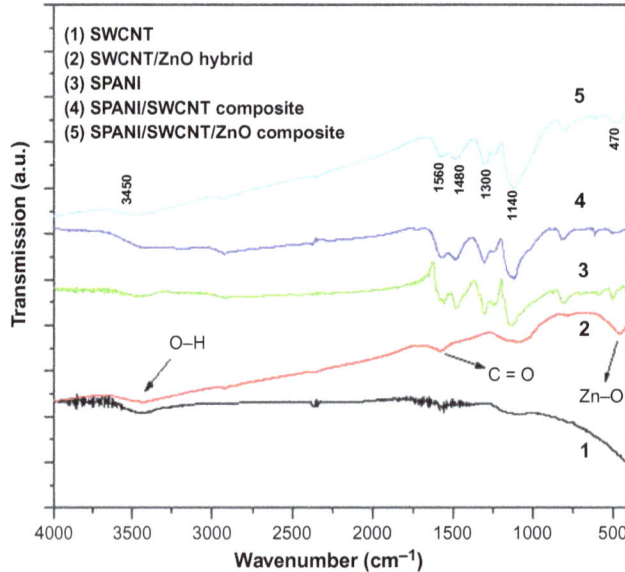

Figure 4.   FTIR spectra of SWCNT, SWCNT/ZnO hybrid, SPANI, SPANI/SWCNT composite, and SPANI/SWCNT/ZnO composite.

stretches at 1560 cm^{-1}, 1480 cm^{-1}, 1300 cm^{-1}, and 1140 cm^{-1}. The stretches at 1560 cm^{-1} and 1480 cm^{-1} arise due to the stretching vibrations of quinoid ring (–N = quinoid = N–) and the benzenoid ring (–N–benzenoid-N–), respectively. The stretches at 1300 cm^{-1} and 1140 cm^{-1} are due to C–N stretching and C = N stretching, respectively. The prominent stretch at 1140 cm^{-1} represents the characteristic stretch of conductivity of SPANI and it measures the degree of delocalization of electrons [31]. The FTIR spectra of SPANI/ SWCNT and SPANI/SWCNT/ZnO composites almost match with that of SPANI, which indicates good coating of SWCNTs and SWCNT/ZnO hybrid with SPANI. But an increase in the characteristic conductivity stretch, i.e., C = N stretching intensity is observed for SPANI/SWCNT composite and it further increases for SPANI/SWCNT/

ZnO composite. This indicates gradual enhancement in the delocalization of electrons due to charge transfer between SPANI and SWCNTs or hybrids. This is also supported by enhanced electrical conductivity of the composites. We have not observed any shift in the stretches of the polymer on reinforcing nanotubes or hybrids, so no conclusions on chemical bonding between the components can be drawn on this basis. Earlier, some researchers reported an interaction between the nanotubes and the polymer from the shifts in FTIR spectra [8], while some reported no significant interaction between them [10].

3.4. Raman spectra

The Figure 5 shows the Raman spectra of SWCNT, SWCNT/ZnO hybrid, and SPANI/SWCNT/ZnO composite. The Raman spectrum of SWCNTs shows the normal characteristic peaks, the G-band at 1589 cm^{-1}, the D-band at 1314 cm^{-1}, and the Radial Breathing Mode (RBM) band at 265 cm^{-1}. The G-band originates from in-plane vibrations of the graphitic wall and the D-band originates from defects in the graphitic structure. The position of G-band of pristine SWCNT is red shifted from 1589 to 1579 cm^{-1} on decorating ZnO nanoparticles, which indicates charge transfer from ZnO to SWCNTs [25]. Further the I_D/I_G factor decreases on decorating SWCNT with the ZnO nanoparticles, which indicates bonding of ZnO nanoparticles with SWCNT surfaces and transfer of charge. For pristine SWCNT, the ratio between the intensity of D-band and that of G-band i.e., I_D/I_G is 0.51, while for SWCNT/ZnO hybrid, it has been found to be 0.47. After coating of the SWCNT/ZnO hybrid with SPANI, the spectrum of the polymer dominates. For SPANI/SWCNT/ZnO, the peak at 1617 cm^{-1} arises from C–C stretching of the benzenoid ring and the peak at 1516 cm^{-1} comes from C=N stretching of the quinonoid ring. The peak at 1350 cm^{-1} arises from the C–N$^+$ stretching and that of 1185 cm^{-1} arises from C–H bending of the quinoid/benzenoid ring. This observation indicates that a good coating of the hybrid nanostructure has been achieved. This is also supported by images of HRTEM and FESEM. No chemical interaction between the polymer and the nanotubes could be proved on the basis of Raman spectra [8].

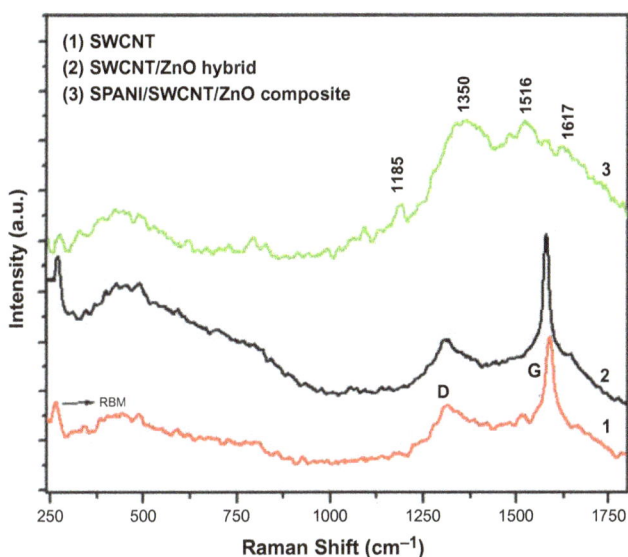

Figure 5.   Raman spectra of SWCNT, SWCNT/ZnO hybrid, and SPANI/SWCNT/ZnO composite.

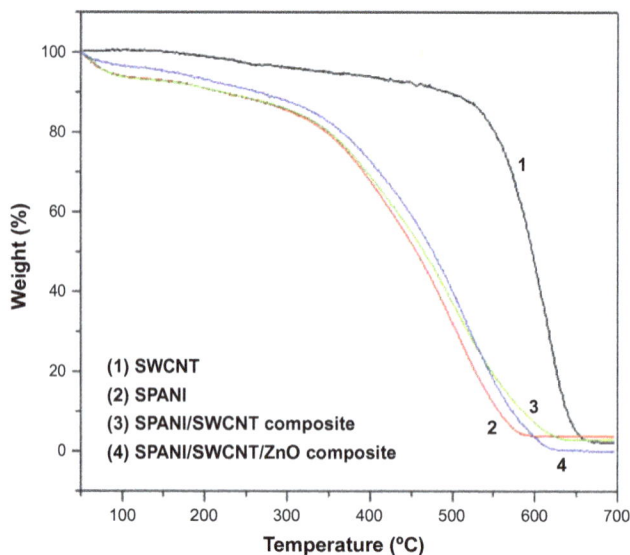

Figure 6. TGA thermograms of SWCNT, SPANI, SPANI/SWCNT composite, and SPANI/ SWCNT/ZnO composite.

3.5. *Thermogravimetric analysis*

TGA was done to investigate the thermal stability of the samples. The Figure 6 shows the TGA thermograms of SWCNT, SPANI, SPANI/SWCNT composite, and SPANI/SWCNT/ZnO composite. The SWCNT has a higher thermal stability than other samples and it shows no appreciable mass loss up to about 550°C. The thermograms of SPANI, SPANI/SWCNT, and SPANI/SWCNT/ZnO composites are almost similar. However, the mass loss for SPANI/ SWCNT/ZnO composite is slightly less than that of SPANI/SWCNT till 550°C, which indicates higher thermal stability for SPANI/SWCNT/ZnO composite in this temperature region.

3.6. *UV-visible absorbance spectra*

The UV-visible absorption spectra of SPANI, SPANI/SWCNT, and SPANI/SWCNT/ZnO are compared and shown in Figure 7. The spectrum of SPANI consists of humps at 360 nm corresponding to π–π^* transition as well as at 445 nm and 750 nm corresponding to polaron transitions. For SPANI/SWCNT composite, a red shift of π–π^* transition from 360 to 365 nm and polaron transition from 445 to 450 nm is observed. A further red shift in these transitions is observed in the spectrum of SPANI/SWCNT/ZnO composite, where we observe transition bands with peak emissions at 370 nm, 455 nm, and 750 nm. An additional peak at 340 nm is also observed for this composite, which is the characteristic absorption peak of ZnO nanoparticles.

Calculation of optical band gap:

The optical band gap E_g is related to the absorption coefficient α by the relation

$$\alpha = \frac{B\left(hv - E_g\right)^{1/2}}{hv} \tag{1}$$

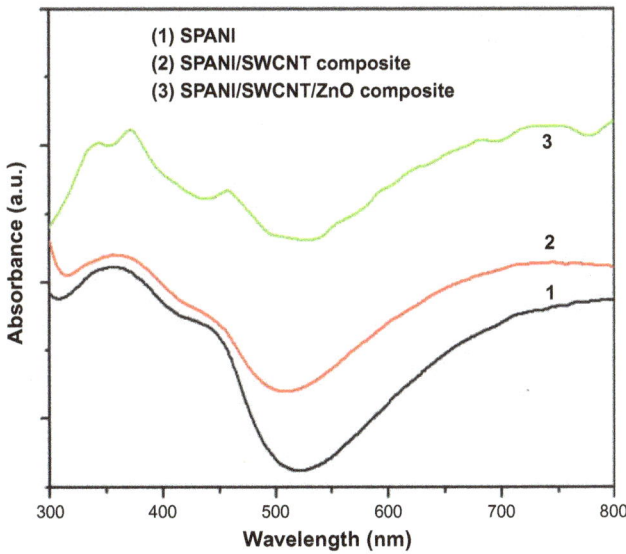

Figure 7. UV-visible absorbance spectra of SPANI, SPANI/SWCNT composite, and SPANI/SWCNT/ZnO composite.

where B is the absorption constant for a direct transition. We have plotted $(\alpha h\nu)^2$ versus $h\nu$ for SPANI, SPANI/SWCNT, and SPANI/SWCNT/ZnO in Figure 8 and extrapolated the linear portion of each to $\alpha = 0$ value and obtained the corresponding band gaps. The optical band gap of pure SPANI and SPANI/SWCNT composite was estimated to be 3.75 eV and 3.6 eV, respectively. For SPANI/SWCNT/ZnO nanocomposite, there is a further decrease in the band gap energy, the value being 3.5 eV. We observe that the

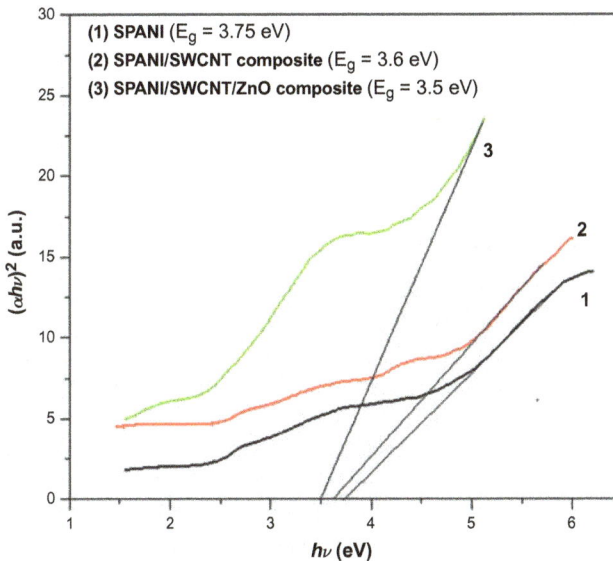

Figure 8. Plot of $(\alpha h\nu)^2$ vs. $h\nu$ for SPANI, SPANI/SWCNT, and SPANI/SWCNT/ZnO composite for calculation of optical band gap.

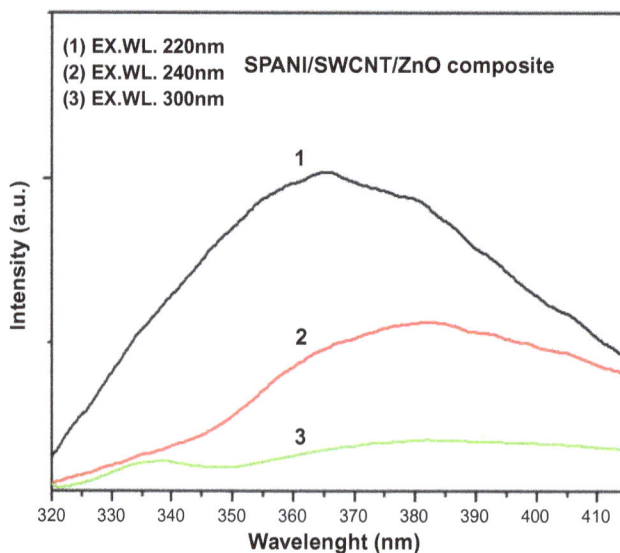

Figure 9.   Emission curves for SPANI/SWCNT/ZnO for four different excitation wavelengths.

decrease in band gap is more when the polymer is reinforced with SWCNT/ZnO nanohybrid rather with SWCNTs, the direct band gap of ZnO nanoparticles being 3.3 eV. These observations indicate creation of new exciton energy levels below the regular band edge due to charge transfer from ZnO to SWCNT, and subsequently from SWCNT to SPANI.

3.7.   Photoluminescence spectra

Figure 9 shows the PL spectra of SPANI/SWCNT/ZnO nanocomposite obtained at different excitation wavelengths from 220 nm to 300 nm. We obtained emission in the UV-visible region with maxima at 365 nm, 380 nm, and 385 nm for excitation wavelengths 220 nm, 240 nm, and 300 nm, respectively. We observed a red shift in the emission maxima with increase in excitation wavelength from 220 to 300 nm. When SPANI/SWCNT/ZnO composite is excited by radiation of wavelengths 220–300 nm, the higher energy photons of smaller wavelength cause emission with enhanced intensity.

3.8.   Electrical conductivity

The Figure 10 shows the DC electrical conductivity of SPANI and the two composites (SPANI/SWCNT and SPANI/SWCNT/ZnO) at temperatures ranging from 317 to 353 K. We observe an increase in the conductivity with increase in temperature for all the samples, thus showing semiconductor like behavior. The addition of SWCNTs or SWCNT/ZnO hybrids to SPANI increases the conductivity of the polymer. The SPANI has conductivity of 0.042 S/m at 323 K, whereas the SPANI/SWCNT and SPANI/SWCNT/ZnO composite samples show the conductivity values of 0.23 S/m and 0.88 S/m, respectively, at that temperature. The value for SPANI/SWCNT/ZnO nanocomposite is 20.8 times and 3.7 times higher than SPANI and SPANI/SWCNT composite, respectively. At 353 K, the conductivity values for SPANI, SPANI/SWCNT and SPANI/SWCNT/ZnO samples are 0.082 S/m, 0.58 S/m, and 1.54 S/m, respectively. For the

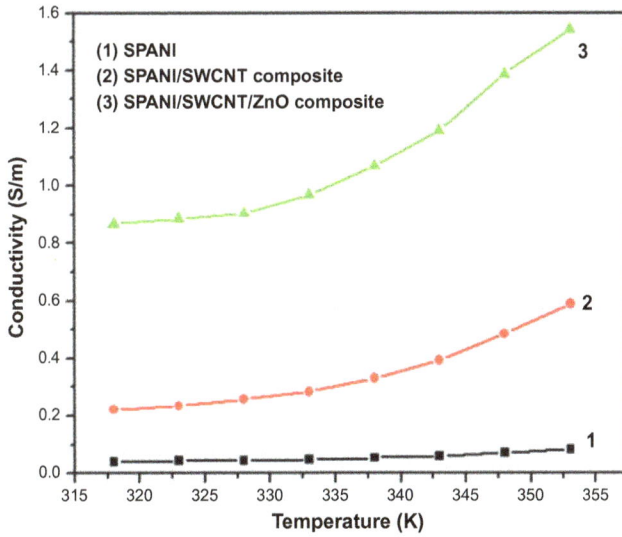

Figure 10.   Comparison of conductivity of SPANI/SWCNT/ZnO nanocomposite with pure SPANI and SPANI/SWCNT composite samples with varying temperature.

sample SPANI/SWCNT/ZnO, the value increases 18.7 times compared to SPANI and 2.6 times compared to SPANI/SWCNT composite.

The Mott's variable range hopping (VRH) model is suitable for explaining the temperature dependence of conductivity. The variation of conductivity as per this theory is given by the relation

$$\sigma = \sigma_0 \exp(-T_0/T)^\gamma \tag{2}$$

where $\gamma = 1/(1 + d)$, where $d = 1$, 2, and 3 for one-, two, and three-dimensional conduction, respectively, and σ_0 is the conductivity when T tends to infinity. T_0 is the characteristic Mott temperature, which depends on the electronic structure and the energy distribution of the localized states as follows

$$T_0 = 16/k_B N(E_F) L_{\text{loc}}^3 \tag{3}$$

where k_B is Boltzmann constant, $N(E_F)$ is the density of states at the Fermi level, and L_{loc} is the localization length.

We find that the log of DC conductivity of SPANI/SWCNT/ZnO composite sample satisfies Mott's VRH model for $\gamma = 1/3$, indicating a two-dimensional conduction, as shown in Figure 11.

We obtained the values of T_0 for the three samples (SPANI, SPANI/SWCNT composite, and SPANI/SWCNT/ZnO composite) from the slopes of the linear variation of log σ with $T^{-1/3}$. The estimated values of T_0 were 180,362 K, 926,859 K, and 112,329 K for SPANI, SPANI/SWCNT composite, and SPANI/SWCNT/ZnO composite, respectively. We see that T_0 for SPANI/SWCNT/ZnO composite is much lower than that for other samples.

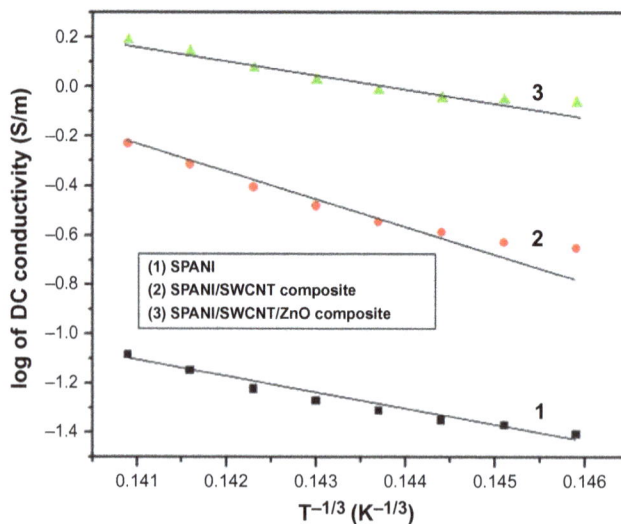

Figure 11. Variation of log of DC conductivity of SPANI, SPANI/SWCNT composite, and SPANI/SWCNT/ZnO composite with $T^{-1/3}$ ($K^{-1/3}$) satisfying VRH model with $\gamma = 1/3$.

4. Conclusions

We reported the synthesis of a novel luminescent and conductive material containing SPANI and SWCNT/ZnO hybrid nanostructures by a simple chemical process. A good coating of the hybrid by the polymer is observed, but a chemical bonding between them is not confirmed. Characterization established the desired nanostructure of the composite. The optical band gap energy of 3.75 eV for pure SPANI was found to decrease to 3.6 eV for the SPANI/SWCNT two-phase composite, which was further reduced to 3.5 eV for the SPANI/SWCNT/ZnO three-phase composite. This was supported by electrical conductivity measurement. The SPANI/SWCNT/ZnO nanocomposite showed strong absorbance peaks in the UV-visible region, including the characteristic peak of ZnO nanocrystals at 340 nm. The PL study showed a broad emission range in the UV-visible region between 340 and 420 nm. Such a ternary polymer nanocomposite may find application as a tailored conducting luminescent material in different electro-optical and energy devices, particularly as a photovoltaic material in dye-sensitized solar cells.

References

[1] X.-G. Li, H. Feng, M.-R. Huang, G.-L. Gu, and M.-G. Moloney, *Ultrasensitive Pb(II) potentiometric sensor based on copolyaniline nanoparticles in a plasticizer-free membrane with a long lifetime*, Anal. Chem. 84 (2012), pp. 134–140. doi:10.1021/ac2028886.

[2] X.-G. Li, H. Feng, and M.-R. Huang, *Strong adsorbability of mercury ions on aniline/ sulfoanisidine copolymer nanosorbents*, Chem. Eur. J. 15 (2009), pp. 4573–4581. doi:10.1002/chem.200802431.

[3] M.-R. Huang, H.-J. Lu, and X.-G. Li, *Synthesis and strong heavy-metal ion sorption of copolymer microparticles from phenylenediamine and its sulfonate*, J. Mater. Chem. 22 (2012), pp. 17685–17699. doi:10.1039/c2jm32361c.

[4] L. Q-F, H. M-R, and L. X-G, *Synthesis and heavy-metal-ion sorption of pure sulfophenylene-diamine copolymer nanoparticles with intrinsic conductivity and stability*, Chem. Eur. J. 13 (2007), pp. 6009–6018. doi:10.1002/chem.200700233.

[5] X.-G. Li, Q.-F. Lu, and M.-R. Huang, *Self-stabilized nanoparticles of intrinsically conducting copolymers from 5-sulfonic-2-anisidine*, Small. 4 (2008), pp. 1201–1209. doi:10.1002/smll.200701002.

[6] P. Kar and A. Choudhury, *Carboxylic acid functionalized multi-walled carbon nanotube doped polyaniline for chloroform sensors*, Sens. Actuators B. 183 (2013), pp. 25–33. doi:10.1016/j.snb.2013.03.093.

[7] E. Lafuente, M.A. Callejas, R. Sainz, A.M. Benito, W.K. Maser, M.L. Sanjuan, D. Saurel, J.M. Teresa, and M.T. Martinez, *The influence of single-walled carbon nanotube functionalization on the electronic properties of their polyaniline composites*, Carbon. 46 (2008), pp. 1909–1917. doi:10.1016/j.carbon.2008.07.039.

[8] E.N. Konyushenko, J. Stejskal, M. Trchova, J. Hradil, J. Kovarova, J. Prokes, M. Cieslar, J.Y. Hwang, K.H. Chen, and I. Sapurina, *Multi-wall carbon nanotubes coated with polyaniline*, Polymer. 47 (2006), pp. 5715–5723. doi:10.1016/j.polymer.2006.05.059.

[9] C. Dhand, P.R. Solanki, M. Datta, and B.D. Malhotra, *Polyaniline/single-walled carbon nanotubes composite based triglyceride biosensor*, Electroanalysis. 22 (2010), pp. 2683–2693. doi:10.1002/elan.201000269.

[10] M.R. Karim, C.J. Lee, Y.T. Park, and M.S. Lee, *SWNTs coated by conducting polyaniline: Synthesis and modified properties*, Synth. Met. 151 (2005), pp. 131–135. doi:10.1016/j.synthmet.2005.03.012.

[11] Y.W. Lin and T.M. Wu, *Synthesis and characterization of externally doped sulfonated polyaniline/multi-walled carbon nanotube composites*, Compo. Sci. Technol. 69 (2009), pp. 2559–2565. doi:10.1016/j.compscitech.2009.07.013.

[12] H. Zhang, H.X. Li, and H.M. Cheng, *Water-soluble multiwalled carbon nanotubes functionalized with sulfonated polyaniline*, J. Phys. Chem B. 110 (2006), pp. 9095–9099. doi:10.1021/jp060193y.

[13] V. Gupta and N. Miura, *Polyaniline/single-wall carbon nanotube (PANI/SWCNT) composites for high performance supercapacitors*, Electrochim. Acta. 52 (2006), pp. 1721–1726. doi:10.1016/j.electacta.2006.01.074.

[14] P.C. Ramamurthy, A.M. Malshe, W.R. Harrell, R.V. Gregory, K. McGuire, and A.M. Rao, *Polyaniline/single-walled carbon nanotube composite electronic devices*, Solid-State Electron. 48 (2004), pp. 2019–2024. doi:10.1016/j.sse.2004.05.051.

[15] M. Tahhan, V.T. Truong, G.M. Spinks, and G.G. Wallace, *Carbon nanotube and polyaniline composite actuators*, Smart Mater. Struct. 12 (2003), pp. 626–632. doi:10.1088/0964-1726/12/4/313.

[16] L. Li-Xia, T. Qin-Xin, S. Chang-Lu, and L. Yi-Chun, *Structure and photoluminescence of nano-ZnO films grown on a Si (100) substrate by oxygen- and argon-plasma-assisted thermal evaporation of metallic Zn*, Chin. Phys. Lett. 22 (2005), pp. 998–1001. doi:10.1088/0256-307X/22/4/061.

[17] D.H. Zhang, Z.Y. Xue, and Q.P. Wang, *The mechanisms of blue emission from ZnO films deposited on glass substrate by R.F. Magnetron sputtering*, J. Phys. D: Appl. Phys. 35 (2002), pp. 2837–2840. doi:10.1088/0022-3727/35/21/321.

[18] H. Hayashi, A. Ishizaka, M. Haemori, and H. Koinuma, *Bright blue phosphors in ZnO-WO₃ binary system discovered through combinatorial methodology*, Appl. Phys. Lett. 82 (2003), pp. 1365–1367. doi:10.1063/1.1554767.

[19] P. Sharma, K. Sreenivas, and K.V. Rao, *Analysis of ultraviolet photoconductivity in ZnO films prepared by unbalanced magnetron sputtering*, J. Appl. Phys. 93 (2003), pp. 3963–3970. doi:10.1063/1.1558994.

[20] S.B. Kondawar, S.D. Bompilwar, V.S. Khati, S.R. Thakre, V.A. Tabhane, and D.K. Burghate, *Characterizations of zinc oxide nanoparticles reinforced conducting polyaniline composites*, Arch. Appl. Sci. Res. 2 (2010), pp. 247–253.

[21] S.B. Kondawar, S.A. Acharya, and S.R. Dhakate, *Microwave assisted hydrothermally synthesized nanostructure zinc oxide reinforced polyaniline nanocomposites*, Adv. Mater. Lett. 2 (2011), pp. 362–367.

[22] M. Dhingra, S. Shrivastava, P.S. Kumar, and S. Annapoorni, *Polyaniline mediated enhancement in band gap emission of zinc oxide*, Composites Part B: Eng. 45 (2013), pp. 1515–1520. doi:10.1016/j.compositesb.2012.09.020.

[23] J. Khanderi, R.C. Hoffmann, A. Gurlo, and J.J. Schneider, *Synthesis and sensoric response of ZnO decorated carbon nanotubes*, J. Mater. Chem. 19 (2009), pp. 5039–5046. doi:10.1039/b904822g.

[24] I. Sameera, R. Bhatia, and V. Prasad, *Preparation, characterization and electrical conductivity studies of MWCNT/ZnO nanoparticles hybrid*, Physica B: Condens. Matter. 405 (2010), pp. 1709–1714. doi:10.1016/j.physb.2009.12.074.

[25] R. Paul, P. Kumbhakar, and A.K. Mitra, *Blue-green luminescence by SWCNT/Zno hybrid nanostructure synthesized by a simple chemical route*, Physica E Low-Dimens Syst. Nanostruct. 43 (2010), pp. 279–284. doi:10.1016/j.physe.2010.07.067.

[26] K. Dai, G. Dawson, S. Yang, Z. Chen, and L. Lu, *Large scale preparing carbon nanotube/zinc oxide hybrid and its application for highly reusable photocatalyst*, Chem. Eng. J. 191 (2012), pp. 571–578. doi:10.1016/j.cej.2012.03.008.

[27] S. Berson, R. De Bettignies, S. Bailly, S. Guillerez, and B. Jousselme, *Elaboration of P3HT/CNT/PCBM composites for organic photovoltaic cells*, Adv. Funct. Mater. 17 (2007), pp. 3363–3370. doi:10.1002/adfm.200700438.

[28] S. Abdulalmohsin, Z. Li, M. Mohammed, K. Wu, and J. Cui, *Electrodeposited polyaniline/multi-walled carbon nanotube composites for solar cell applications*, Syntheticmetals. 162 (2012), pp. 931–935.

[29] G. Wang, W. Xing, and S. Zhuo, *The production of polyaniline/graphene hybrids for use as a counter electrode in dye-sensitized solar cells*, Electrochim. Acta. 66 (2012), pp. 151–157. doi:10.1016/j.electacta.2012.01.088.

[30] S. Ibrahim, M. Soliman, M. Anas, M. Hatez, and T.M. Abdel-Fattah, *Dye-sensitized solar cell based on polyaniline/multiwalled carbon nanotubes counter electrode*, Int. J. Photoenerg. (2013) Article ID 906820 doi:10.1155/2013/906820.

[31] J.-E. Huang, X.-H. Li, J.-C. Xu, and H.-L. Li, *Well-dispersed single-walled carbon nanotube/polyaniline composite films*, Carbon. 41 (2003), pp. 2731–2736. doi:10.1016/S0008-6223(03)00359-2.

2

Shungite as the natural pantry of nanoscale reduced graphene oxide

Elena F. Sheka[a]* and Natalia N. Rozhkova[b]

[a]*General Physics Department, Peoples' Friendship University of Russia, Moscow 117198, Russian Federation;* [b]*Institute of Geology, Karelian Research Centre RAS, Petrozavodsk, Russian Federation*

Shungite is presented as a natural carbon allotrope of a multilevel fractal structure that is formed by a successive aggregation of ~1 nm reduced graphene oxide nanosheets. Turbostratic stacks of the sheets of ~1.5 nm in thickness and globular composition of the stacks of ~6 nm in size determine the secondary and tertiary levels of the structure. Aggregates of globules of tens of nanometers complete the structure. Molecular theory of graphene oxide, supported by large experience gained by the modern graphene science, has led to the foundation of the suggested presentation. The microscopic view has found a definite confirmation when analyzing the available empirical appearance of shungite. To our knowledge, this is the first time a geological process is described at quantum level.

Keywords: shungite; reduced graphene oxide; molecular theory

1. Introduction

Carbon is an undisputed favorite of Nature that has been working on it for billions of years thus creating a number of natural carbon allotropes. Representatively, we know nowadays diamond, graphite, amorphous carbon (coal and soot), and lonsdaleite. For the last three decades, the allotrope list has been expanded over artificially made species such as fullerenes, single-walled and multi-walled carbon nanotubes, glassy carbon, linear acetylenic carbon, and carbon foam. The list should be completed by nanodiamonds and nanographites as well as one-layer and multi-layer graphenes. Evidently, one-to-few carbon layers adsorbed on different surfaces should be attributed to this cohort as well.

In spite of high abundance of the carbon allotropes, the above list remains incomplete until shungite is added to the group of natural allotropes. As has been known, this natural carbon deposit cannot be attributed to either diamond, or graphite and amorphous carbon. A lot of efforts have been undertaken to exhibit that the material, once pure carbon by content, presents a fractal structure of agglomerates consisting of nanosize globules [1], each of which presents a cluster of ~1 nm graphene-like sheets [2]. The current article presents a summarized view on shungite as carbon allotrope and suggests a microscopic vision of the shungite derivation supplemented by the presentation of its structure as a multistage fractal net of reduced graphene oxide (rGO) nanosheets.

---

*Corresponding author. Email: sheka@icp.ac.ru

2.   Shungite as we know it

Shungite rocks are widely known and are in a large consumer's demand due to its unique physicochemical [3] and biomedical properties [4]. For a long period, thorough and systematic studies, aimed at clarification of the reasons of such uniqueness, have been performing. As has been gained [1], shungite carbon of natural deposits is a densely packed porous structure of agglomerates with a large variety of pore size from units to hundreds nanometer. Such a large dispersion of the pore size evidences a multistage structure of agglomerates that points to the fractal accommodation [5]. The shungite fractal structure has been clearly evidenced by small angle neutron scattering (SANS) [6] that showed two types of shungite pores, namely, small pores with linear dimensions of 2–10 nm and large pores of more than 100 nm in size. The findings allowed suggesting that the shungite fractal structure is provided with aggregates of globular particles of ~6 nm in average size. Within the framework of the first ideas in the 90s, the globules were seen as fullerene-like structures [7]. However, thorough studies, particularly, high-resolution electron microscopy, X-Ray diffraction, and Raman scattering showed that the globules are clusters (stacks) of graphene-based fragments of ≤1 nm in size [8,9]. X-ray [3] and neutron [8] diffraction showed the graphite-like packing of the latter characterized by the interfacial distance of 3.40(3) Å [3] and 3.50(3) Å [10]. The length of the coherent scattering region along c-axis constitutes 2.2(1) nm [3] and 1.5(4) nm [10]. The data show that shungite globules present aggregates of six- and five-layered graphene-like stacks.

As has often been observed [11], solid fractals have increased solubility with respect to dense solids. In full accordance with this, shungite can be quite easily dispersed in water in contrast to solid fullerenes, nanographites, and other sp^2 nanocarbons. Shungite aqueous dispersions, with maximum achieved concentration of ~0.1 mg/ml, are described in details elsewhere [1,8]. If the water evaporation is blocked, the dispersions are quite stable and conserve properties during a long period of time (for several years). A complete drying of the dispersions results in the formation of densely packed powdered colloid condensate. The structural and physicochemical characteristics of the latter are similar to those of the pristine shungite [9].

The close link between the pristine shungite, its aqueous dispersions, and the condensate is clearly visible through Raman scattering. Figure 1 shows Raman spectra of the

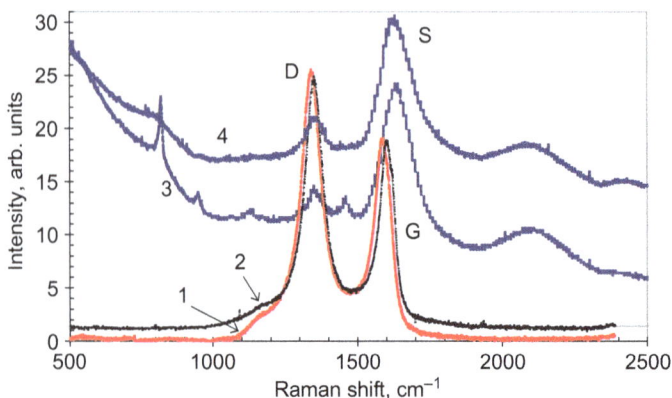

Figure 1.   Raman spectra of pristine shungite powder (1), colloid condensate (2), and aqueous dispersions of shungite carbon nanoparticles with concentration of 0.06 mg/ml (3), and 0.12 mg/ml (4). T = 293°C, laser excitation at 532 nm.

two solids and two aqueous dispersions recorded on a Nicolet Almega XR (Thermo Scientific) spectrometer at laser excitation of 532 nm. The instrument spectral resolution is 1 cm^{-1} and the laser power is 15 mW. As seen in the figure, practically identical dublets of G and D bands present the main features of the spectra of both solids indicating their close similarity. However, both bands of the condensate are slightly up shifted from 1341 to 1348 cm^{-1} and from 1586 to 1596 cm^{-1} for D and G bands, respectively. The data show that the solids are pretty similar, albeit not absolutely identical due to a certain structure rearrangement on the way from pristine shungite to colloid condensation from shungite water dispersions.

In the current graphene science, the relative intensity of the two bands I_D/I_G is usually considered as a measure of the perfectness of graphene regular structure [12]. However, in relation to such irregular structure as shungite, better to speak differently and to attribute the bands to stretching vibrations of the sp^3 and sp^2 C–C bonds. Evidently, the vibrational frequency of non-distorted benzenoid unit, described by G band, should be close to that of benzene molecule of 1599 cm^{-1}. Once practically inactive in the Raman scattering of the molecule, the vibration is drastically enhanced for the extended set of condensed benzenoid units thus transferring to intense G band that corresponds to the Γ-point $k = 0$ phonon mode of the graphene crystal. Similarly, D band is characteristic of the distorted benzenoid units on their way to cyclohexanoid ones. The relevant frequencies of stretching vibrations of a set of benzene-to-cyclohexane molecules cover a quite extended interval 1330–1390 cm^{-1} depending on how many additional hydrogen atoms are added to the benzene ring or how many sp^2 C–C bonds are transformed into sp^3 ones. The position of D bands of both solids fits the interval pretty well indicating that a part of the pristine benzenoid units are distorted toward the cyclohexanoid ones. Similarly to the former, inactive stretching of the pristine cyclohexane is enhanced for the extended set of partial or complete cyclohexanoid units (see the intense Γ-point $k = 0$ phonon mode at 1342 cm^{-1} in the experimental Raman spectrum of graphane [13]). Accordingly, the corresponding I_D/I_G ratio determines the extent of the total sp$^2 \rightarrow$ sp^3 transformation that is quite similar in the studied solids.

Stretching O–H vibrations of ~3400 cm^{-1} dominate in the Raman spectra of aqueous dispersions that is why recording of their low-frequency parts presented in Figure 1 required 30-min accumulation. Raman spectra of both dispersions bear the clear imprint of the benzenoid-to-cyclohexanoid transformation presented by D bands located at 1350 cm^{-1} (curve 3) and 1353 cm^{-1} (curve 4). It should be noted that the band positions are vividly up shifted with respect to that of the pristine shungite while practically coincide with the position of D band of the colloid condensate. Since the benzenoid-to-cyclohexanoid transformation is quite alike in the two solids, it hardly differs for the dispersion colloids, due to which the shift and relative intensity of their G bands should be similar to that observed for the condensate. Consequently, the relevant G bands of the dispersions, much lower by intensity than D bands, should be located at 1596 ± 4 cm^{-1} and falls on the low-frequency wing of intense S bands far from their maxima located at 1635 cm^{-1} (curve 3) and 1627 cm^{-1} (curve 4). Therefore, the bands should be attributed to Raman scattering from non-carbon scatterer. It is reasonable to suggest the scissor deformational vibrations of water play the role. A shift from the frequency of 1585 cm^{-1} related to free water molecules to 1627–1635 cm^{-1} in the dispersions points to the hydrated state of water. The appearance of the vibration, once symmetrically forbidden for free water, with up shifted frequency was observed in the Raman spectra of water retained in clay minerals [14].

The aqueous dispersions exhibit a large number of peculiar properties that, on the one hand, have a direct connection with the unique properties of shungite while on the other hand, are pretty similar to those characteristic for aqueous dispersions of such quantum dots as either Ag and Au nanoparticles or CdS and CdSe nanocompositions, on the one hand, and synthetic graphene quantum dots (see exhausted review [15]), on the other hand. Similarly to the former, shungite dispersions reveal high activity toward enhancing nonlinear [16–18] and spectral [19] optical properties. Analogously to the latter, shungite dispersions exhibit a close similarity in the appearance of a high inhomogeneity of both morphological and spectral properties. A particular attention should be given to their biomedical behavior [4,20,21]. Thus, the study of the dispersion effect on the behavior of serum albumin showed that shungite globules and proteins form stable bioconjugates. The latter do not change the protein secondary structure, but causes a drastic lowering of the compactness of the protein tertiary structure that might promote various biomedical applications.

3. Shungite in view of graphene molecular chemistry

The graphene-based basic structure of shungite provides a good reason to consider the latter at the microscopic level by using high power of the modern empirical and theoretical molecular science of graphene. This approach allows us not only to explain all the peculiarities of the shungite behavior, but also to lift the veil on the mystery of its origin. To consider shungite from the viewpoint of the graphene molecular science, in fact, is to find answers to the following questions.

(1) Where did graphene-based basic units of shungite come from?
(2) Why is the unit linear size limited to ~1 nm?
(3) What did this size stabilize during the geological time of life?
(4) What is the chemical composition of the basic units?
(5) Why and how do the units aggregate?
(6) Why are there two sets of pores in the shungite carbon?

Knowledge that the molecular graphene science has accumulated for the last years is so vast that it allows considering the totality of issues simultaneously. Obviously, not all the answers to the above questions have been so far fully exhaustive. However, they present the first attempt of seeing the problem as a whole leaving details of the subsequent refinements for future investigations. Currently, the following answers can be suggested.

Answer 1. To answer the first question, we have to address the geological story of shungite. Although shungite is about 2 billions years old, its origin has been still under discussion [22]. The available hypothesizes are quite controversial. According to the biogenic concept, it is formed of organic-carbon-rich sediments. Following the others, shungite is of either volcanic endogenous or even extraterrestrial origin. In contrast to graphite, which is largely distributed over the Earth, the shungite deposits are space limited, and the Onega Lake basin of Karelia is the main area for the rock mining.

Two distinct peculiarities are characteristic for the geological Karelian region, namely: (1) shungite deposits around the Onega Lake neighbor with graphites in the vicinity of the Ladoga Lake and (2) there is abundance of water, both open and mineral one, in the former case. The first feature gives clear evidence that the Karelian region as a whole is favorable for graphene-layer deposition of carbon which might imply the presence of a common framework of the two geological processes. The second forces to draw a particular attention to the aquatic environment of the deposition.

Geology of graphite is well developed by now. According to the modern concept [23], graphite can be (i) either syngenetic, formed through the metamorphic evolution of carbonaceous matter dispersed in the sediments or (ii) epigenetic, originating from precipitation of solid carbon from carbon-saturated C–O–H fluids. Currently, the privilege has been given to the first one. The transformation of carbonaceous matter involves structural and compositional changes of basic structural units in graphite in the form of aromatic lamellae (graphene sheets) and occurs in the nature in the framework of thermal or regional metamorphism that apart from temperature, involves shear strain and strain energy. Temperature from 380°C to ~450°C and pressure between 2×10^8 Pa (2 kbar) and 3×10^8 Pa (3 kbar) efficiently govern the graphenization [24,25].

Accepting the syngenetic graphenization (the definition [26] exactly suits the processes occurred) to be a common process for the derivation of both graphite and shungite in the Karelian region, we can suggest the following answer to the first question.

The graphenization is a longtime complicated process that, occurring during a geological scale of time, can be subjected to various chemical reactions. Tempo and character of the reactions are obviously dictated by the environment. It is quite reasonable to suggest that the aqueous environment at 300–400°C, under which the metamorphic evolution of carbonaceous matter occurs, is a dynamically changeable mixture of water molecules, hydrogen and oxygen atoms, hydroxyl and carboxyl radicals as well. The interaction of the carbonaceous matter subjected to structural and compositional changes in the course of alignment of graphene lamellae and pore coalescence with this mixture accompanies the process. The most expected reactions concern hydration, hydrogenation, oxidation, hydroxylation, and carboxylation of the formed lamellae. At this point, it is important to note that, according to the molecular theory of graphene [27–32], any reaction of these kinds primarily involves edge carbon atoms of the sheets. Actually, Figure 2 presents a typical image map of the atomic chemical susceptibility (ACS) distribution over (5, 5) nanographene (NGr) atoms (the latter is presented by a rectangular graphene fragment, below (5, 5) NGr molecule, containing $n_a = 5$ and $n_z = 5$ benzenoid units along armchair and zigzag edges, respectively). The ACS map shape is characteristic for a graphene fragment with bare edge atoms of any size and shape. Presented in Figure 2 shows that the chemical reactivity of the edge atoms exceeds that of the basal plane atoms by approximately four times. Consequently, any addition reaction will start by involving these atoms first. The addition, obviously, terminates the lamellae growth thus limiting the lateral dimensions of the formed graphene layers. Empirically, it has been repeatedly observed in the case of graphene oxide (GO) [33–36]. Therefore, since the above mentioned reactions start simultaneously with the deposit formation, their efficiency determines if either the formed graphene lamellae will increase in size (low efficient reactions) or the lamellae size will be terminated (reactions of high efficiency). Since large graphite deposits are widely distributed over the Earth, it might be accepted that aqueous environment of the organic-carbon-rich sediments generally does not provide suitable conditions for the effective termination of graphene lamellae in the course of the deposit graphitization. Obviously, particular reasons may change the situation that can be achieved in some places of the Earth. Apparently, this occurred in the Onega Lake basin, which caused the replacement of the graphite derivation by the shungite formation. Some geologists

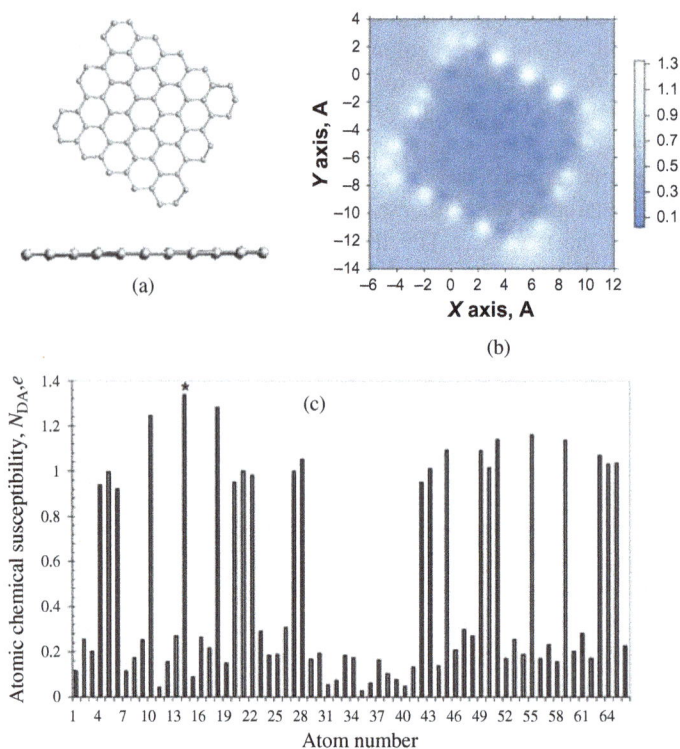

Figure 2. Top and side views of the equilibrium structure of the (5, 5) NGr molecule (a), and ACS distribution over atoms in the real space (b), and according to atom numbers in the output file (c) (UHF calculations [27]).

reported on the correlation of shungite formation with the increase of oxygen concentration in atmosphere that took place in 1.9–2.1 Ga [37]. This fits well the geochemical boundary of the Earth history of 2.2 Ga.

Answer 2. If chemical modification of graphene lamellae is responsible for limiting their size, the answer to the question about the size limitation to ~1 nm should be sought in the relevant reaction peculiarities. First of all, one must choose among those reactions that are preferable under the pristine graphenization conditions. Including hydroxylation and carboxylation into oxidation reaction, we must make choice among three of them, namely, hydration, hydrogenation, and oxidation. All the three reactions are well studied for graphene at molecular level both empirically and theoretically.

The pristine graphene lamella is hydrophobic so that its interaction with water molecules is weak. Chemical coupling of a water molecule with graphene sheets can rarely occur at the zigzag edge and is characterized by small coupling energy. According to this, water cannot be considered as a serious chemical reactant responsible for the chemical modification of the pristine graphene lamellae. Nevertheless, water plays extremely important role in the shungite fortune that will be discussed later.

Graphene hydrogenation has been actively studied both computationally and experimentally. At molecular level, depending on external conditions concerning the fixation of the sheet perimeter and the accessibility of its basal plane to hydrogen atoms either from

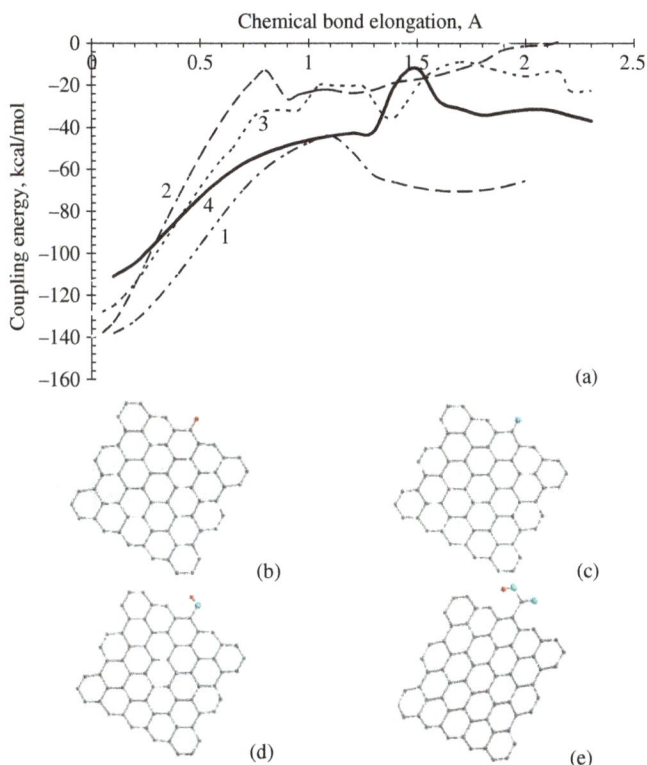

Figure 3. Barrier profiles (a) for desorption of hydrogen (1) and oxygen (2) atoms as well as hydroxyl (3) and carboxyl (4) units from the (5, 5) NGr molecule monohydride and monoxides shown in panels (b)–(e). Here and in all the subsequent figures, dark gray, blue, and red balls mark carbon, oxygen, and hydrogen atoms, respectively (UHF calculations).

one- or two-sides, different graphene hydrides (GHs) are formed [30]. Empirically (see [38] and references therein), the graphene hydrogenation is a difficult task, and the process usually involves such severe conditions as high temperatures and high pressure or employs special devices, plasma ignition, electron irradiation, and so forth. One of the explanations can be connected with the necessity in overcoming a barrier at each addition of the hydrogen atom to the graphene body. Figure 3 demonstrates the dependence of coupling energies of different addends on their distance from the targeted carbon atom at the zigzag edge of the (5, 5) NGr molecule. In the case of hydrogen, the plotting clearly reveals the barrier that constitutes ~13 kcal/mol.

In contrast, graphene oxidation can apply for a role. The reaction is studied thoroughly at different conditions (see reviews [33–35] and references therein) and the achieved level of its understanding is very high. The latter has led the foundation of massive fabrication of a particular 'graphene' that, in fact, is rGO. The oxidation may occur under conditions that provide the shungite derivation in spite of low acidity of the aqueous surrounding but due to long geological time and practically barrierless character of the reaction concerning additions of either oxygen atoms or hydroxyls to the graphene body as seen in Figure 3. As shown, oxidation causes a destruction of both pristine graphite and graphene sheets just cutting them into small pieces [34,39]. Thus, 900 s of continuous oxidation cut a large graphene sheet into pieces of ~1 nm in size [34]. Important that further prolongation of the oxidation does not cause decreasing the size thus stabilizing them at the 1 nm level. This finding allows suggesting that shungite sheets of ~1 nm in size have been formed in the course of geologically prolong oxidation of graphene lamellae derived from the graphe-nization of carbon sediments.

   Answer 3. Numerous experimental studies (see [34–36,40–42]) and recent detailed consideration of GOs from the viewpoint of molecular theory [31] showed that the latter are products of the hetero-oxidant reaction. Three oxidants, among which there are oxygen atoms O, hydroxyls OH, and carboxyls COOH, are the main partners of the process albeit in different ways by participating in the formation of the final product. Figure 4 presents final products of the (5, 5) NGr molecule oxidation that follow from the graphene molecular theory. GO molecules I and II were obtained in the course of the stepwise addition of the above oxidants to the molecule under the conditions that the pristine molecule basal plane is accessible for the oxidants either from the top only (Figure 4(a)) or from both sides (Figure 4(b)). The choice of the preferable oxidant was made following the criterion of the largest per step coupling energy [31]. Plottings in

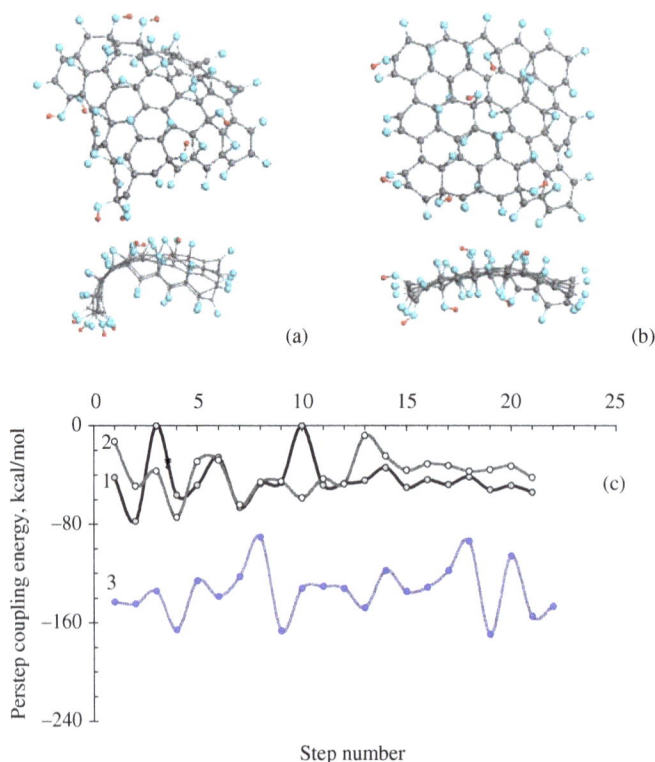

(a)   (b)

(c)

Per step coupling energy, kcal/mol

Step number

Figure 4.   Based on the (5, 5) NGr molecule, oxygen-saturated GOs under one-side (top-down exfoliated (5, 5) GO molecule I) (a) and two-side (GO molecule II) (b) oxidation. (c) Per step coupling energy versus step number for the oxides family covering by (5, 5) GO molecules I under subsequent O- and OH-additions to carbon atoms at either the molecule basal plane (curves 1 and 2) or edges (curve 3) [31] (UHF calculations).

Figure 4(c) correspond to the per step coupling energies that describe the energetics of the oxidants attachment in the course of oxidation of GO molecule I (analogous dependences take place for GO molecule II). Contrary to the common opinion, affirmed in the GO science, carboxyl units, located both at the edge of the molecule and at its basal plane, do not meet the criterion and lose a competition to two other oxidants. Once currently investigated, increasing the molecule size allows for revealing a small fraction of the units in the molecule edge area only (see the next subsection below).

Hydrothermal conditions of the shungite derivation present serious arguments in favor of a hypothesis about the GO origin of the deposits. However, the hypothesis strongly contradicts the atomic percentage of oxygen in the carbon-richest shungite rocks that only constitutes a few percents [43] instead of several tens of percents expected from the saturated GOs. This contradiction forces to think about full or partial reduction of the preliminary formed GO occurred during geological process.

As follows from the plottings in Figure 4(c), GO is characterized by two regions of chemical bonding of oxidants with the graphene body. While the edge region should be attributed to that one of strong chemical bonding, the basal plane, for which the average coupling energy is three times less, is evidently related to the area of a weak coupling. The finding is crucial for the GO reduction showing that the latter concerns primarily basal plane while oxidants located in the framing area may not be removed under conditions of the convenient reduction without destruction of the carbon skeleton. This finding explains the residual oxygen content in reduced rGOs of 8–10% [34–36]. As seen, shungite is no exception to this pattern and with low content of oxygen fits the overall picture related to rGOs.

Usually a synthetic GO reduction occurred when using strong reducing agents that are not available in the natural environment of shungite. However, as has been recently shown, the GO reduction can take place just in water, which only requires a much longer time for the process completion [44]. Evidently, the geological time of the shungite derivation is quite enough for the reduction of pristine GOs in water to take place.

Answer 4. Figure 5 presents the equilibrium structures of the (5, 5) rGO and (11, 11) rGO molecules with 1.3×1.4 nm^2 and 10×10 nm^2 cross-section, respectively. The (5, 5) rGO molecule was obtained in the course of the structure optimization after removing all epoxy and hydroxy groups from the basal plane of the (5, 5) GO molecules shown in Figure 4(a) and (b). The (11, 11) rGO molecule was computationally synthesized in the

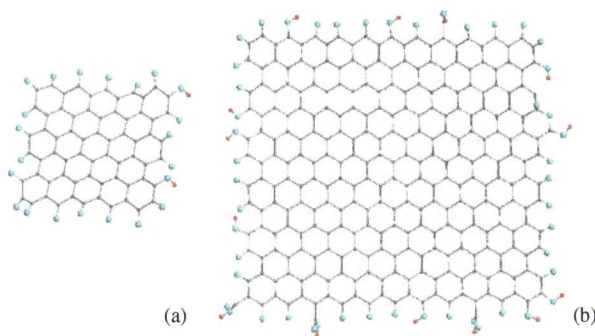

(a)   (b)

Figure 5.   Based on the (5, 5) (a) and (11, 11) (b) NGr molecules, reduced graphene oxides ((5, 5) rGO [31] and (11, 11) rGO, respectively) (UHF calculations).

course of per step oxidation described in detail in Ref. [31]. As seen in Figure 5b, the framing shell of the (11, 11) rGO is replenished with four carboxyl units additionally to carbonyls and hydroxyls as was mentioned earlier.

Due to recovering sp^2 configuration for carbon atoms at basal plane, the (5, 5) rGO molecule noticeably regenerates its planarity, although impaired, especially in the corner areas. Basing on empirical estimation of ~1 nm for a basic shungite graphene-based sheet, this molecule, better rGO nanosheet, can be considered as a reliable configuration of the basic shungite structural unit. However, the atomic percentage of oxygen in the case constitutes 24 at% that is far from the empirical contamination. The controversy may mean that the actual oxygen framing of the rGO nanosheets is not fully saturated. A large oscillation amplitude of the per step coupling energy related to the framing area (see curve 3 in Figure 4(c)) may be one of possible reasons. Actually, the atoms, which correspond to the top part of the plotting, may be removed during the reduction addition-ally to the basal ones. Another reason can be connected with the stability of the rGO sheets that depends on the sheet shape and corner structure, on the one hand, and thermodynamic conditions of the oxygen reservoir, on the other hand [45]. At any rates, the problem needs a further examination.

Answer 5. Assuming that rGO nanosheets generated in aqueous media present the first stage of shungite derivation, let us trace their path from individual molecules to densely packed shungite carbon. Obviously, the path is through successive stages of the sheets aggregation. Empirically was proven that aggregation is characteristic for synthetic GO and rGO sheets as well. Thus, infrared absorption [46] and inelastic incoherent neutron scattering (IINS) [47,48] showed that synthetic GO forms stacked turbostratic structures that confine water. Just recently, a similar picture was obtained for synthetic rGO [49] and shungite [10,49]. Neutron diffraction showed therewith [10,48,49] that the characteristic graphite interfacial distance d_{002} constitutes, in average, ~6.9 Å in the case of GO and reduces to ~3.5 Å for rGO of both synthetic and natural origin, evidently indicating the recovery of the GO carbon carcass planarity due to its reduction. Computationally, it was confirmed [48] that water molecules are comfortably located between the neighboring layers of the GO stacked structure while none of the water molecules can be retained near the rGO basal plane (see Figure 6). Regardless of the molecule starting position, each of them is displayed outside the rGO basal plane area once kept in the vicinity of framing atoms. The finding is well correlated with the rGO short-packed stacked structure leaving the place for water molecules confinement in pores formed by the stacks. The IINS study [10] has proven the pore location of retained water in shungite and confirms that the rGO nanosheets present the basic structural units of shungite thus allowing for suggesting a vision of the next stages of the shungite structure toward shungite rocks.

The characteristic d_{002} diffraction peaks of shungite are considerably broadened in comparison with those of graphite, which allows for estimating approximate thickness of the stacks of ~1.5 nm [10] that corresponds to five- and six-layered rGO sheets. The stacks form the secondary structure of shungite. Aggregating, the stacks are combined in globules, a planar view of one of possible models of which is presented in Figure 7. The globules of a few nanometer in size present the third stage of the shungite fractal structure [1]. The inner surface of the pores is mainly carpeted with oxygen atoms that can willingly hold water molecules in their vicinity as follows from Figure 6. Thus retained water is the main content of the registered IINS spectra [10]. Further aggregation of the

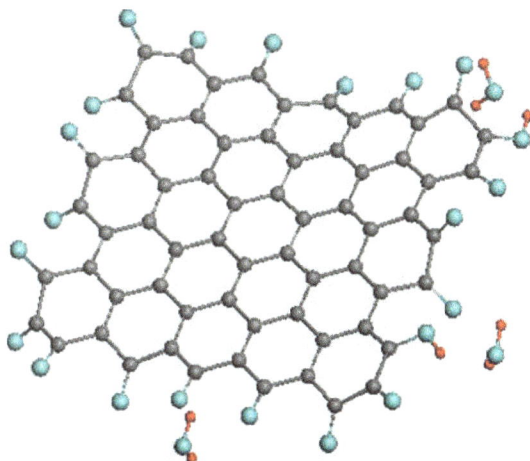

Figure 6.   Equilibrium structure of a complex that involves (5, 5) rGO molecule and three water molecules. Initially, water molecules were accommodated within the basal plane region at 1.8 Å above the plane. The total coupling energy constitutes −27.15 kcal/mol (UHF calculations).

globules leads to the formation of bigger aggregates with lateral dimensions of 20–100 nm. The aggregate agglomeration completes the formation of the fractal structure of shungite.

Answer 6. In fractal structures, which are rich in pores, the pore size is usually tightly connected with the lateral dimension of structural elements involved in the pore formation [50]. Moreover, the larger is variety of the elements size and structure, the bigger is distribution of the pores over size. In view of this, the different size of the multilevel shungite fractal structural elements evidently predetermines different sizes of shungite pores. Thus, as seen in Figure 7, the irregular distribution of stacks in space causes the formation of different interglobular pores, linear dimensions of which are comparable with those of stacks. Actually, one of the linear sizes of the pores formed by rGO nanosheets is determined by linear dimensions of the latter while two others are defined by the thickness of the sheet stacks. Therefore, basic rGO nanosheets and their stacks are responsible for shungite pores of 2–5 nm in size. Following this line, globules may form pores up to 10 nm while extended aggregates of globules obviously form pores of a few tens nanometer and bigger. This presentation well correlates with the SANS experimental data that evidence the presence of two sets of pores in shungite in the range of 2–10 nm and above 100 nm [6]. Taking together, the multitier of structural elements and various porosity make the fractal structure of shungite good self-consistent.

4.   Conclusive remarks

The main idea discussed in the current article is that molecular chemistry at the graphenization stage lays the foundation of the difference in graphite and shungite derivation under natural conditions during the geological time. The concept showed the way of checking this suggestion by exhibiting chemical reactions that are responsible for the deposits derivation as well as by simulating final products of the relevant reactions. Based

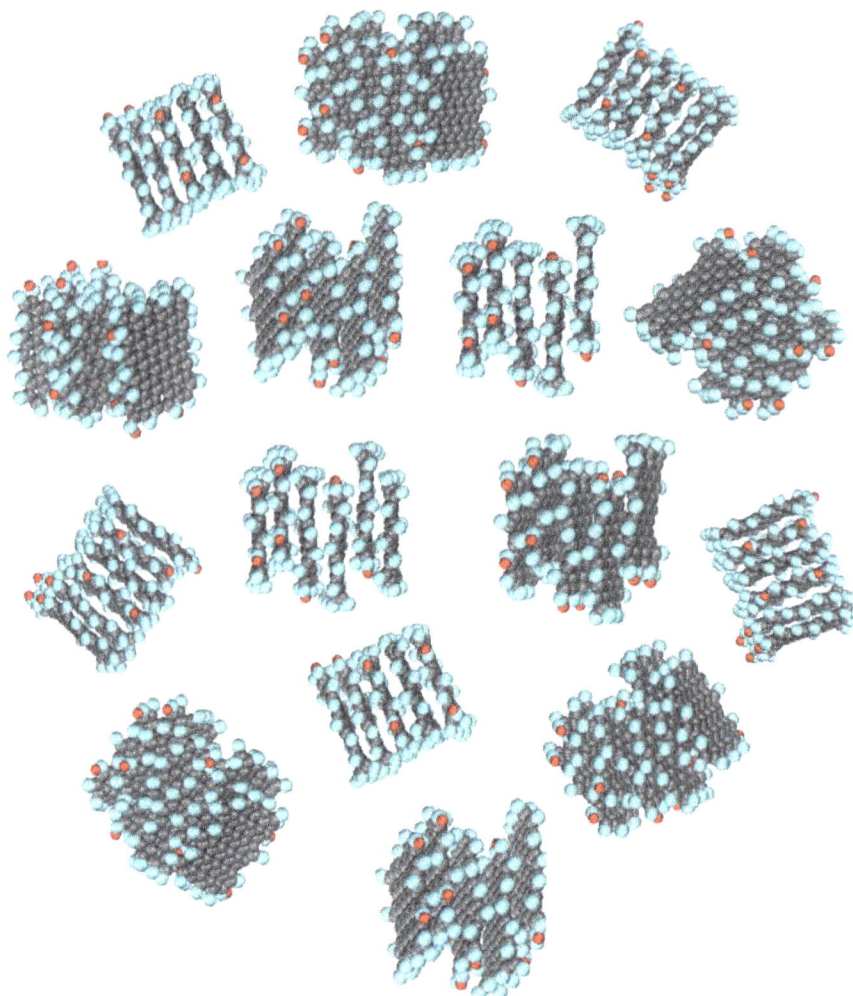

Figure 7.   A planar view of a randomly designed shungite globe formed by five- and six-layered rGO stacks. Vertical and horizontal dimensions are of ~7 nm. The average interfacial distance d_{002} in the stacks and the stack length constitute ~0.35 nm, ~1.4 nm, and ~1.7 nm, respectively.

on the theory of graphite genesis [23], the reaction participants involve molecular objects simulating fragments of polycondensed benzenoid molecules (carbon substrate, or naked graphene lamellae), on the one hand, and molecular (water, carboxyl, hydroxyl) as well as atomic (hydrogen, oxygen) species (chemical reactants), on the other hand. As follows from the molecular theory of graphene [32] and general grounds of the chemistry for nanoscience [51], the naked graphene lamellae are kinetically unstable since covalent bonds are cut at their edges. Such species will try to heal themselves and external molecules may stabilize them. In full agreement with our suggestion, Hoffmann continues [51] that 'too great stabilization will inhibit growth; too little stabilization will not prevent from collapse to the solid'. The difference in the graphene lamellae stabilization was the second basic idea of our approach. Thus, the great stabilization of the lamellae results in the shungite formation while the little one provides derivation of graphite deposits. The stabilizing reactions are controlled by the reactants coming on and off the pristine naked graphene lamellae while both thermodynamic (Gibbs energies) and kinetic (activation energies) factors matter in the dynamic process. Shungite is suggested to be formed as a result of a balance of a number of multi-reactant processes, each governed by its own thermodynamics and kinetics. The presence of other elements such as silicon and metals undoubtedly influences the deposit formation. Actually, the Karelian deposits of shungite are non-uniform by the carbon content, value of which changes from 3% to 98 wt% [1]. Silica is the main partner of the mixed depositions. However, speaking about shungite as carbon allotrope, we imply a particular shungite rock from the Shun'ga deposit with the highest carbon content up to 98.8% [43] for which the presence of other Earth element is negligible [52].

The two main concepts have been considered in the article addressing oxidation/reduction reactions that govern chemical modification of the pristine graphene lamellae. Basing on a wide experience gained for graphene chemistry in the laboratory conditions and an extended computational system experiment performed earlier [31], the oxidation/reduction reactions are shown to have a big privilege against hydration and hydrogenation of graphene. The two reactions work simultaneously but serving different purposes: oxidation stabilizes the growth of pristine graphene lamellae thus determining their size, while reduction releases the oxygenated nanosheets from weakly bound reactants located through over basal plane and partially at the sheets perimeter leaving remaining in the sheet circumference thus preserving their stabilization. This conclusion is in full agreement with shungite empirical data related to exhibiting (1) ~1 nm planar-like sheets of rGO as the basic structural element of the macroscopic shungite structure and (2) remaining low content of oxygen in the most carbon-pure shungite samples.

Shungite is formed in aqueous surrounding and although water molecules do not act as active chemical reactants at the oxidation stage, they play a very important role at the reduction [44] as well as in composing shungite as a solid. First, the slow rate of reduction evidently favors the accumulation of rGO nanosheets during a long shungite geological story. Should not to exclude also a possible chemical modification of the sheets framing due to their long stay in hot water. Second, water molecules fill the pores, helping to strengthen the framework of fractal shungite carbon. These processes, when taken together, have led to the creation of a unique natural pantry of nanoscale rGO.

Acknowledgments

The work was supported by Basic Research Program, RAS, Earth Sciences Section-5 and grant RFBI 13-03-00422. The authors are grateful to N. Popova for assisting in the barrier calculations.

References

[1] N.N. Rozhkova, *Shungite Nanocarbon (in Russian)*, Karelian Research Centre of RAS, Petrozavodsk, 2011.

[2] N.N. Rozhkova, A.V. Gribanov, and M.A. Khodorkovskii, *Water mediated modification of structure and physical chemical properties of nanocarbons*, Diamond Relat. Mater. 16 (2007), pp. 2104–2108.

[3] N.N. Rozhkova, G.I. Emel'yanova, L.E. Gorlenko, A. Jankowska, M.V. Korobov, and V.V. Lunin, *Structural and physico-chemical characteristics of shungite nanocarbon as revealed through modification*, Smart Nanocomposites 1 (2010), pp. 71–90.

[4] S.P. Rozhkov, N.N. Rozhkova, G.A. Suhanova, A.G. Borisova, and A.G. Goryunov, *DSC data on interaction of carbon nanoparticles with protein molecules*, in *Nanoparticles in Condensed Media*, P.A. Vityaz, ed., Publishing Center BSU, Minsk, 2008, pp. 134–139.

[5] N.N. Rozhkova, L.E. Gorlenko, G.I. Emel'yanova, M.V. Korobov, V.V. Lunin, and E. Osawa, *The effect of ozone on the structure and physico-chemical properties of ultradisperse diamond and shungite nanocarbon elements*, Pure Appl. Chem. 81 (2009), pp. 2093–2105.

[6] M.V. Avdeev, T.V. Tropin, V.L. Aksenov, L. Rosta, V.M. Garamus, and N.N. Rozhkova, *Pore structures in shungites as revealed by small-angle neutron scattering*, Carbon 44 (2006), pp. 954–961.

[7] V.V. Kovalevski, N.N. Rozhkova, AZ. Zaidenberg, and A.N. Yermolin, *Fullerene-like structures in shungite and their physical properties*, Mol. Mat. 4 (1994), pp. 77–80.

[8] N.N. Rozhkova, G.I. Yemel'yanova, L.E Gorlenko, A.V. Gribanov, and V.V. Lunin, *From stable aqueous dispersion of carbon nanoparticles to the clusters of metastable carbon of shungites*, Glass Phys. Chem. 37 (2011), pp. 621–626.

[9] N.N. Rozhkova, *Aggregation and stabilization of shungite carbon nanoparticles*, Ecol. Chem. 4 (2012), pp. 240–251.

[10] E.F. Sheka, N.N. Rozhkova, I. Natkaniec, and K. Holderna-Natkaniec, *Inelastic neutron scattering study of reduced graphene oxide of natural origin*, JETP Lett. 99 (2014).

[11] F.K. Smidt, *Fractals in Physical Chemistry of Heterogeneous Systems and Processes (in Russian)*, Irkutsk University, Irkutsk, 2000.

[12] M.A. Pimental, G. Dresselhaus, M.S. Dresselhaus, L.A. Cancado, A. Jorio, and R. Sato, *Studying disorder in graphite-based systems by Raman spectroscopy*, Phys. Chem. Chem. Phys. 9 (2007), pp. 1276–1290.

[13] D.C. Elias, R.R. Nair, T.M.G. Mohiuddin, S.V. Morozov, P. Blake, M.P. Halsall, A.C. Ferrari, D.W. Boukhvalov, M.I. Katsnelson, A.K. Geim, and K.S. Novoselov, *Control of graphene's properties by reversible hydrogenation: Evidence for graphane*, Science 323 (2009), pp. 610–613.

[14] A.B.H. Amara, *X-ray diffraction, infrared and TGA/DTG analysis of hydrated nacrite*, Clay Minerals 32 (1997), pp. 463–470.

[15] L. Li, G. Wu, G. Yang, J. Peng, J. Zhao, and J.-J. Zhu, *Focusing on luminescent graphene quantum dots: current status and future perspectives*, Nanoscale 5 (2013), pp. 4015–4039.

[16] I.M. Belousova, I.M. Kislyakov, D.A. Videnichev, N.N. Rozhkova, and A.G. Tupolev. *Shungite carbon as a material for optical limiting of high intensity laser radiation in the visible and near infrared region*. Abstracts, 9th Biennal Int Workshop on Fullerenes and Atomic Clusters, (St. Petersburg, Russia) Ioffe Phys-Techn Inst. RAS. (2009), p. 222.

[17] N.V. Kamanina, S.V. Serov, N.A. Shurpo, and N.N. Rozhkova, *Photoinduced changes in refractive index of nanostructured shungite-containing polyimide systems*, Tech. Phys. Lett. 37 (2011), pp. 949–951.

[18] N.V. Kamanina, S.V. Serov, N.A. Shurpo, S.V. Likhomanova, D.N. Timonin, P.V. Kuzhakov, N.N. Rozhkova, I.V. Kityk, K.J. Plucinski, and D.P. Uskokovic, *Polyimide-fullerene nanostructured materials for nonlinear optics and solar energy applications*, J. Mater. Sci.: Mater. Electron 23 (2012), pp. 1538–1542.

[19] B.S. Razbirin, N.N. Rozhkova, E.F. Sheka, D.K. Nelson, A.N. Starukhin, and A.S. Goryunov. *Fractals of graphene quantum dots in photoluminescence of shungite*. arXiv:1308.2569v2 [cond-mat.mes-hall] 2013.

[20] A.S. Goryunov, A.G. Borisova, and S.P. Rozhkov. *Raman spectroscopy of bioconjugates of bovine serum albumin and shungite nanocarbon*. Proc. Karelian Res. Center RAS, Exp. Biol. Ser. 2 (2012), pp. 154–158.

[21] S. Rozhkov, G. Sukhanova, A. Borisova, N. Rozhkova, and A. Goryunov, *Effects of carbon nanoparticles on protein thermostability revealed by DSC and ESR spin-labelling methods*, Ann. World Conf. Carbon Biarritz (France) (2009), p. 201.

[22] V.A. Melezhik, A.E. Fallick, M.M. Filippov, A. Lepland, D.V. Rychanchik, Y.E. Deines, P.V. Medvedev, A.E. Romashkin, and H. Strauss, *Petroleum surface oil seeps from a Paleoproterozoic petrified giant oilfield*, Terra Nova. 21 (2009), pp. 119–126.

[23] B. Kwiecińskaa and H.I. Petersen, *Graphite, semi-graphite, natural coke, and natural char classification – ICCP system*, Int. J. Coal. Geol. 57 (2004), pp. 99–116.

[24] C.A. Landis, *Graphitization of dispersed carbonaceous material in metamorphic rocks*, Contrib. Mineral Petrol. 30 (1971), pp. 34–45.

[25] C.F.K. Diessel and R. Offler, *Change in physical properties of coalified and graphitized phytoclasts with grade of metamorphism*, Neues. Jahrb. Mineral Monatsh H. 1 (1975), pp. 11–26.

[26] A. Bianco, H.-M. Cheng, T. Enoki, G. Yu, R.H. Hurt, and N. Koratkar, *All in the graphene family – a recommended nomenclature for two-dimensional carbon materials*, Carbon 65 (2013), pp. 1–6.

[27] E.F. Sheka and L.A. Chernozatonskii, *Chemical reactivity and magnetism of graphene*, Int. J. Quant. Chem. 110 (2010), pp. 1938–1946.

[28] E.F. Sheka and L.A. Chernozatonskii, *Broken spin symmetry approach to chemical susceptibility and magnetism of graphenium species*, J. Exp. Theor. Phys. 110 (2010), pp. 121–132.

[29] E.F. Sheka, *Fullerenes: Nanochemistry, Nanomagnetism, Nanomedicine, Nanophotonics*, CRC Press, Taylor and Francis Group, Boca Raton, FL, 2011.

[30] E.F. Sheka and N.A. Popova, *Odd-electron molecular theory of the graphene hydrogenation*, J. Mol. Mod. 18 (2012), pp. 3751–3768.

[31] E.F. Sheka and N.A. Popova, *Molecular theory of graphene oxide*, Phys. Chem. Chem. Phys. 15 (2013), pp. 13304–13322.

[32] E.F. Sheka, *Computational strategy for graphene: Insight from odd electrons correlation*, Int. J. Quant. Chem. 112 (2012), pp. 3076–3090.

[33] D.R. Dreyer, S. Park, CW. Bielawski, and R.S. Ruoff, *The chemistry of graphene oxide*, Chem. Soc. Rev. 39 (2010), pp. 228–240.

[34] W. Hui and H.H. Yun, *Effect of oxygen content on structures of graphite oxides*, Ind. Eng. Chem. Res. 50 (2011), pp. 6132–6137.

[35] T. Kuila, S. Bose, A.K. Mishra, P. Khanra, N.H. Kim, and J.H. Lee, *Chemical functionalization of graphene and its applications*, Prog. Mat. Sci. 57 (2012), pp. 1061–1105.

[36] A.M. Dimiev, L.B. Alemany, and J.M. Tour, *Graphene oxide. Origin of acidity, its instability in water, and a new dynamic structural model*, ACS Nano. 7 (2013), pp. 576–584.

[37] AI. Golubev, AE Romashkin, and D.V. Rychanchik, *Relation of carbon accumulation to Paleoproterozoic basic volcanism in Karelia (Jatulian-Ludicovian transition)*, Geol. Useful Minerals Karelia 13 (2010), pp. 73–79.

[38] W. Chen, Z. Zhu, S. Li, C. Chen, and L. Yan, *Efficient preparation of highly hydrogenated graphene and its application as a high-performance anode material for lithium ion batteries*, Nanoscale 4 (2012), pp. 2124–2129.

[39] S. Pan and I.A. Aksay, *Factors controlling the size of graphene oxide sheets produced via the graphite oxide route*, ACS Nano. 5 (2011), pp. 4073–4083.

[40] V. Singh, D. Joung, L. Zhai, S. Das, S.I. Khondaker, and S. Seal, *Graphene based materials: Past, present and future*, Prog. Mat. Sci. 56 (2011), pp. 1178–1271.

[41] A.L. Ivanovskii, *Graphene-based and graphene-like materials*, Russ. Chem. Rev. 81 (2012), pp. 571–605.

[42] Y.Z. By, M. Shanthi, C. Weiwei, L. Xuesong, W.S. Ji, J.R. Potts, and R.S. Ruoff, *Graphene and graphene oxide: Synthesis, properties, and applications*, Adv. Mater. 22 (2010), pp. 3906–3924.

[43] M.M. Filippov, *Shungitonosnyje porody Onezhskoj struktury (Shungite Rocks of Onega's Structure) (Russian)*, Karelian Research Centre of RAS, Petrozavodsk, 2002.

[44] K.-H. Liao, A. Mittal, S. Bose, C. Leighton, K.A. Mkhoyan, and C.W. Macosko, *Aqueous only route toward graphene from graphite oxide*, ACS Nano. 5 (2011), pp. 1253–1258.

[45] H. Shi, L. Lai, I.K. Snook, and A.S. Barnard, *Relative stability of graphene nanosheets under environmentally relevant conditions*, J. Phys. Chem. C. 117 (2013), pp. 15375–15382.

[46] M. Acik, C. Mattevi, C. Gong, G. Lee, K. Cho, M. Chhowalla, and Y.J. Chabal, *The role of intercalated water in multilayered graphene oxide*, ACS Nano. 4 (2010), pp. 5861–5868.

[47] A. Buchsteiner, A. Lerf, and J. Pieper, *Water dynamics in graphite oxide investigated with neutron scattering*, J. Phys. Chem. B. 110 (2006), pp. 22328–22338.

[48] I. Natkaniec, K. Druzbicki, S.P. Gubin, K. Holderna-Natkaniec, S.V. Tkachev, and E.F. Sheka, *IINS and DFT studies of vibrational spectra of water retained in graphene oxide.* 2nd satellite

workshop of ICNS 2013 on Dynamics of Molecules and Materials, University of Glasgow, Glasgow, Scottland, 2013, p. 21.

[49] E.F. Sheka, N.N. Rozhkova, I. Natkaniec, K. Holderna-Natkaniec, and K. Druzbicki, Water dynamics in shungite with inelastic neutron scattering, International Conference on Advanced Carbon Nanostructure, Ioffe Phys-Techn Inst RAS, St. Petersburg, Russia, 2013, p. 68.

[50] J.-F. Gouyet, *Physics and Fractal Structures*, Masson Springer, Paris/New York, 1996.

[51] R. Hoffmann, *Small but strong lessons from chemistry for nanoscience*, Ang. Chem. Int. Ed. 52 (2013), pp. 93–103.

[52] N.N. Rozhkova, A.Z. Zaidenberg, and A.I. Golubev. *Challenges to chemical geology. Refereed papers from MAEGS-10. Prague. Czech Geological Survey 1998; MAEGS-10.* Prague, pp. 137–143.

3

Design of double negativity elastic metamaterial

Y.C. Su and C.T. Sun*

School of Aeronautics and Astronautics, Purdue University, West Lafayette, IN 47907, USA

A metamaterial model that possesses simultaneously negative effective mass density and negative effective Young's modulus is proposed in this study. Dispersion curves and dynamic responses of the model are investigated. In the double negative frequency region, it is demonstrated that the phase velocity is negative. In addition, it was found that the band gap region of the metamaterial can be predicted accurately by taking parts of single-unit cell to analyze the steady-state response. The design is also fabricated by a 3D printer.

Keywords: acoustic metamaterial; double negativity; local resonance; negative mass; negative modulus

1. Introduction

Materials with periodic structures are well known for their use in filtering waves. However, since its underlying principle for blocking waves is to apply impedance variation, the wavelength of the gap must be the same order as the lattice constant of the periodic structure. In the field of mechanics, blocking low-frequency waves is generally more desired than blocking those with high frequencies, thus rendering the periodic structure impractical due to its large lattice constant. To overcome the limitation of low-frequency wave mitigation in periodic structures [1], Liu et al. first applied local resonance to fabricate an acoustic metamaterial with negative effective mass [2]. Using lead balls coated with silicone rubber, they found that the metamaterial could filter waves with wavelengths two orders larger than its lattice constant. Since then, acoustic metamaterials have become a hot topic among researchers [3–5]. In the area of elastic metamaterials, Huang and Sun [6] connected resonators using rigid and massless trusses to create a negative effective Young's modulus (NYM) metamaterial.

By combing negative effective mass and negative effective modulus, one can obtain 'double negativity metamaterials'. Although a number of double negativity (DN) acoustic metamaterials have been investigated [7–15], there are few reports in literature on DN elastic metamaterials. Among the elastic metamaterials that have been published, Wang theoretically investigated the mechanism of negative mass and/or negative modulus of lumped mass models [16]. Jin et al. proposed a broadband DN metamaterial using active control [17]. Pope proposed the designs of DN elastic metamaterials using electrical–mechanical circuit analogies [18]. Soon after, Pope and Laalej performed the first active DN elastic metamaterial

---

*Corresponding author. Email: sun@purdue.edu

experiment [19]. Using composite materials to make elastic metamaterials, Wu et al. [20] and Lai et al. [21] numerically validated the DN. Liu et al. [22] proposed the idea of chiral metamaterials and validated it through theoretical analysis and numerical simulation. Expanding upon that concept, Zhu et al. [23] created a single-phase solid based on the lumped mass model. The present work is based on the lumped mass model proposed in [24], and the metamaterial was fabricated as a single-phase solid using a 3D printer.

2. Metamaterial models and design process

The metamaterial design in this study is based on the lumped mass model [24]. Since the model is still in the theoretical stage, this study realizes it using an Eden 350 3D printer (Stratasys Ltd., Eden Prairie, MN, USA).

Figure 1 shows the design process for determining the dimensions of NYM resonator. To change the lumped mass model into a single-phase solid, we used beam bending to replace the vertical spring. After performing parametric studies on the length and thickness of the horizontal beam, we found that a beam with a length of 30 mm and width of 1 mm can provide resonance at the frequencies of interest. It is noted that due to geometric complexity it is impractical to use theoretical analysis to calculate local resonance. Instead, we numerically estimate the local resonances by the finite element analysis. Since the inclined frame is only used for transmitting force, the width of the frame is designed as 1.5 mm to make the mode of frame rotation/stretching occur at higher frequencies. For the mass of the NYM resonators, we found that the arrow shape gives the best spatial usage.

After determining the geometry, we proceeded to adjust the material properties in each part Furthermore, since the thickness is converted into effective material properties in the simulations, the range of effective material properties is set to be $E = 5$–20 GPa,

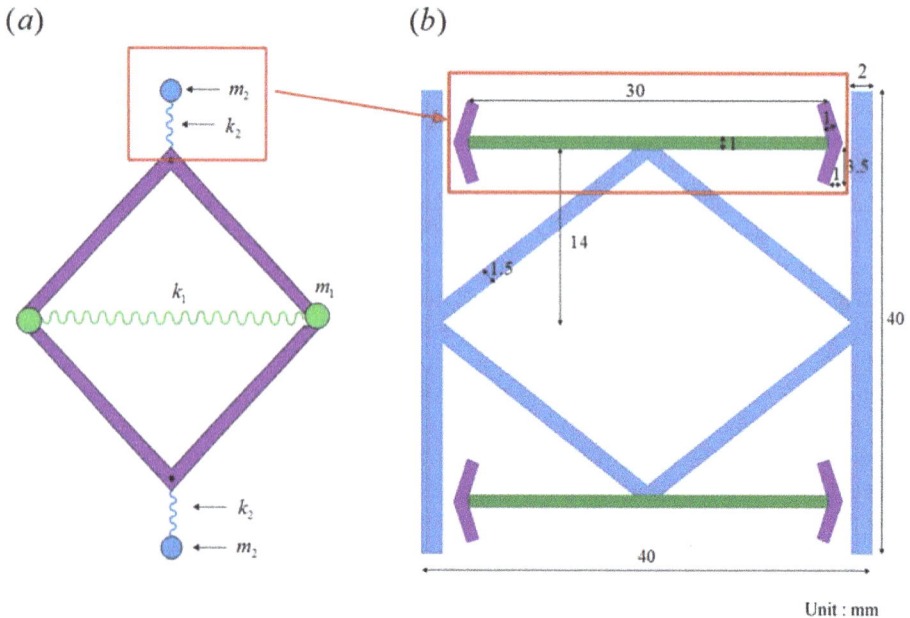

Unit : mm

Figure 1.   Design of the NYM resonator based on the lumped mass model: (a) NYM lumped mass model; and (b) NYM submodel.

Table 1.   Material parameters for the NYM submodel.

Marked area	Mass density (kg/m^2)	Young's modulus (GPa)	Poisson's ratio
Blue	1300	5.0000	0.49
Purple	2350	9.0385	0.49
Green	1742	6.7000	0.49

ρ = 1300–5200 kg/m^3 (E = 5 GPa and ρ = 1300 kg/m^3 are the material properties of the high-temperature photopolymer ink that was used for the 3D printing). Figure 1b and Table 1 show the finalized parts of the unit cell of the metamaterial that produces NYM. The characteristic length of the unit cell, L, is 40 mm.

In the design of the negative effective mass density (NMD) resonator, as shown in Figure 2, we use the same procedure to investigate the length of the vertical beam, which serves as the horizontal spring of the lumped mass model. The shape of the NMD resonator is also determined according to the efficiency of spatial usage. Figure 2b and Table 2 show the finalized parts of the unit that produces NMD. The vertical beams supporting two triangular-shaped masses serve as two horizontal springs. The top and

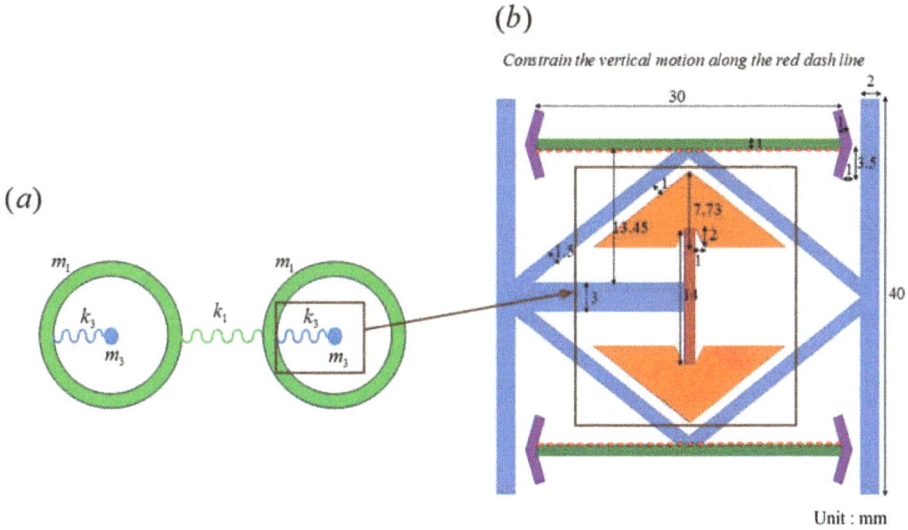

Figure 2.   Design of the NMD resonator based on the lumped mass model: (a) NMD lumped mass model; (b) NMD submodel.

Table 2.   Material parameters for the DN metamaterial and NMD submodel.

Marked area	Mass density (kg/m^2)	Young's modulus (GPa)	Poisson's ratio
Blue	1300	5.0000	0.49
Purple	2350	9.0385	0.49
Green	1742	6.7000	0.49
Orange	4000	15.3846	0.49
Red	3354	12.9000	0.49

bottom horizontal beams are restrained from vertical motion. The characteristic length of the unit cell, L, is also 40 mm.

The DN model is the same as the NMD model except that the horizontal beams are free to move vertically. The material parameters for the DN model are also listed in Table 2. In order to have a DN frequency region, the dimensions for the NYM and NMD resonators are designed to have overlapping band gaps.

3. Dispersion curves for the DN model

Dispersion curves show the frequencies at which harmonic waves can propagate without amplitude attenuation. Frequencies between two dispersion curves locate the stopping band.

Considering 1D longitudinal harmonic wave propagation, the displacement can be expressed as

$$u(x,t) = Ae^{iq(x-ct)} = Ae^{iqx}e^{-i\omega t} \tag{1}$$

where c is the phase velocity and q is the wave number. By selecting two locations as nodes (zero displacement) in simulation, Equation (1) can be treated as a free vibration problem of a finite length body with fixed end conditions. As demonstrated in Figure 3, if the harmonic wave denoted by the blue line is the longest wavelength propagating within the infinite number of metamaterial unit cells, it can be modeled by setting the harmonic waves with fixed end condition at the left and right edges of either 400 unit cell, 200 unit cell or 100 unit cell metamaterial strip. In this study, we found that the 100 unit cell strip is long enough to model the dispersion curves of the infinite metamaterial chain.

To model the infinite number of unit cells metamaterial by the 100 unit cells, we constrained the horizontal movements of the left and right boundaries of the 100 unit cell metamaterial strip, so that these two boundaries became nodal points. The result of the free vibration analysis yielded the natural frequencies and mode shapes, from which the corresponding wave numbers were obtained.

We used ABAQUS commercial software (finite element software by Abaqus Inc., Providence, RI, USA) for finite element analysis. Since beam bending is used as the

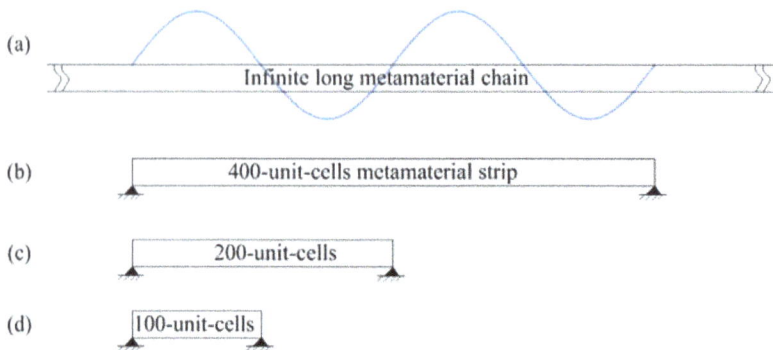

Figure 3. (a) Infinite long metamaterial chain with a harmonic wave (blue line); (b) 400 unit cell metamaterial strip modeling; (c) 200 unit cell metamaterial strip modeling; (d) 100 unit cell metamaterial strip modeling.

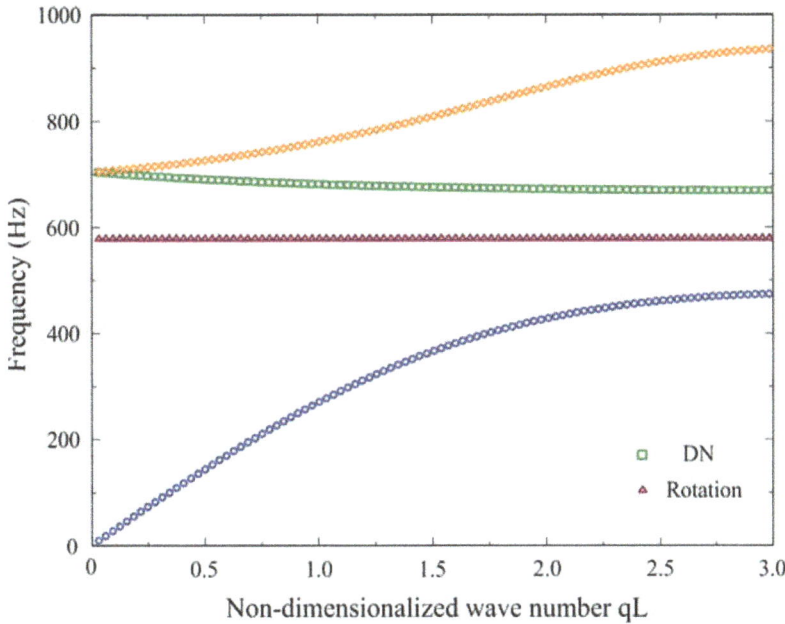

Figure 4.   Dispersion curves for the DN metamaterial.

resonator's stiffness, accuracy for the beam bending mode estimation is crucial. For element-type selection, the incompatible mode element (CPS4I) was used owing to its efficiency for bending calculation. Four CPS4I elements are placed along the thickness of the beam to achieve element number convergence.

Figure 4 shows the dispersion curve of the DN metamaterial. Each point on the dispersion curve represents an eigenmode (calculated by ABAQUS) of the 100 unit cell long metamaterial strip.

As shown in Figure 4, the frequencies in the DN region are 668–703 Hz. The DN region is recognized by negative phase velocity [24]. It is noted that the narrow passing band, 579.28–579.37 Hz, is owing to the rotational mode of the horizontal beam about the joint.

4.   Wave propagation in the DN model

The negative phase velocity is one of the characteristics of DN metamaterials. This phenomenon can be observed from wave propagation. Hence, we also investigated the dynamic behavior of the DN metamaterial to demonstrate the negative phase velocity.

Since a large number of finite elements are used for beam bending calculation in each unit cell, accumulated numerical error is an issue for ABAQUS Dynamic Explicit packet. To deal with this numerical problem, Dynamic Implicit version is used instead. A long metamaterial strip (200 unit cells) was built to delay wave reflection, and a sinusoidal displacement with a frequency of 685 Hz was applied at the first unit cell (metamaterial strip with wave speed of 67.7966 m/s and wavelength of 340.4255 mm at 685 Hz). As before, the vertical displacements of the upper and lower boundaries of the unit cells are constrained because of the plane wave assumption. The negative phase velocity is observed by tracking the responses of three proximal unit cells.

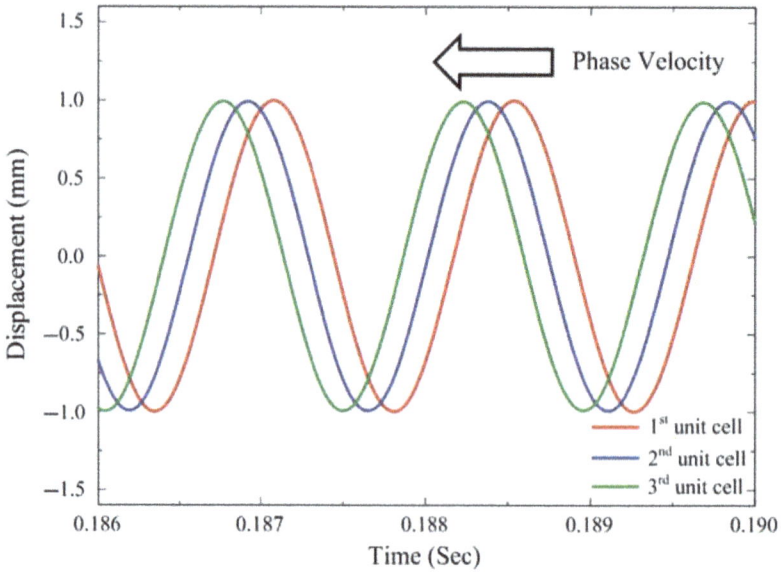

Figure 5.   Demonstration of negative phase velocity.

Since the initial condition is quiescent before the application of the sinusoidal motion at the frequency of 685 Hz, the steady-state frequency is reached after a period of time. As shown in Figure 5, the negative phase velocity is observed after 0.186 s. It is noted that the wave amplitude has no decay since 685 Hz is within the DN frequency range and there is no damping assumption in the simulation.

5.   Band gap region determination by single-unit cell

To ensure that the negative phase velocity of the DN model is caused by overlapping NYM and NMD band gaps, we present a procedure for band gap prediction using single-unit cells (NYM submodel and NMD submodel).

The NYM submodel is shown in Figure 1b and Table 1. The frequency-dependent effective modulus of a metamaterial can be obtained by applying symmetrical loading $F_0 \cos \omega t$ on the representative unit cell, as shown in Figure 6. The corresponding displacements at the right and left edges of the NYM unit cell are $u = +u_0 \cos \omega t$ and $u = -u_0 \cos \omega t$, respectively. The change of the unit cell dimension is $2u_0$, and the dynamic stiffness of the unit cell is given by

$$k = \frac{F_0}{2u_0} \tag{2}$$

It is noted that the stiffness of the unit cell defined by Equation (2) is proportional to the effective Young's modulus of the metamaterial. The NYM band gap region is defined as the frequency range in which k is negative. By using this procedure, the band gap region is 588.6–704.0 Hz (see Figure 7).

For comparison, we also used 100 unit cells to calculate the dispersion curves for the NYM submodel, as shown in Figure 8, from which the NYM band gap region is

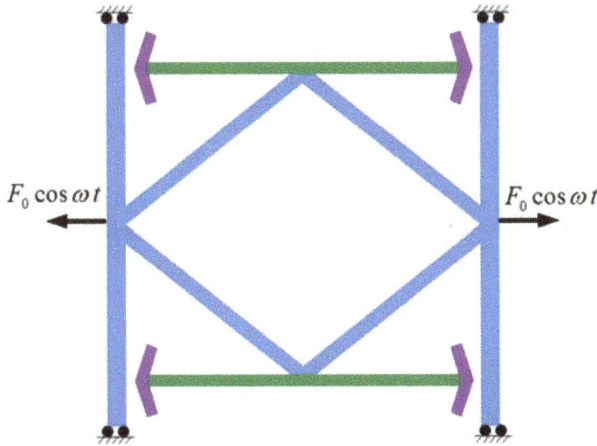

Figure 6.   NYM unit cell subjected to symmetrical loading condition for testing dynamic stiffness (effective modulus).

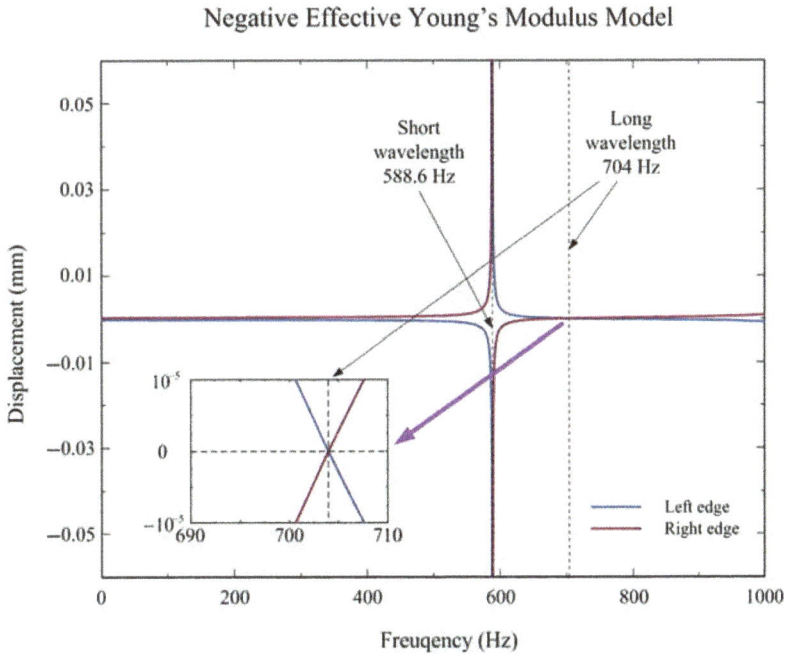

Figure 7.   Displacement frequency spectrum.

588–704 Hz, which agrees with the band gap obtained from the unit cell. The narrow passing band, 576.01–576.61 Hz, is the horizontal beam rotational mode. In Figure 7, the lower bound of the NYM band gap shows out-of-phase motion on the left and right edges of unit cell, which indicates short wavelength corresponding to dispersion curve (the shortest wavelength in dispersion curve is 2L). On the contrary, the upper bound of the NYM band gap shows in-phase motion (zero displacement) on the left and right edges of unit cell, which corresponds to the longest wavelength of optical branch in dispersion curve.

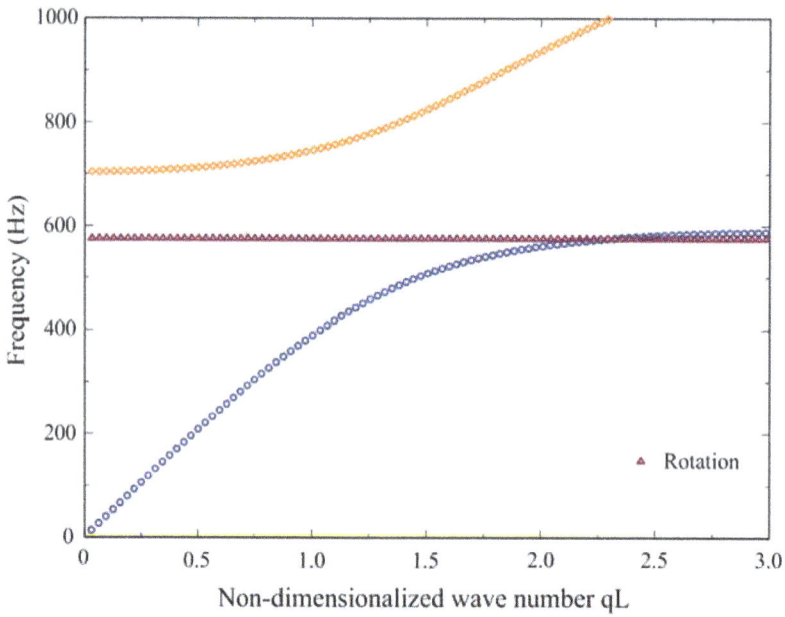

Figure 8.   Dispersion curves for the NYM submodel.

The NMD submodel is made by suppressing the bending motions of the two green horizontal beams, as shown in Figure 2b. The effective mass density of the NMD submodel can be obtained from the dynamic response of the representative unit cell loaded as shown in Figure 9. We also set uniform displacement as an unknown on the left and right edges of the NMD unit cell.

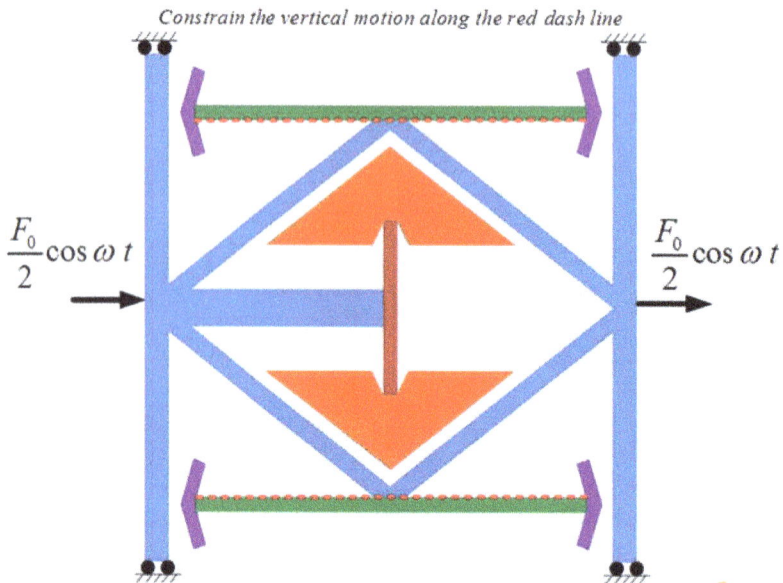

Figure 9.   NMD unit cell subjected to antisymmetrical loading for testing effective mass.

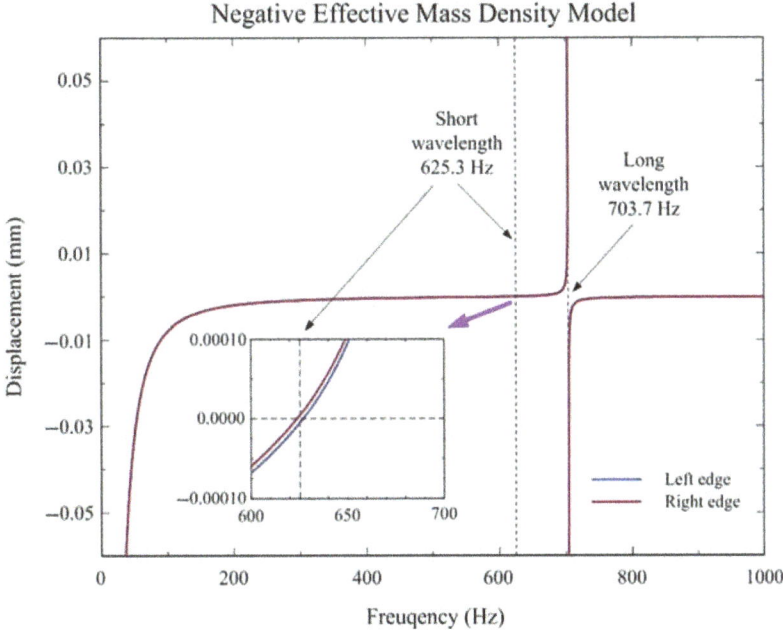

Figure 10.   Displacement frequency spectrum.

The effective mass of the unit cell is given by

$$m_{\text{eff}} = -\frac{F_0}{u_0 \omega^2} \tag{3}$$

From the result of the finite element simulation, Figure 10 shows the displacement spectrum. The band gap region is defined as the frequency range where u_0 becomes positive, 625.3–703.7 Hz, in which m_{eff} has negative value.

Similarly, for validation, we also use the 100 unit cells of the NMD submodel to evaluate the band gap region from the dispersion curves, as shown in Figure 11. The NMD band gap region is 625–705 Hz. Again, this agrees with that obtained from the unit cell of the NMD submodel. In Figure 10, the lower bound of NMD band gap shows out-of-phase motion on the left and right edges of NMD unit cell, which indicates shortest wavelength in the acoustic branch of dispersion curve. On the contrary, the upper bound of NMD band gap shows in-phase motion on the left and right ends of unit cell, which corresponds to the longest wavelength in the optical branch of dispersion curve.

A similar numerical method of effective medium theory can be found in [21,22]. The only variation is the calculation of average stress and strain in the unit cell, and their unit cell predictions have not been investigated quantitatively in the studies.

6.   3D printing fabrication

The metamaterial proposed in this study can be fabricated by a 3D printer. Since 3D printing is three-dimensional with varying thicknesses, we use the following principles to convert the 2D metamaterial model with a uniform thickness to a 3D specimen.

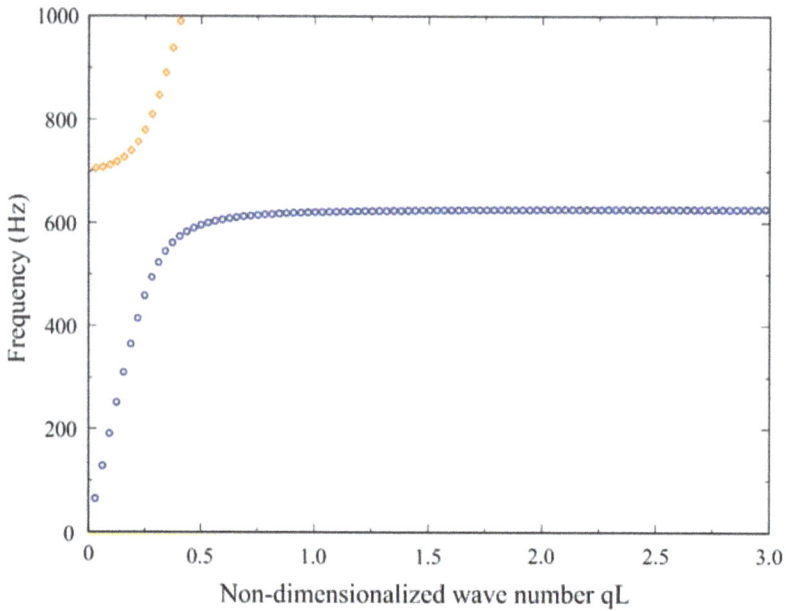

Figure 11.   Dispersion curve for the NMD submodel.

(1) The effective Young's modulus in a 2D model is proportional to the thickness of the specimen.
(2) Mass density is proportional to the thickness of the specimen.

Figure 12 shows the DN metamaterial specimen, which was fabricated using an Eden 350 3D printer, based on the calculations in this study. The fabrication process can be divided into three steps: preprocessing, production, and support removal.

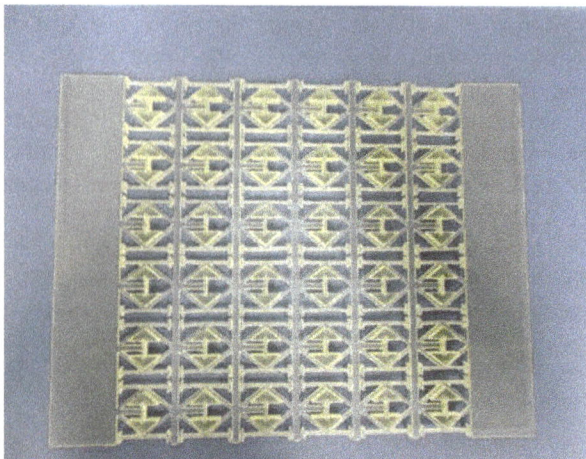

Figure 12.   A DN metamaterial made with a 3D printer.

In the preprocessing step, we built the specimen into a 3D form of various thicknesses using the CAD software, CATIA, and input the CAD file into the Eden 350. The 3D printing jet then dripped liquid photopolymer layer-by-layer to build the 3D model. We used high-temperature material RGD 525 as the photopolymer ink to produce the specimen. The liquid photopolymer was then cured by UV light during processing. It is noted that the 3D printer also dripped removable gel-like supporting materials on the parts with weaker structures to stabilize them. Finally, we used a water jet to remove the supporting materials and obtain a metamaterial specimen with a 16 μm layer resolution and 0.1 mm accuracy.

7. Conclusion

A DN elastic metamaterial has been designed. The dynamic behavior of the elastic metamaterials was studied. The negative phase velocity of the DN model is demonstrated by both the dispersion curve and wave propagation. The metamaterial proposed in this study is fabricated using a 3D printer by converting distinct material properties into different thicknesses in the individual parts of the metamaterial.

Acknowledgment

Dr. Les Lee was the program manager.

Disclosure statement

No potential conflict of interest was reported by the authors.

Funding

This work was supported by AFOSR [grant number #FA9550-10-1-0061].

References

[1] R. Martinez-Sala, J. Sancho, J.V. Sanchez, V. Gómez, J. Llinares, and F. Meseguer, *Sound attenuation by sculptures*, Nature 378 (1995), pp. 241. doi:10.1038/378241a0.

[2] Z. Liu, X. Zhang, Y. Mao, Y.Y. Zhu, Z. Yang, C.T. Chan, and P. Sheng, *Locally resonant sonic materials*, Science. 289 (2000), pp. 1734–1736. doi:10.1126/science.289.5485.1734.

[3] Z. Yang, J. Mei, M. Yang, N.H. Chan, and P. Sheng, *Membrane-type acoustic metamaterial with negative dynamic mass*, Phys. Rev. Lett. 101 (2008), pp. 204301. doi:10.1103/PhysRevLett.101.204301.

[4] N. Fang, D. Xi, J. Xu, M. Ambati, W. Srituravanich, C. Sun, and X. Zhang, *Ultrasonic metamaterials with negative modulus*, Nat. Mater. 5 (2006), pp. 452–456. doi:10.1038/nmat1644.

[5] S.H. Lee, C.M. Park, Y.M. Seo, Z.G. Wang, and C.K. Kim, *Acoustic metameterial with negative modulus*, J. Phys. Condens. Matter. 21 (2009), pp. 175704. doi:10.1088/0953-8984/21/17/175704.

[6] H.H. Huang and C.T. Sun, *Theoretical investigation of the behavior of an acoustic metamaterial with extreme Young's modulus*, J. Mech. Phys. Solids. 59 (2011), pp. 2070–2081. doi:10.1016/j.jmps.2011.07.002.

[7] J. Li and C.T. Chan, *Double-negative acoustic metamaterial*, Phys. Rev. E. 70 (2004), pp. 055602(R). doi:10.1103/PhysRevE.70.055602.

[8] Y.Q. Ding, Z.Y. Liu, C.Y. Qiu, and J. Shi, *Metamaterial with simultaneously negative bulk modulus and mass density*, Phys. Rev. Lett. 99 (2007), pp. 093904. doi:10.1103/PhysRevLett.99.093904.

[9] S.H. Lee, C.M. Park, Y.M. Seo, Z.G. Wang, and C.K. Kim, *Composite acoustic medium with simultaneously negative density and modulus*, Phys. Rev. Lett. 104 (2010), pp. 054301. doi:10.1103/PhysRevLett.104.054301.

[10] L. Fok and X. Zhang, *Negative acoustic index metamaterial*, Phys. Rev. B. 83 (2011), pp. 214304. doi:10.1103/PhysRevB.83.214304.

[11] Z. Liang and J. Li, *Extreme acoustic metamaterial by coiling up space*, Phys. Rev. Lett. 108 (2012), pp. 114301. doi:10.1103/PhysRevLett.108.114301.

[12] Z. Liang, T. Feng, S. Lok, F. Liu, K.B. Ng, C.H. Chan, J. Wang, S. Han, S. Lee, and J. Li, *Space-coiling metamaterials with double negativity and conical dispersion*, Sci. Rep. 3 (2013), pp. 1614. doi:10.1038/srep01614.

[13] K.H. Lee, T.G. McRae, G.I. Harris, J. Knittel, and W.P. Bowen, *Cooling and control of a cavity optoelectromechanical system*, Phys. Rev. Lett. 104 (2010), pp. 123604. doi:10.1103/PhysRevLett.104.123604.

[14] Y. Xie, B.-I. Popa, L. Zigoneanu, and S.A. Cummer, *Measurement of a broadband negative index with space-coiling acoustic metamaterials*, Phys. Rev. Lett. 110 (2013), pp. 175501. doi:10.1103/PhysRevLett.110.175501.

[15] H. Chen, H. Zeng, C. Ding, C. Luo, and X. Zhao, *Double-negative acoustic metamaterial based on hollow steel tube meta-atom*, J. Appl. Phys. 113 (2013), pp. 104902. doi:10.1063/1.4790312.

[16] X. Wang, *Dynamic behaviour of a metamaterial system with negative mass and modulus*, Int. J. Solids Struct. 51 (2014), pp. 1534–1541. doi:10.1016/j.ijsolstr.2014.01.004.

[17] Y. Jin, B. Bonello, and Y. Pan, *Acoustic metamaterials with piezoelectric resonant structures*, J. Phys. D: Appl. Phys. 47 (2014), pp. 245301. doi:10.1088/0022-3727/47/24/245301.

[18] S.A. Pope, *Double negative elastic metamaterial design through electrical-mechanical circuit analogies*, IEEE Trans. Ultrason., Ferroelect., Freq. Control. 60 (2013), pp. 1467–1474. doi:10.1109/TUFFC.2013.2718.

[19] S.A. Pope and H. Laalej, *A multi-layer active elastic metamaterial with tuneable and simultaneously negative mass and stiffness*, Smart Mater. Struct. 23 (2014), pp. 075020. doi:10.1088/0964-1726/23/7/075020.

[20] Y. Wu, Y. Lai, and Z.-Q. Zhang, *Elastic metamaterials with simultaneously negative effective shear modulus and mass density*, Phys. Rev. Lett. 107 (2011), pp. 105506. doi:10.1103/PhysRevLett.107.105506.

[21] Y. Lai, Y. Wu, P. Sheng, and Z.-Q. Zhang, *Hybrid elastic solids*, Nat. Mater. 10 (2011), pp. 620–624. doi:10.1038/nmat3043.

[22] X.N. Liu, G.K. Hu, G.L. Huang, and C.T. Sun, *An elastic metamaterial with simultaneously negative mass density and bulk modulus*, Appl. Phys. Lett. 98 (2011), pp. 251907. doi:10.1063/1.3597651.

[23] R. Zhu, X.N. Liu, G.K. Hu, C.T. Sun, and G.L. Huang, *Negative refraction of elastic waves at the deep-subwavelength scale in a single-phase metamaterial*, Nat. Commun. 5 (2014), pp. 1–8. doi:10.1038/ncomms6510.

[24] H.H. Huang and C.T. Sun, *Anomalous wave propagation in a one-dimensional acoustic metamaterial having simultaneously negative mass density and Young's modulus*, J. Acoust. Soc. Am. 132 (2012), pp. 2887–2895. doi:10.1121/1.4744977.

4

Filtration of sodium chloride from seawater using carbon hollow tube composed of carbon nanotubes

Chaudhary Ravi Prakash Patel, Prashant Tripathi, O.N. Srivastava* and T.P. Yadav

Department of Physics (Centre of Advanced Studies), Nanoscience Centre, Banaras Hindu University, Varanasi 220005, India

The present article deals with filtration of seawater to remove sodium chloride (NaCl) using filter made from organized structures of carbon nanotubes (CNTs). The filter consists of hollow carbon cylinder (length ~10 cm, diameter ~1 cm), which is composed of radially aligned CNTs. This carbon hollow cylinder has been synthesized by continuous spray pyrolysis of ferrocene–benzene solution in argon atmosphere. The hollow cylinder has been turned into a water filter by closing one end and keeping a small funnel at the other. Filtration of seawater (Marina Beach, Chennai, India) has been obtained both under the self pressure of seawater column in the hollow cylinder and under the difference of pressure created by enclosing the filter in a vacuum tight container. It has been found that the efficiency of filtration is about two times higher under partial vacuum (~10^{-2} torr) created on the filtrate (water) side. After filtration of seawater, a deposit in the inner surface of hollow cylinder has been found. This deposit has been characterized by X-ray diffraction, transmission electron microscopy and energy dispersive X-ray analysis, and it has been found that the deposit was NaCl. The filtration leads to almost complete removal of NaCl from the seawater.

Keywords: carbon nanotube; filtration; chemical vapor deposition; spray pyrolysis

1. Introduction

Problems with drinking water are expected to become more serious world over in the coming decades [1]. Seawater is a good source of water but its salinity does not allow it for human consumption. As is known, the major component making seawater saline is NaCl [2]. Mass wise chlorine is about 16 times higher than Mg, ~22 times than sulfur, ~48 times than Br and K. Similarly sodium is 9 times higher than Mg, 12 times to sulfur, 17 times to K and 180 times higher than Br and C. Addressing these problems calls out for research to be conducted in order to find new methods of removing NaCl at lower cost and input energy while minimizing the use of chemicals and impact on the environment at the same time [3]. Therefore, in order to utilize seawater for domestic purposes, removal of sodium chloride is required. Filtration, i.e. separation of sodium chloride from water, is the simplest process for removal of NaCl from seawater [4]. Unlike desalination, the energy requirement in filtration is very small and it can even work under gravity [5].

The advent of nanoscience and technology manifested by nanomaterials including such nanostructures that possess membrane like pores has led to exploration of

*Corresponding author. Email: heponsphy@gmail.com

nanomaterials for filtration [6]. Recently, instead of pores embodied in polymer membrane, the native pores of carbon nanostructures, particularly carbon nanotubes (CNTs), have caught attention for filtration of water [7]. Polymeric membranes have flexible chains, and hence, well-defined pores necessary for filtration cannot be invariably present [8]. The use of CNT as an additional entity in conventional filters, or as filters on their own, has increased in recent times [9]. This is due to the various pore sizes and configurations, which may be obtained from CNTs by tailoring the growth parameters. Some studies in this direction have been made recently [10]. Based on extensive calculations, it has been shown that membranes comprising CNTs can provide an efficient means of water filtration [11]. The narrow pores of CNTs are capable of filtering water while rejecting ions (Na^+/Cl^-) [12]. It has been shown that CNT-based filters are capable of removing bacteria from water and heavy hydrocarbon from petroleum [13]. It has been reported that the basic filtration action partially depends on the design of the CNT-based filter [14]. In addition to simple flow through pores embodied in CNTs, electrochemical filters for removing bacteria using CNTs as anodes have also been developed [15]. Polymer filters cannot be repeatedly used through several cycles because removal of fouling ingredients is not easy. However, the CNT-based filter can be re-used for several cycles, by autoclaving them after each use. Even though CNT-based filters have marked advantages due to hydrophobic nature, high porosity and specific area, there are several aspects, which are yet to be studied and optimized [8–11]. For example, investigations are required to study decline in water flux through the CNT due to polarization-induced formation of stagnant layer on the surface of CNT base [16–19].

Keeping the abovementioned aspects in view, we have developed hollow carbon cylindrical filter consisting of radially aligned CNTs. This has been used for water filtration.

2. Experiment

2.1. Synthesis of carbon hollow cylinder consisting of radially aligned CNTs

It has been suggested that vertically aligned arrays of CNTs could form a unique membrane for water filtration [20]. Generally, the aligned CNT arrays get synthesized on a substrate [21,22]. Such configurations of CNTs do not have both ends open. They are not suitable to act like a membrane for filtration. Here, we have adopted a modified version of our earlier synthesis protocol [23], where radially aligned CNTs lead to the formation of long hollow carbon cylinder as shown in Figure 1(a). The deposition of hollow cylinder was done by spray pyrolysis of ferrocene ($C_{10}H_{10}Fe$) and benzene (C_6H_6) solution. An optimum concentration of ferrocene in benzene, 30 mg/ml, was utilized. Argon gas was used as carrier gas for spray pyrolysis and the temperature employed was 900°C. The CNTs get deposited on Fe nanoparticle catalyst coming from ferrocene. Both benzene and ferrocene provided carbon sources. There was a thick homogeneous carbon film-like deposition along the total heating zone of the furnace on the walls of the quartz tube, which worked like template. This thick carbon deposit was in the form of replica of inner surface of quartz tube. It was in the form of hollow cylinder (tube) and it corresponds to carbon hollow cylinder comprising radially aligned CNTs (see inset of Figure 1(a); the arrows indicate radially aligned CNTs). In order to achieve the longer carbon hollow cylinder, we employed a furnace having longer hot zone (15 × 4 cm diameter). One difficulty faced was in regard to taking the long (~10 cm) hollow carbon cylinder out of the quartz tube. To accomplish this, various parameters including the

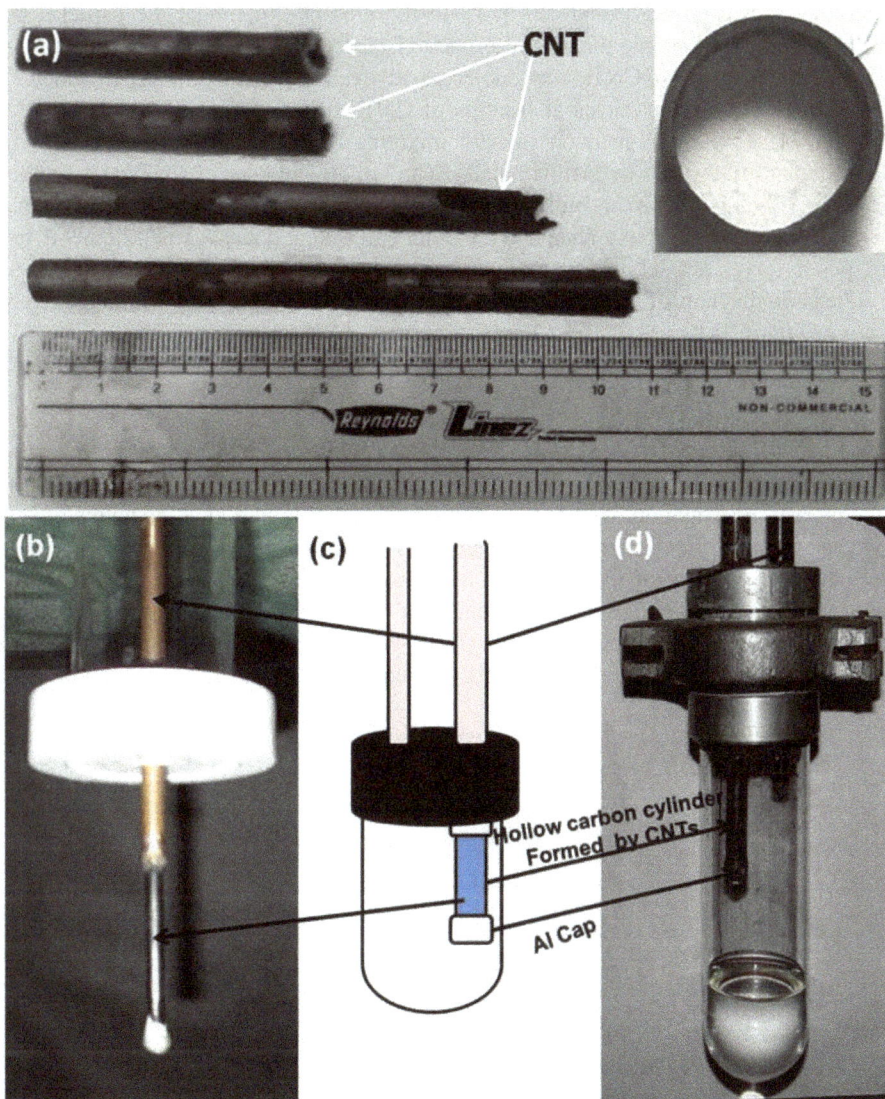

Figure 1.   (a) Hollow cylinder of carbon nanotubes, (b) filtration setup of hollow cylinder of carbon nanotubes, (c) the schematic diagram of the filtration setup and (d) optical photograph of the filtration setup.

treatment of quartz tube before deposition of CNTs were varied. It was found that the hollow carbon cylinder could easily come out by mild etching using dilute HNO_3, if the quartz tube was repeatedly flushed (three to four times) at 500°C for 2 hours by gas mixture (90% Ar + 10% H_2, volume wise).

2.2.   Fabrication of filtration setup

Since ferrocene ($FeC_{10}H_{10}$) has been used as catalyst, Fe particles would be present at CNTs tip. However, in our earlier studies, we have found that most of the Fe gets removed by prolonged acid (HNO_3) treatment [24]. In the present experiment, Fe particles have been removed by the above mentioned process. Before fabrication of filtration setup, the bulk hollow cylinder of CNTs was subjected to overnight acid (HNO_3) treatment for removal of Fe catalyst particles at the tips of CNTs that acted as catalyst followed by ultrasonic cleaning by immersion in acid mixtures. The hollow carbon cylinder was flushed four to five times by distilled water. The hollow cylinder of CNTs was mounted as a filter (Figure 1(b)). The bulk hollow cylinder tube is closed at one end with an aluminum cap fixed by epoxy resin and the other end is kept open. A small funnel is fixed here, which serves as inlet port for entry of seawater with copper pipes. Figure 1(c) shows the schematic diagram of the filtration setup. Optical photograph of the filtration setup is shown in Figure 1(d).

2.3.   Filtration process

Filtration process will become cost-effective if the utilized pressure for water flow is low. In the present investigation, we have utilized filtration employing only the native pressure of water column in CNT-based hollow carbon cylinder as shown in Figure 1(b) and (c). The schematic diagram of the working mechanism of the CNT-based filtration setup is shown in Figure 2. A hollow carbon cylinder of diameter 1 cm and length ~10 cm was used, where C_1, C_2, C_3, C_4 etc. represent CNTs with diameter ~10–12 nm. The length of CNTs, which corresponds to thickness of the hollow cylinder, is 300 μm. The drawing is

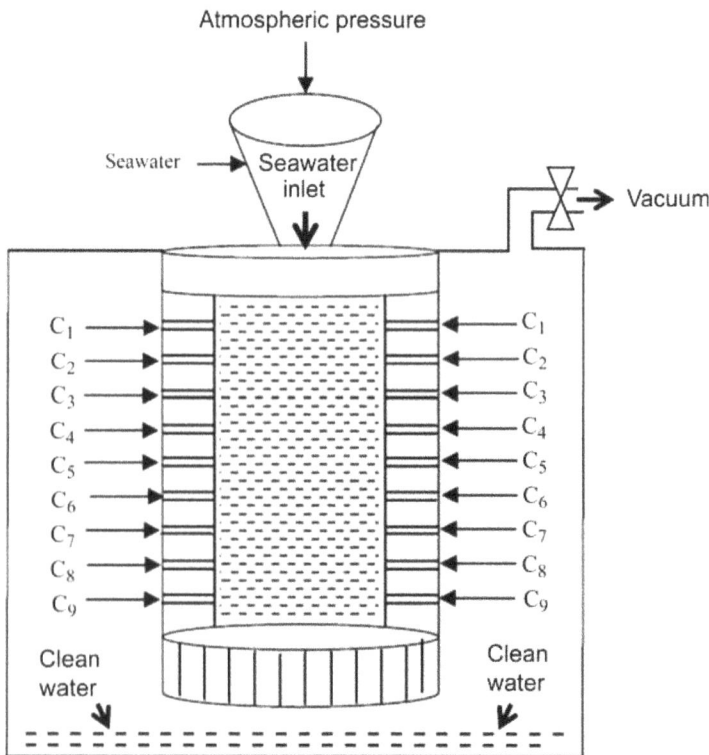

Figure 2.   The schematic diagram of the working mechanism of the CNT-based filtration setup.

not to the scale and also the separation between labels has been exaggerated for clarity. The CNTs are nearly in contact with each other. The arrows depict flow of water through the CNTs. Here, the water used was the seawater from Marina Beach Chennai (India).

2.4. *Characterization*

The structural characterization of the as-grown CNTs before and after filtration have been carried out using X-ray diffraction (XRD) employing X'Pert PRO PANalytical diffract-ometer equipped with a graphite monochromator with Cu Kα source (λ = 1.5402 Å, operating at 45 kV and 40 mA). The microstructural characterizations were carried out using a scanning electron microscope (SEM: QUANTA 200). The detailed structural and microstructural characterization was carried out using transmission electron microscopy (TEM) (TECNAI-20G^2 at 200 kV) in the diffraction and imaging modes. Compositional analysis was performed by an energy-dispersive X-ray (EDX) analysis coupled with SEM. Raman spectra of the CNTs sample were recorded using Horiba Jobin Yuon Raman spectrometer model no. H 45517 using an argon ion laser source λ = 514 nm.

3.   **Result and discussions**

3.1. *Seawater filtration through CNT-based filter*

The efficiency of the CNT-based filtration setup was monitored at different pressures exerted by the seawater column in the hollow tube of the filter. It may be pointed out that the top of the seawater column filled with CNT-based hollow tube is at atmospheric pressure. The open ends of CNTs (as shown by arrow in Figure 2) are at atmospheric pressure. Thus, the flow of water from the CNT in filtration will only be due to water column in the CNT-based hollow cylinder. Initially, we utilized a 10-cm water column. It was found that the flow rate of water outside the CNT exit end was negligible. However, when an additional seawater column of 100 cm (pressure 10 Pa) above the top end of CNT-based filter was used, there was a flow of 3 cc of water in 12 hours (Figure 3(a)). Here, D1 and D5 are five different runs under identical conditions (100-cc water column added to the top of the initial 10-cm water column). The five runs were made to check the reproducibility of the results. It was found that the rate becomes negligibly small after 12 hours and stopped completely after ~20 hours. It may be pointed out that even though the filtration rate from CNT-based carbon hollow cylinder is very low, this mode of filtration is still interesting if we can use a liquid column immiscible with water and having a density significantly higher than water at the top of the water column. With the increase in the liquid column pressure, the filtration rate also increases. It may possibly become nearly competitive with other filtration processes where external pressure is applied.

We next proceeded to carry out filtration by evacuating the container where filter configuration was mounted. This was achieved by enclosing the CNT-based filter assem-bly in a vacuum tight enclosure. The top end is the filter from where seawater is poured, which was kept outside the vacuum enclosure. The seawater was filled in the hollow cylinder and the enclosure was evacuated by a vacuum system. After rotary evacuation, the vacuum valve was closed. It is apparent now that there will be a small air pressure at the end of CNT terminal from where water flows out (Figure 3(b)). Similar to the case of D1–D5, VF1–VF3 are three different experiments under vacuum (up to 10^{-2} torr) to monitor the filtration rate of water under vacuum. The filtration will now be under a

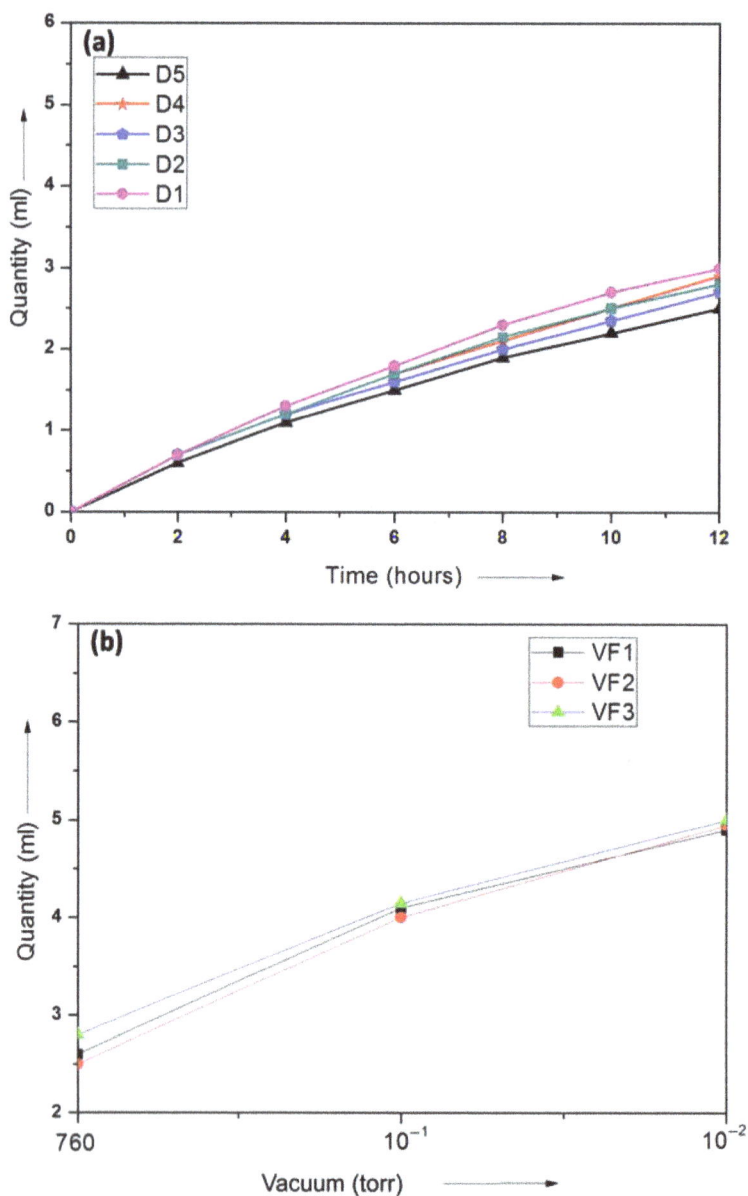

Figure 3.   Efficiency of seawater filtration with (a) 100-cm water column of (pressure 10 Pa) and (b) water collected as a function of vacuum level in 12 hours.

pressure of about 776 torr, which gets applied from the top of water column. When water flows into the container, the vacuum will deteriorate. The saturation vapor pressure of water is 17.5 torr. Thus, a pressure difference of ~742 torr gets applied at seawater entering points. The amount of filtrated water was measured after an interval of 12 hours. The amount of water with a pressure difference ~742 torr was found to be nearly two times higher. Thus, creating vacuum in the filtrate side improves the rate of filtration.

3.2.   Characterization of CNT filter before and after filtration

Detailed microstructural characterization using SEM of the bulk hollow cylinder of CNTs before and after filtration of seawater is shown in Figure 4. Figure 4(a) shows the SEM image of a broken piece from thick bulk hollow cylinder. This figure clearly shows that hollow cylinder consists of radially aligned CNTs. The length of the multiwall carbon nanotubes (MWCNTs) in bulk hollow cylinder corresponds to the wall thickness (~300 μm) of the bulk structure. Figure 4(b) shows the magnified SEM image of the bulk hollow cylinder. This figure shows the dense packing of aligned MWCNTs. Figure 4 (c) shows the SEM image of a broken piece of bulk hollow cylinder of CNTs after filtration of seawater. The deposition of particles, presumably NaCl crystallites, can be seen easily. The SEM image of a broken piece of inner wall of the filter clearly shows the presence of tiny crystals of NaCl (Figure 4(d)). These tiny crystals get nucleated as filtration proceeds with water flowing through CNTs and the seawater becomes super-saturated with NaCl. The purity of seawater was also checked by EDX (Figure 5). The 10 ml water (before and after filtration) was heated at 100°C and the dry product was subjected to EDX. Based on the analysis of the EDX data (Figure 5 and inset table), it was

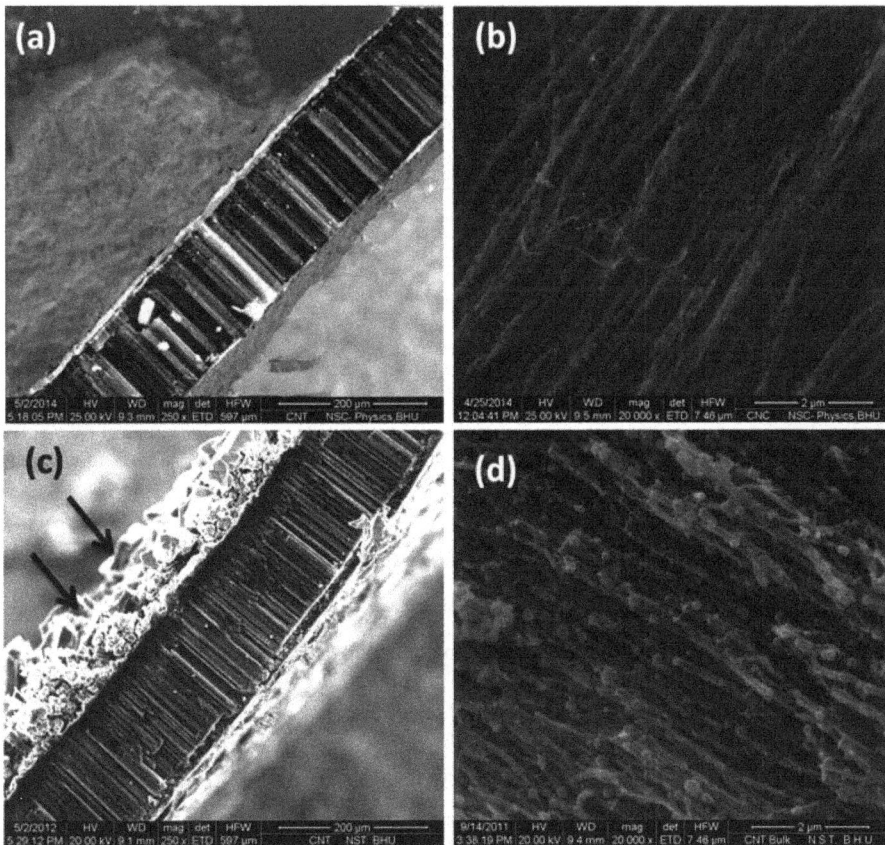

Figure 4.   SEM of the bulk hollow cylinder of CNTs before (a–b) and after (c–d) filtration of seawater. The SEM image of a broken piece of inner wall of the filter clearly shows the presence of tiny crystals of NaCl.

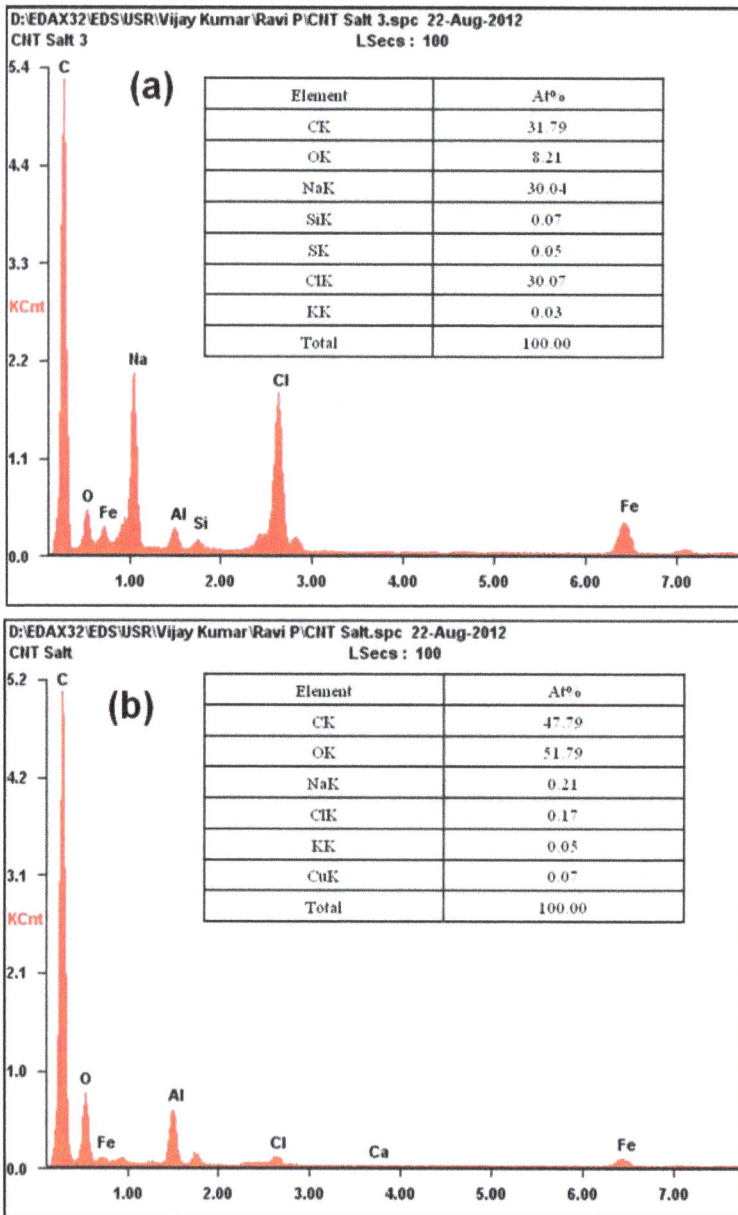

Figure 5.   (a–b) EDX data of salt of seawater, before and after filtration of seawater.

found that the quantity of sodium decreased from 30.06 wt% to 0.04 wt%, whereas, after filtration, the quantity of chlorine decreased from 11.44 wt% to 0.17 wt%. It may be noticed that after filtration, some Al and O peaks have appeared. These peaks are presumably arising due to Al cap on the filter and its oxidation on heating to 100°C.

The characterization of bulk hollow CNT cylinder before and after filtration was done by XRD. Figure 6(a) shows the XRD pattern of the as-grown bulk hollow cylinder/tube of CNTs where the (00.2), (10.0), (10.1) and (00.4) diffraction spots of CNT have been

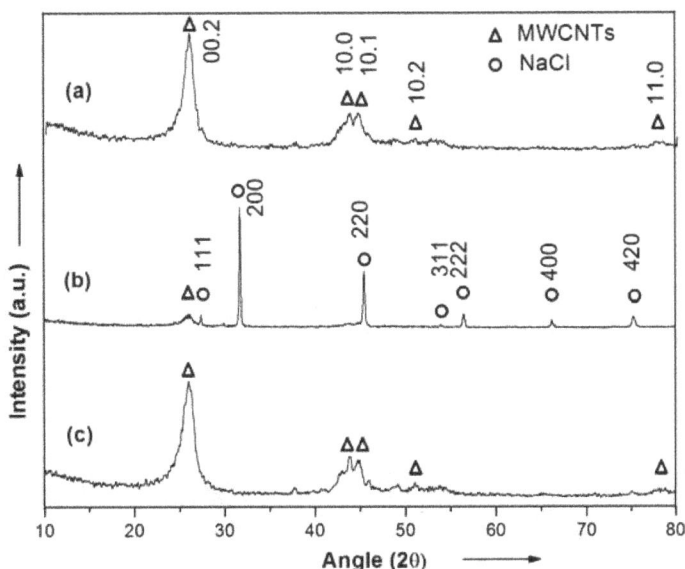

Figure 6. (a) XRD pattern of the as-grown bulk hollow cylinder/tube of CNTs, (b) the XRD pattern of inner surface of bulk hollow cylinder of CNTs after filtration of seawater and (c) XRD of carbon cylinder after filtration and then washed with distilled water.

depicted. The higher intensity of (00.2) peaks at ~26° corresponds to that of CNT [23]. Figure 6(b) shows the XRD pattern of inner surface of bulk hollow cylinder of CNTs after filtration of seawater. After filtration of water, the inner surface of hollow tube contained whitish deposit consisting of small crystallites. Detailed analysis of the XRD peaks showed that after filtration besides CNT peaks, the other peaks that appear can be successfully indexed based on NaCl lattice structure with lattice parameter a = 5.4 Å. The peaks correspond to (111), (200), (220), (311), (222) and (400) diffraction spots from NaCl. The XRD studies of the inner portion of the hollow carbon cylinders from five filtration runs invariably revealed that the deposit obtained after filtration corresponded to NaCl. This was further confirmed from the EDX analysis to be described later. Figure 6(c) shows the XRD pattern of inner surface of the bulk hollow cylinder of CNTs after few cycle of filtration of seawater and then washed with ordinary water. About 100 ml of ordinary tap water was required to wash the tube. Also, the salt that would get dissolved in this water can be removed by the process described presently. It can be seen that XRD pattern of the hollow cylinder after filtration is quite similar to the as grown bulk hollow CNT cylinder. It is thus noted that there is no significant change in the hollow CNT cylinder after filtration.

Figure 7(a) shows TEM images of the CNTs forming bulk hollow cylinder. Figure 7(b) shows the high-resolution TEM image of typical CNTs in the bulk hollow cylinder assembly after filtration of seawater; it shows the well-graphitized walls of the MWCNTs along with tiny NaCl crystals on the tip of CNTs. In Figure 7(b), as can be seen from the border, the CNT extends only up to A'. The particle at B is NaCl. This was verified through diffraction pattern taken from particle B. It is to be mentioned here that the formation of NaCl inside the hollow carbon cylinder is not expected to be uniform during water passage. It is rather expected to be discrete. The NaCl crystal nuclei may get formed

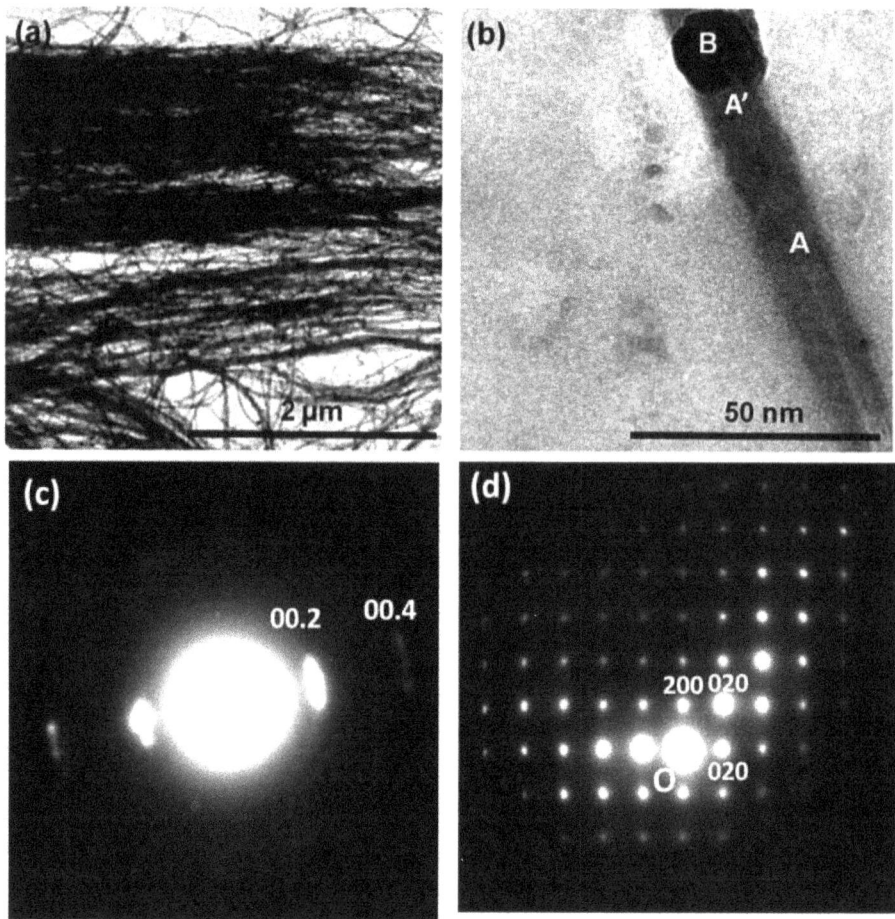

Figure 7. (a) TEM images of the CNTs forming bulk hollow cylinder, (b) high-resolution TEM image of typical CNTs in the bulk hollow cylinder assembly after filtration of seawater and (c–d) selected area diffraction pattern taken from regions A and B, respectively.

at localized defects on the inner surface of hollow carbon tube. These nuclei would grow through addition of NaCl molecules from surrounding regions. Thus, although NaCl deposit would block some CNTs, the surrounding CNT channels may still remain open for filtration. However, after a comparatively longer time (~20 hours), the NaCl crystallite would eventually become large enough to cover most of the CNT pores leading to stoppage of filtration process altogether. This is in keeping with experimental observations. The inner and outer diameters of these CNTs were found to be ~10 nm from TEM analysis. It may be mentioned that the size of the hydrated Na ion is 76 nm [9]; thus, it can be filtered through CNTs. Figure 7(c–d) shows selected area diffraction (SAD) pattern taken from regions A and B of Figure 7(b). A representative SAD pattern from CNT is shown in Figure 7(c). The indexing of diffraction spots in Figure 7(c) has now been done based on the evaluation of d spacing. The d spacing of 3.37 Å corresponds to 00.2 spot of graphite. This has been outlined in Figure 7(c) and similarly 00.4 peaks with $d = 1.687$ Å, which clearly represents the SAD pattern of MWCNTs. Figure 7(d) shows a typical SAD pattern from tiny crystals, which gets formed on the inner surface of hollow cylindrical filter. Similar procedure has been employed for indexing diffraction spots in Figure 7(d).

It may be pointed out that for CNT-based filters, the recyclability is quite good, unlike the polymer membrane filters. For re-using the CNT-based filters, the NaCl deposits have to be removed by flushing it with distilled water. The CNT-based filter can be heated to ~300°C to remove leftover water and also volatile impurities. The treatment can be repeatedly done without affecting the functioning of the CNT-based filter. This feature also makes CNT-based filters attractive in comparison to polymer membrane filter. Test for recyclability has been tested based on Raman Spectroscopy before and after filtration through several cycles. This is described in the following. Raman spectroscopy is widely used to investigate both vibrational and electronic properties of CNTs. Figure 8 shows Raman spectra of the bulk hollow cylinder consisting of CNTs, before and after seawater filtration. The three peaks at 1345, 1578 and 2697 cm^{-1} have been observed before filtration, and these correspond to D, G and 2D known peaks of CNT. After filtration, these three peaks have been found to be shifted upward and observed at 1353, 1583 and 2702 cm^{-1}. Because of the small wave vector transfer in first-order scattering, only phonon modes with a nearly zero wave vector are expected in first-order Raman spectra due to the breathing modes in ring of sp^2 carbon atoms. In particular, two of the characteristic peaks in the first-order spectra of CNTs, the disorder-induced band mode at ~1330 cm^{-1} assumed to be D-point modes and the high-energy graphite-like mode at ~1600 cm^{-1} assumed to be G-point modes, have been observed due to bond stretching of all pairs of sp^2 atoms in both ring and chains. Third peak is known as second order of D peak and arises due to two phonons with opposite momentum. The I_{2D}/I_G and I_D/I_G ratios of CNTs cylinder before filtration were found to be 0.57 and 0.37, respectively, and after filtration, the I_{2D}/I_G and I_D/I_G ratios were found to be changed as 0.44 and 0.73, which indicate the nature of ordered MWCNTs before filtration and after 5th cycle of filtration.

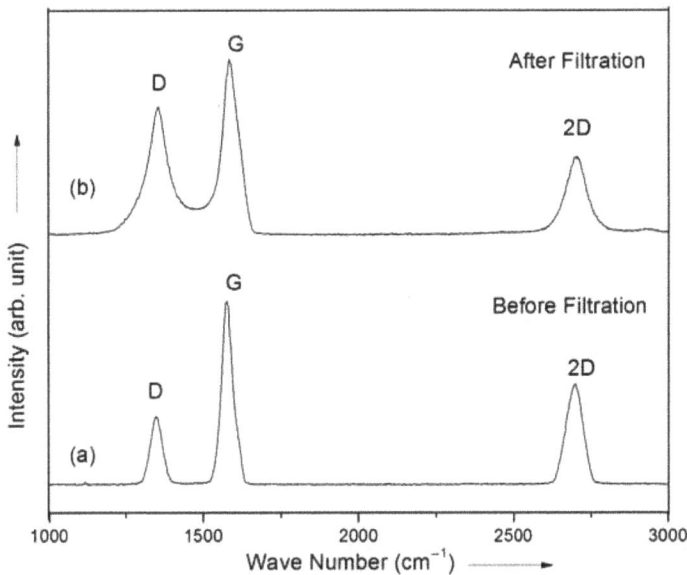

Figure 8. Raman spectra (a) of MWCNTs of carbon cylinder (before filtration) and of (b) MWCNTs of carbon cylinder (after filtration and washed with distilled water).

However, after five cycles of filtration, a small disorder was induced. The Raman spectra after filtration (but before removing the filtered NaCl) look similar to that which is obtained after filtration and removal of NaCl. This is expected since NaCl will not produce any Raman peak in the spectrum range of interest, i.e. $1000–3000$ cm^{-1}. This testifies the recyclability of CNT-based filters. I_D/I_G ratio increases after filtration since the process of filtration may lead to creation of some disorder.

Pore size and diameter play important roles in NaCl filtration. The nanoporous surfaces of CNTs are suitable for rejecting salt (NaCl). The hydrophobic nature of CNT encourages nearly frictionless movement of water molecules without the need of any energy-driven force. The microscopic characterization shows that NaCl is mostly on the tip of the CNTs. This implies that rather than adsorption of NaCl on the membrane walls, it flows through the CNT, which leads to filtration. It is true that efficiency of filtration is low. It may be pointed out that unlike desalination where huge consumption of electricity is involved, the filtration process discussed here is not energy-intensive [10]. However, the removal of NaCl through CNT membrane by filtration does not seem to have been tried so far [9,10]. However, CNTs have been used for increasing the efficiency of metal filters. Apparently compared to this, our CNT filter that does not employ metal filter will be more viable. The efficiency can be improved by applying pressure on the water entry.

4. Conclusions

A stand-alone CNT configuration forming a carbon hollow cylinder has been fabricated through spray pyrolysis. The pyrolysis process has been tailored to synthesize CNTs in order to work as CNT-based filters for filtration of seawater. The filtration has been done under pressure created by seawater column (~100 cm) and also under vacuum created in the container housing the CNT-based hollow cylinder. With vacuum pressure difference of ~742 torr gets applied. This makes the rate of filtration two times higher. The rate of filtration is comparatively lower but holds promise for improvement. Nevertheless, the efficiency of filtration relating to removal of NaCl from seawater is good. The Na and Cl concentration has been found to decrease from 30.06 wt% to 0.4 wt% for Na and for Cl from 11.44 wt% to 0.17 wt%. It has been shown that CNT-based filters are capable of withstanding recyclability in regard to repeated filtration.

Acknowledgments

We would like to thank Prof. C.N.R. Rao, Prof. P. M. Ajayan, Dr M.A. Shaz, Dr Kalpana Awasthi and Dr R.R. Shahi for helpful discussions. The authors are grateful to the Department of Science and Technology (DST), Nano Science Mission scheme and Ministry of New and Renewable Energy (MNRE) India for financial support.

References

[1] R. Sheikholeslami, *Strategies for future research and development in desalination – Challenges ahead*, Desalination. 248 (2009), pp. 218–224. doi:10.1016/j.desal.2008.05.058

[2] H.P. Eugster, C.E. Harvie, and J.H. Weare, *Mineral equilibria in a six-component seawater system, Na–K–Mg–Ca–SO₄–Cl–H₂O, at 25°C*, Geochimica Cosmochimica Acta. 44 (1980), pp. 1335–1347. doi:10.1016/0016-7037(80)90093-9

[3] Y. Oren, *Capacitive deionization (CDI) for desalination and water treatment – Past, present and future (a review)*, Desalination. 228 (2008), pp. 10–29. doi:10.1016/j.desal.2007.08.005

[4]   M.C. Duke, S. Mee, and J.C.D. Da Costa, *Performance of porous inorganic membranes in non-osmotic desalination*, Water Res. 41 (2007), pp. 3998–4004. doi:10.1016/j. watres.2007.05.028

[5]   M.A. Shannon, P.W. Bohn, M. Elimelech, J.G. Georgiadis, B.J. Mariñas, and A.M. Mayes, *Science and technology for water purification in the coming decades*, Nature. 452 (2008), pp. 301–310. doi:10.1038/nature06599

[6]   Y. An, K. Zhang, F. Wang, L. Lin, and H. Guo, *Removal of organic matters and bacteria by nano-MgO/GAC system*, Desalination. 281 (2011), pp. 30–34. doi:10.1016/j.desal.2011.07.035

[7]   P.S. Goh, A.F. Ismail, and B.C. Ng, *Carbon nanotubes for desalination: performance evaluation and current hurdles*, Desalination. 308 (2013), pp. 2–14. doi:10.1016/j.desal.2012.07.040

[8]   S. Subramanian and R. Seeram, *New directions in nanofiltration applications – Are nanofibers the right materials as membranes in desalination?* Desalination. 308 (2013), pp. 198–208. doi:10.1016/j.desal.2012.08.014

[9]   M.S. Mauter and M. Elimelech, *Environmental applications of carbon-based nanomaterials*, Environ. Sci. Technol. 42 (2008), pp. 5843–5859. doi:10.1021/es8006904

[10] Q. Zaib and H. Fath, *Application of carbon nano-materials in desalination processes*, Desalin. Water Treat. 51 (2013), pp. 627–636. doi:10.1080/19443994.2012.722772

[11] S.T. Mostafavi, M.R. Mehrnia, and A.M. Rashidi, *Preparation of nanofilter from carbon nanotubes for application in virus removal from water*, Desalination. 238 (2009), pp. 271–280. doi:10.1016/j.desal.2008.02.018

[12] M.G. Buonomenna, *Nano-enhanced reverse osmosis membranes*, Desalination. 314 (2013), pp. 73–88. doi:10.1016/j.desal.2013.01.006

[13] A. Srivastava, O.N. Srivastava, S. Talapatra, R. Vajtai, and P.M. Ajayan, *Carbon nanotube filters*, Nat Mater. 3 (2004), pp. 610–614. doi:10.1038/nmat1192

[14] M.A. Tofighy and T. Mohammadi, *Salty water desalination using carbon nanotube sheets*, Desalination. 258 (2010), pp. 182–186. doi:10.1016/j.desal.2010.03.017

[15] A.S. Brady-Estévez, S. Kang, and M. Elimelech, *A single-walled-carbon-nanotube filter for removal of viral and bacterial pathogens*, Small. 4 (2008), pp. 481–484. doi:10.1002/ smll.200700863

[16] B. Corry, *Designing carbon nanotube membranes for efficient water desalination*, J. Phys. Chem. B. 112 (2008), pp. 1427–1434. doi:10.1021/jp709845u

[17] K. Gethard, O. Sae-Khow, and S. Mitra, *Carbon nanotube enhanced membrane distillation for simultaneous generation of pure water and concentrating pharmaceutical waste*, Separation Purif. Technol. 90 (2012), pp. 239–245. doi:10.1016/j.seppur.2012.02.042

[18] S.J. Park and D.G. Lee, *Performance improvement of micron-sized fibrous metal filters by direct growth of carbon nanotubes*, Carbon. 44 (2006), pp. 1930–1935. doi:10.1016/j. carbon.2006.02.005

[19] A. Subramani, M. Badruzzaman, J. Oppenheimer, and J.G. Jacangelo, *Energy minimization strategies and renewable energy utilization for desalination: A review*, Water Res. 45 (2011), pp. 1907–1920. doi:10.1016/j.watres.2010.12.032

[20] R. Das, M. Eaqub Ali, S.B.A. Hamid, S. Ramakrishna, and Z.Z. Chowdhury, *Carbon nanotube membranes for water purification: A bright future in water desalination*, Desalination. 336 (2014), pp. 97–109. doi:10.1016/j.desal.2013.12.026

[21] P.M. Ajayan, *Nanotubes from carbon*, Chem. Rev. 99 (1999), pp. 1787–1800. doi:10.1021/ cr970102g

[22] Y.J. Jung, B. Wei, R. Vajtai, P.M. Ajayan, Y. Homma, K. Prabhakaran, and T. Ogino, *Mechanism of selective growth of carbon nanotubes on SiO₂/Si patterns*, Nano Lett. 3 (2003), pp. 561–564. doi:10.1021/nl034075n

[23] K. Awasthi, R. Kumar, H. Raghubanshi, S. Awasthi, R. Pandey, D. Singh, T.P. Yadav, and O. N. Srivastava, *Synthesis of nano-carbon (nanotubes, nanofibres, graphene) materials*, Bull Mater Sci. 34 (2011), pp. 607–614. doi:10.1007/s12034-011-0170-9

[24] K. Awasthi and O.N. Srivastava, *Synthesis of carbon nanotubes*, in *Chemistry of Carbon Nanotubes*, V.A. Basiuk and E.V. Basiuk, eds., Vol. 1, American Scientific Publishers (ASP), Valencia, CA, 2008, pp. 1–26.

5

Graphene–platinum nanocomposite as a sensitive and selective voltammetric sensor for trace level arsenic quantification

R. Kempegowda[a], D. Antony[b] and P. Malingappa[a]*

[a]Department of Chemistry, Bangalore University, Central College Campus, Bangalore 560001, India; [b]Raman Research Institute, C.V. Raman Avenue, Bangalore 560080, India

A simple protocol for the chemical modification of graphene with platinum nanoparticles and its subsequent electroanalytical application toward sensitive and selective determination of arsenic has been described. Chemical modification was carried out by the simultaneous and sequential chemical reduction of graphene oxide and hexachloroplatinic acid in the presence of ethylene glycol as a mild reducing agent. The synthesized graphene–platinum nanocomposite (Gr–nPt) has been characterized through infrared spectroscopy, x-ray diffraction study, field emission scanning electron microscopy and cyclic voltammetry (CV) techniques. CV and square-wave anodic stripping voltammetry have been used to quantify arsenic. The proposed nanostructure showed linearity in the concentration range 10–100 nM with a detection limit of 1.1 nM. The proposed sensor has been successfully applied to measure trace levels of arsenic present in natural sample matrices like borewell water, polluted lake water, agricultural soil, tomato and spinach leaves.

Keywords: arsenic; graphene–platinum nanocomposite; chemical modification; square-wave anodic stripping voltammetry; spinach leaves

1. Introduction

The presence of heavy metal ions even at trace level in the environment has been considered as a major concern in recent years due to their toxicity and prolonged distribution [1]. Hence, quantification of these ions at trace level becomes a major area of research in the field of environmental science and technology. Arsenic is considered to be one of the naturally occurring highly toxic elements which are widely distributed in nature and cause serious threat to both mankind and aquatic systems even at ultra trace level [2]. Therefore, the World Health Organization (WHO) has recommended a maximum threshold limit value of 10 ppb for arsenic in drinking water [3]. Generally, arsenic exists in +3 and +5 oxidation states, former is approximately 60 times more toxic than the latter, and hence its concentration measurement is essentially required toward environmental concern [4]. Several methods have been employed for the determination of arsenic including spectroscopic methods like atomic absorption spectroscopy (AAS), inductively coupled plasma atomic emission spectroscopy (ICPAES) and UV-Vis spectroscopy. These

*Corresponding author. Email: mprangachem@gmail.com
Present address for R. Kempegowda: Department of Chemistry, Siddaganga Institute of Technology, Tumkur, Karnataka, India.

methods can be successfully applied in the determination of arsenic well below the guideline levels prescribed by WHO. However, these methods are time intensive and require high capital cost, skilled personnel to operate the instruments, prolonged sampling procedures and expensive day-to-day maintenance. In contrast, electrochemical methods offer comparably superior merits over these methods. The superiority of the electrochemical methods is due to their high sensitivity, field applicability and possibility of determining multiple analytes in single measurement [5]. Therefore, significant focus has been made in recent years to develop electrochemical sensors to quantify arsenic at trace level. In all these methods, chemical modification of working electrode toward the arsenic plays a key role in achieving the required selectivity, sensitivity and desired detection limit [6]. Modification of electrode surface using metal nanoparticles is one of the fascinating approaches in designing electrochemical interface for the electrocatalysis. The nanosized transition metal particles are of great interest due to their unique catalytic, electronic properties and noble performance over their bulk counterparts [7]. The catalytic properties of the metal nanoparticles are associated with the quantum scale dimensions and large surface-to-volume ratio, and availability of specific binding sites on the surface of the particles constitutes the driving force in the nanoparticle-based electroanalysis [8]. Gold, iridium oxide and cobalt oxide nanoparticles have been widely used in the determination of arsenic, whereas very few reports are available on the use of platinum [9–13]. Therefore, it is worthwhile to develop a sensing platform using platinum nanoparticles. It is generally accepted that the shape, size, surface structure of the nanoparticles and the catalyst support have tremendous influence on their catalytic properties [7]. Graphene-modified electrodes have been widely used as substrate material for wide variety of applications due to its fast electron transfer sites. In order to improve its analytical performance toward particular applications, they have been modified with nanoparticles. Various carbon-based supports have been used to anchor the nanoparticles, which include graphite, glassy carbon, carbon nanotubes and thin diamond films, among which graphene became a significant substrate material to anchor the metal nanoparticles due to the availability of large surface area to accommodate the nanoparticles. These nanoparticle-modified graphene composites were used as sensing and biosensing platforms [14–17]. In this report, we have made an attempt in designing chemically modified graphene sheets with platinum nanoparticles in the presence of ethylene glycol as a chemical reducing agent. The fabricated nanocomposite was used in the electrochemical determination of arsenic and subsequently applied to wide variety of environmental sample matrices using square-wave anodic stripping voltammetry (SWASV), with least interference from most of the common cations and anions.

2. Materials and methods

2.1. Apparatus

All electrochemical measurements were carried out using electrochemical analyzer (CH Instruments, 3700, Tennison Hill Drive, Austin, TX, USA, Model: CHI 619B) at room temperature in an electrochemical cell of 10 mL volume with a standard three-electrode configuration. Graphene–platinum nanoparticle-modified glassy carbon electrode (GCE) (diameter = 5 mm) acted as the working electrode, a Pt (purity 99.99%) wire as the counter electrode and Ag/AgCl (3M KCl) as reference electrode (CH Instruments). Before electrochemical measurements, all the solutions were degassed using ultra pure nitrogen gas for about 15 minutes. SWASV was used in the potential range –0.2 to 0.8 V with an amplitude of 0.025 V, frequency of 40 Hz and 0.004 V of potential increment. All pH measurements were carried out using Control Dynamics pH meter (Model: APX 175, Control Dynamics Instruments, Mumbai, India). Infrared spectrometric measurements were recorded using Fourier transform infrared spectroscopy (FTIR) Shimadzu spectrometer (Model: 8400S, Shimadzu Corporation, Kyoto, Japan) in the range 1000–4000 cm^{-1} with a resolution of 4 cm^{-1}. The surface morphology and particle size of platinum nanoparticle-decorated graphene sheets were characterized using field emission scanning electron microscope (FESEM) (Ultra plus, Carl Zeiss, Jena, Germany). The samples for FESEM analysis were prepared by the drop casting of dilute ethanolic solution of Gr–nPt composite on ITO slides. Powder x-ray diffraction (XRD) studies were carried out using Bruker aXS Model D8 Advance powder x-ray diffractometer with a Cu Kα radiation (λ = 0.154 nm). The diffraction data were recorded for 2θ angles between 5° and 85°.

2.2. Chemicals and reagents

All reagents used were analar grade and used without any further purification. Graphite, hexachloroplatinic acid ($H_2PtCl_6 \cdot 6H_2O$), ethylene glycol and potassium bromide were purchased from Sigma-Aldrich (Bangalore, India). Buffer solutions of known pH in the range pH 1–12 were prepared using deionized water from MilliQ water purifier (Denvers, MA, USA) with a resistivity of not less than 18.2 MΩ cm at 27 ± 2°C. Arsenic stock solution (1 mM) was prepared by dissolving 0.012 g of analar grade $NaAsO_2$ in 100 mL distilled water. Working standards were prepared by diluting the standard stock solution on the day of analysis.

2.3. Preparation of graphene oxide and graphene–nanoplatinum composite

Graphene oxide (GO) was synthesized according to the modified Hummer's method by the exfoliation of graphite [18]. Two grams each of graphite and sodium nitrate was taken in 1 L round-bottom flask to which 100 mL of concentrated sulfuric acid was added slowly at 0°C. Twelve grams of solid potassium permanganate was added into the reaction vessel, and the reaction mixture was stirred continuously for 1 hour. Then, 400 mL of distilled water was added and the stirring was continued for another 15 minutes. Then, hydrogen peroxide (30% v/v) was added to the reaction vessel dropwise until the gas evolution completely ceases. The reaction mixture was centrifuged to separate the unexfoliated graphite from the exfoliated one. The residue was separated by decantation and washed several times using small quantities of hydrochloric acid 5% (v/v) solution, until the filtrate solution gave a negative test for the presence of sulfate with barium chloride. Then, the residue was further washed with copious amounts of double-distilled water and dried in vacuum to obtain yellow-brown solid of GO.

The Gr–nPt was prepared by the simultaneous and sequential reduction of hexachloroplatinic acid ($H_2PtCl_6 \cdot 6H_2O$) and GO in one step. In order to convert GO into graphene, 5 mL solution of GO (1 mg mL^{-1} of water) and 1 mM of $H_2PtCl_6 \cdot 6H_2O$ were taken in 100 mL volumetric flask and mixed thoroughly. Then, 40 mL of ethylene glycol was added and sonicated for about 1 hour to obtain homogeneous dispersion. Then, the reduction was carried out at 120°C with constant stirring under reflux conditions. The pale yellow color of the reaction mixture turns to black, which is an indication of formation of platinum nanoparticles [19]. The resulting Gr–nPt composite material was filtered and washed thoroughly with distilled water and dried in vacuum desiccator under ambient conditions. Similarly, the nanocomposite with different loadings of Pt nanoparticles was obtained by varying the concentration of H_2PtCl_6 with the same amount of GO (1 mg mL^{-1} of water).

2.4. Modification of GCE

GCE was used as a conducting substrate for the development of the sensing platform. Prior to its modification, the GCE was polished thoroughly using 1.0, 0.3 and 0.05 μm alumina slurry and then repeatedly washed with double-distilled water followed by sonication in distilled water and ethanol for 10 minutes each. The preparation of modified electrode was carried out by the drop casting of 10 μL of aqueous colloidal dispersion of Gr–nPt (1 mg mL^{-1}) on the surface of pretreated GCE and dried at room temperature (27 ± 2°C).

2.5. Analytical procedure

The electrochemical determination of As(III) was carried out using SWASV in the potential range –0.2 to 0.8 V under optimized conditions. Known amounts of As(III) were taken in an electrochemical cell of 10 mL volume containing 8 mL of hydrochloric acid (1 M) as supporting electrolyte which is fitted with a tab-controlled magnetic stirrer. As(III) was preconcentrated by the electroreduction of As(III) to As(0) at a potential of – 0.4 V for 80 seconds from the bulk of the electrolytic solution onto the surface of modified electrode at the interface and subsequently stripped off from the electrode surface into the bulk of the solution by scanning the potential in a positive direction after 15 seconds of equilibration time.

2.6. Sample preparation

Water samples from borewell and polluted lakes were collected and stored in polyethylene containers as per the standard protocols. The collected water samples were filtered using Whatman filter paper to remove any suspended particulate matter. After filtration, known volumes of the water samples were treated with 0.5 mL each of potassium iodide (5%) and hydrochloric acid (5 M) solutions to reduce any As(V) present in the sample to As(III). The presence of excess of iodine, indicated by pale brown color, was destroyed by adding few drops of ascorbic acid [20]. Known aliquots of these solutions were analyzed for arsenic content according to the procedure described above.

Soil samples were collected from agricultural field, dried and milled to break the lumps into fine powder. The grinded soil powder samples were sieved to get homogeneous powder. One gram of powdered sample was weighed and transferred into 100 mL beaker. To this, 2 mL of water and 0.5 g of potassium hydroxide pellets were added, and the mixture was heated on hot plate until the water evaporates and fuses. Then, it was diluted up to 100 mL, and known aliquot of this sample solution was used to determine arsenic content [21].

The tomato and spinach leaves were dried under sunlight and grinded and sieved in the form of fine powder. About 100 g of the powdered sample was placed in a 250 mL beaker, and 10 mL each of nitric and sulfuric acids was added. This content was heated to 100°C for 20 minutes in fume hood and then cooled to room temperature. To this, 10 mL of perchloric acid was added and heated again in fume hood for 5 minutes, until the dense white fumes of sulfur dioxide cease. The sample was then cooled and 1 mL of concentrated HCl was added to remove any heavy metal ions present in the sample. The solution was heated for 15 minutes and washed with distilled water. Then, As(V) if present was reduced to As(III) by the process described as above. The solution was washed and diluted to 100 mL with distilled water in a calibrated flask. Known aliquots of the diluted solution were used for the determination of arsenic.

3. Results and discussions

3.1. Characterization of the composite

The formation of GO from native graphite and its decoration with platinum nanoparticles was characterized by studying FTIR, XRD and FESEM techniques.

The FTIR spectra of native graphite, GO and Gr–nPt composite are shown in Figure 1. The samples were uniformly mixed with KBr in 1:100 ratio using pestle and mortar and compressed into a thin pellet using pellet press. The spectra of GO (Figure 1b) showed a strong and broad band at 3250 cm^{-1} due to –OH stretching vibration. The carbonyl (–C=O) stretching of carboxylic groups present at the edge planes of the GO sheets was observed at 1728 cm^{-1}. The absorption due to –OH bending, epoxide groups and skeletal ring vibrations were observed at 1616 cm^{-1}. The deformation stretching frequency of –OH groups attached to the aromatic ring was found to be at 1387 cm^{-1} [22]. However, the peaks pertaining to the oxygen-containing functional groups are absent in the spectra of graphite (Figure 1a). All these studies reveal the formation of GO from native graphite and also the generation of oxygen-containing functionalities during oxidation process. However, in case of Gr–nPt composite, the intensity of the peaks pertaining to the oxygen-containing functional groups significantly decreases, and some of the peaks are completely diminished, indicating the absence of oxygen-containing functional groups on the surface of the composite (Figure 1c). All these observations are in good agreement with the reported literature [11]. This is due to the reason that during simultaneous

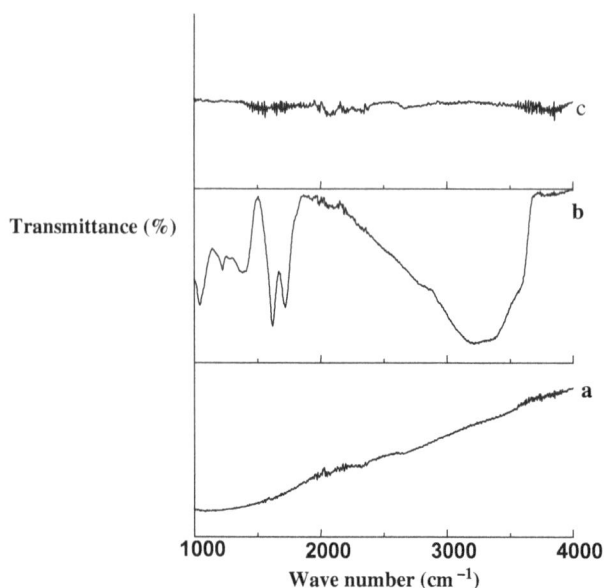

Figure 1. FTIR spectra of (a) graphitic carbon, (b) graphene oxide and (c) graphene–platinum nanocomposite (Gr–nPt).

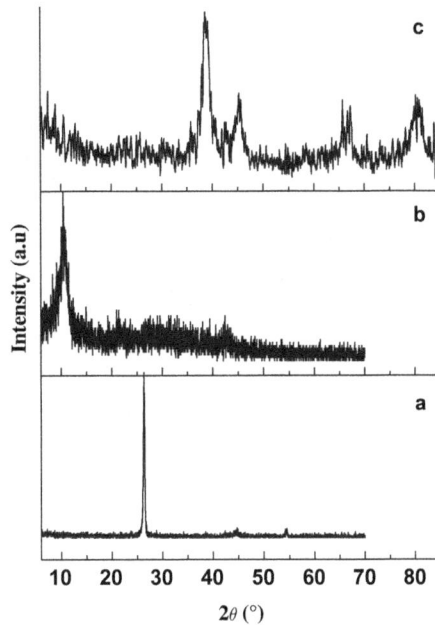

Figure 2. XRD peak patterns of (a) graphite, (b) graphene oxide and (c) graphene–platinum nanocomposite.

reduction of GO and Pt precursor in the presence of ethylene glycol as a reducing agent, all the oxygen-containing functionalities present on the surface of GO get reduced and converted into graphene material [15].

The XRD pattern of graphite, GO and Gr–nPt composite was recorded and compared to confirm the formation of GO as well as the generation of platinum nanoparticles on the layers of graphene (Figure 2). Native graphite showed a sharp peak at $2\theta = 26.34$ with a d spacing of 0.38 nm, which is a characteristic peak of graphitic carbon which corresponds to the diffraction of (002) plane. After chemical oxidation, the (002) peak of graphite has been shifted to 10.5° with a d spacing of 0.94 nm. This shift might be attributed to the introduction of oxygen-containing functional groups like epoxy, hydroxyl, carbonyl and carboxylic groups at both the sides and edges of the graphene sheets, which in turn confirms the formation of GO from natural graphite during oxidation process and also the formation of platinum particles [23].

The XRD pattern of Gr–nPt showed four peaks at 39.8°, 46.6°, 67.7° and 81.4° corresponding to the diffraction of (111), (200), (220) and (311) crystalline planes of face-centered cubic lattice of platinum, respectively [24]. It confirms the presence of platinum particles on the graphene substrate. The size and distribution of Pt nanoparticles on the surface of Gr were studied by recording the images of Pt-containing graphene sheets through FESEM. The selected region of FESEM images revealed that the particles have the size distribution between 1.25 and 6.75 nm with an average size distribution of 4 nm (Figure 3b), which is in good agreement with the XRD measurements. Enlarged portion of FESEM image showed that the Pt particles are uniformly distributed throughout the graphene layers without any agglomeration in the substrate (Figure 3a).

(a)

(b)

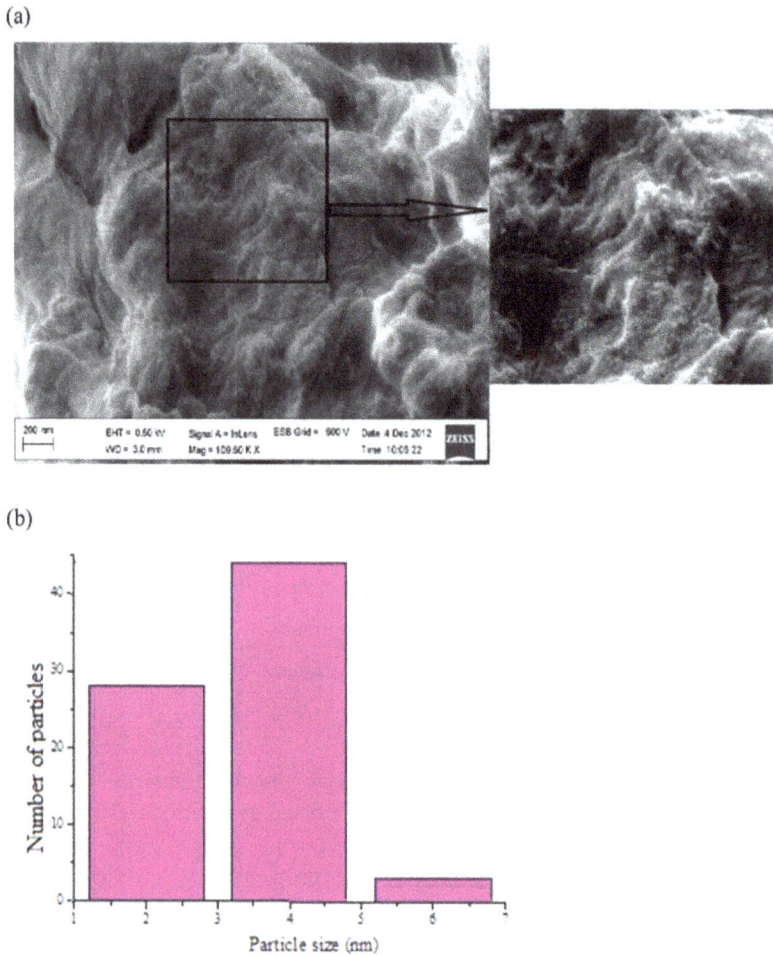

Figure 3.   FESEM images of (a) graphene–platinum composite and (b) particle size profile.

3.2.   Electrochemistry of As(III) at Gr–nPt-modified GCE

The electrochemical behavior of Gr–nPt-modified GCE in presence of As(III) was initially examined by using cyclic voltammetry (CV) in order to decipher the electro-catalytic activity of Gr–nPt composite toward arsenic. Cyclic voltammetric profile of 10 μM of As(III) in 1 M HCl at graphene and Gr–nPt-modified electrode in the potential range from 0.6 to 0.6 V with a scan rate 50 mV sec^{-1} is shown in Figure 4. The graphene-modified GCE did not show any voltammetric signature for the presence of arsenic in the potential window used in the present investigation. This indicates that the graphene alone is not capable of favoring the redox process for arsenic; hence, it is insensitive and not useful material for the construction of sensing platform in arsenic measurement. Whereas Gr–nPt composite-modified GCE showed a voltammetric peak in the forward cathodic sweep at –0.18 V which can be ascribed to the three-electron reduction process of As(III) to As(0), and in the reversal anodic sweep an intense peak at 0.16 V was observed which corresponds to reoxidation (stripping) of As(0) to As(III). The potential required for the reduction of As(III) to As(0) is much less negative than that observed on Pt and Au

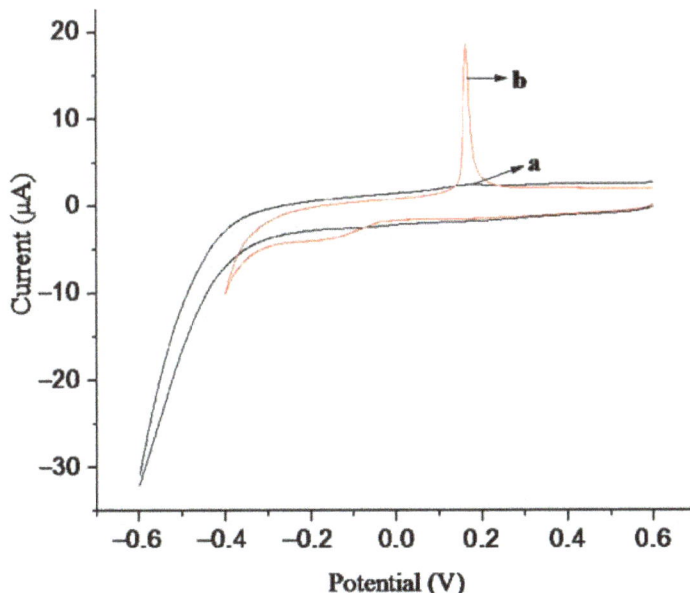

Figure 4.   Overlaid cyclic voltammograms recorded at (a) graphene and (b) Gr–nPt composite-modified glassy carbon electrodes in 1 M HCl in the presence of 10 μM As(III). Scan rate –50 mV sec^{-1}.

nanoparticle-modified GCE [24,25]. These observations revealed that the arsenic reduction process is more favorable at Gr–nPt-modified GCE with low overpotential. Such voltammetric response was not observed in graphene, showing that the observed response at Gr–nPt electrode is due to the presence of Pt nanoparticles. Hence, Gr–nPt composite-modified GCE showed potential affinity toward arsenic, and therefore, the feasible redox reaction of As(III) and the enhancement in the peak current can be attributed to the high electrocatalytic activity of the platinum nanoparticles toward As(III).

It is evident from the Figure 4 that the oxidation peak (stripping) is sharper and more significant than the corresponding reduction peak; hence, the oxidation peak was systematically studied using SWASV in order to achieve the desired detection limit. The stripping voltammetric determination of arsenic proceeds through the following two steps.

(1) Preconcentration of As(III) from the bulk of the electrolytic solution onto the modified electrode surface under the applied potential of –0.4 V for preselected time of 80 seconds under constant stirring.
(2) Stripping (oxidation) of As(0) from the electrode surface into the bulk of the electrolytic solution as As(III). This anodic stripping step generates quantifiable analytical signal, and the signal current has been correlated to the arsenic concentration.

The general scheme for the preconcentration and stripping of arsenic into the electrolytic solution has been shown in the Scheme 1. In a typical stripping voltammogram of arsenic measurement (data not shown), the Gr–nPt composite electrode did not show any voltammetric peak in the absence of arsenic in the potential window used in present study. Hence, it can be applied for the quantification of analytes (metal ions) which are

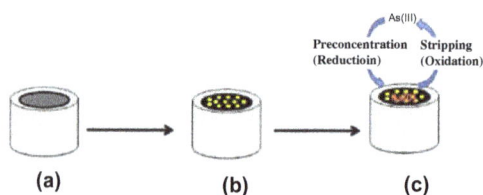

Scheme 1.   Schematic pathway of arsenic detection. (a) Bare glassy carbon electrode. (b) Glassy carbon electrode modified with Gr–nPt composite. (c) Electrocatalysis of As(III) at Gr–nPt interface.

active in this potential region. Whereas in the presence of arsenic, it showed a well-defined and sharp oxidation peak at 0.14 V, indicating that that the Gr–nPt-modified GCE showed potential affinity toward arsenic, and hence this kind of modified electrode can be used for the electrochemical measurement of arsenic at very low concentration level.

3.3.   Optimization study

In order to obtain maximum efficiency of the proposed electrode toward the electrochemical measurement of arsenic, the reaction variables like platinum loading, electrolytic solution, deposition potential and deposition time which controls the analytical response have been optimized.

The amount of platinum nanoparticles present on the graphene layers plays a very important role in the electrochemical determination of arsenic as they behave as catalyst for the redox process of arsenic. Therefore, the Gr–nPt of different loadings of platinum nanoparticles was prepared by varying the concentration of H_2PtCl_6 from 0.2 to 1.4 mM with the same amount of GO (1 mg mL^{-1} of water). The electrochemical performance of each composite has been studied with respect to As(III) using CV. The cyclic voltammetric response increases with increase of platinum loading up to 1 mM; thereafter, there is no significant improvement in the electrochemical performance of arsenic. Therefore, platinum loading of 1 mM of H_2PtCl_6 has been used in the composite preparation.

The shape and sensitivity of the stripping peak in any voltammetric experiment mainly depend on the diffusion of the analyte species from bulk of the electrolytic solution toward the interface and vice versa. The diffusion toward the electrode surface depends on the surface structure and the diffusion from the surface into the bulk of the solution depends on the ions of the electrolytic solution, which show potential interacting ability toward the metal ion present at the interface to form their respective compounds. The Cl$^-$ ions present in the HCl solution show a strong affinity toward ionic arsenic, resulting through oxidation at the interface to form $AsCl_3$ species [26]. Hence the choice of supporting electrolyte governs the magnitude of the stripping current generated at the electrode interface. Hence, the electrochemical response of arsenic at Gr–nPt composite-modified GC electrode has been examined in the presence of some of the commonly used electrolytes like HCl, H_2SO_4 and HNO_3 at different concentration levels (Figure 5) [27]. Among these, HCl showed well-defined sharp and intense peak in comparison with other electrolytes. The concentration of HCl was varied in the concentration range 0.1 to 2 M, and it was found that the peak current increases from 0.1 to 1 M and then it decreases. The increase of peak current might be due to the effective complexing ability of Cl$^-$ ions with As(III) and the decrease of peak current might be due to the adsorption of Cl$^-$ ions on Pt

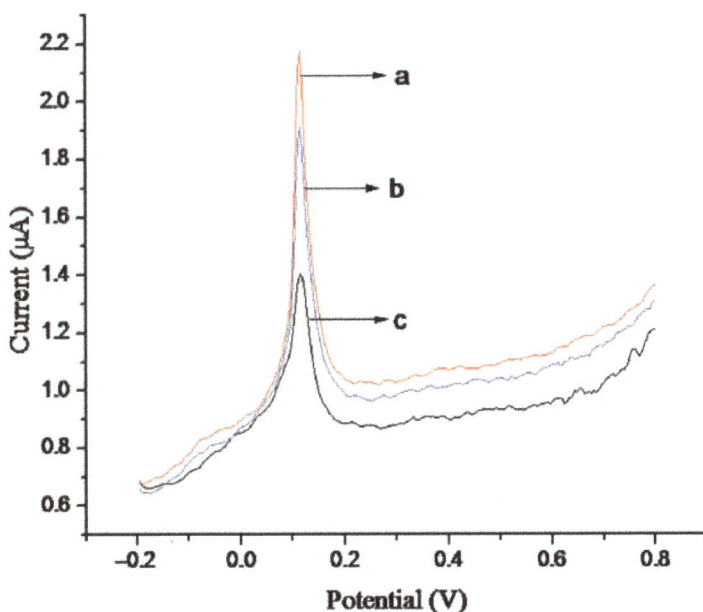

Figure 5. Overlaid stripping voltammograms of 50 nM of As(III) present in 1 M (a) HCl, (b) HNO$_3$ and (c) H$_2$SO$_4$.

surface at higher concentration level [28,29]. In case of H$_2$SO$_4$, least stripping response was noticed which might be due to the bulky and less complexing nature of SO$_4^{2-}$ ions, which results lower coordination number with As(III). Thus, more energy might be required for its dissociation [27]. The peak intensity observed in the presence of hydro-chloric acid as an electrolytic solution was better than nitric acid and sulfuric acids. Hence, hydrochloric acid (1 M) has been used as an optimum electrolyte throughout the study.

The effect of applied potential on the reduction of As(III) to As(0) at the modified interface by holding the electrode at a particular potential was studied. It is well known that higher reduction potential values can be expected to give the maximum current due to the reduction of more and more As(III) ions. Therefore, the dependence of peak current with respect to deposition potential was studied from –0.1 to –0.6 V (Figure 6). The peak current increases with increase of potential from –0.1 to –0.4 V and then decreases. The decrease of peak current is attributed to the inefficient deposition of arsenic. This might be due to the reason that at more cathodic potentials, the water molecules compete with arsenic to undergo reduction and result in the formation of H$_2$ gas which further blocks the surface and decreases the current response. In addition to this, at more cathodic potentials, some of the elemental arsenic can be converted into As^{3-} [25]. Due to this, a deposition potential of –0. 4 V was used as an optimum potential in all further studies. The time required for the accumulation of As(III) from the bulk of the electrolyte solution onto the surface of the Gr–nPt-modified electrode plays a crucial role in trace level estimations. Therefore, the dependence of peak current with respect to the preconcentration time is studied from 20 to 120 seconds (Figure 7). It has been found that the peak current increases with an increase of preconcentration time from 20 to 80 seconds, and then it

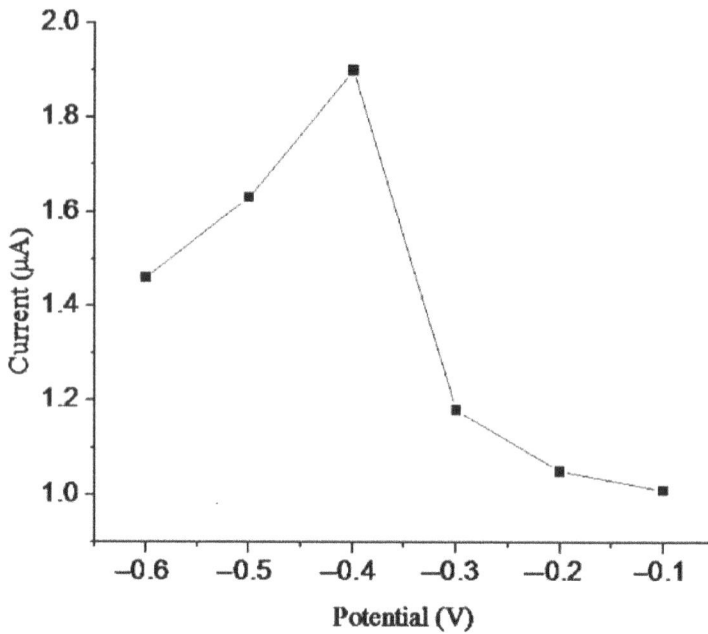

Figure 6. Effect of deposition potential on the stripping peak current obtained for 50 nM of As(III) present in 1 M HCl with a deposition time of 80 seconds.

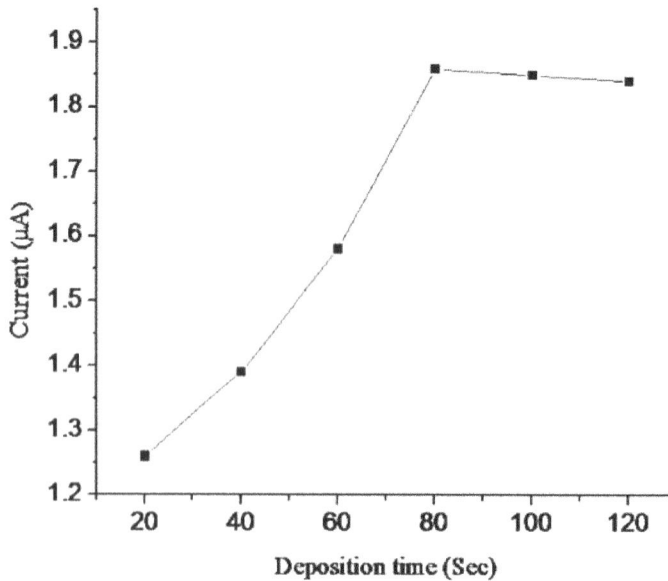

Figure 7. Effect of deposition time on the stripping peak current obtained for 50 nM of As(III) present in 1 M HCl with a deposition potential of –0.4 V.

remained constant. This might be due to the saturation response of arsenic on the electrode at higher concentration as it is a nonconducting semiconductor and does not easily deposit on itself [30].

3.4. Analytical merits

The calibration plot was constructed by measuring the peak currents produced by the stripping of arsenic with the successive addition of 10 nM of As(III) into an electrochemical cell of 10 mL volume under optimized conditions. The Gr–nPt-modified electrode showed a linear response in the concentration range up to 100 nM (Figure 8) with a detection limit of 1.1 nM which is well below the threshold limit value of arsenic in drinking water prescribed by WHO [3].

The proposed sensor has been compared with some of the existing electrodes used for arsenic measurement in terms of detection limits (Table 1). It signifies the importance of the proposed sensor prepared using Gr–nPt composite in the detection of As(III). The proposed sensor provides comparably low detection limits; hence, it can be used as an alternative tool to the existing protocols.

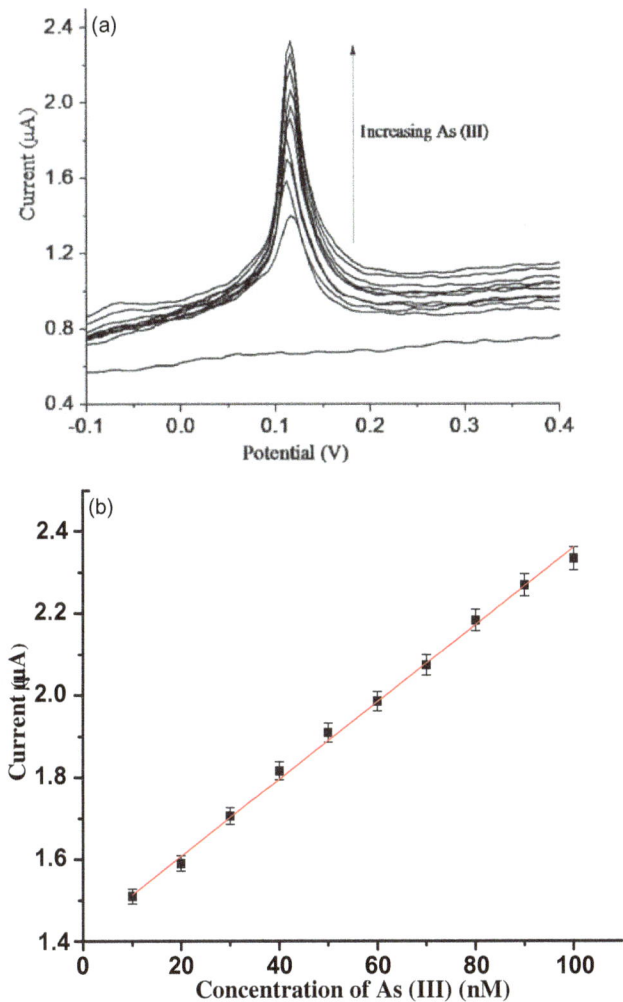

Figure 8. (a) Overlaid square-wave anodic stripping voltammograms of As(III) under optimized conditions and (b) calibration plot.

Table 1. Comparison of few existing electrodes with the proposed electrode in As(III) measurement.

Modified carbon electrodes	Method	Limit of detection (LOD)	Linear range	Reference
Pt–GCE	DPASV	0.48 μM	1–50 μM	[25]
Au–GC/BPPG	CV	0.8 μM	2–12 μM	[31]
Pt-Pd-Polymer/GCE	DPASV	2.4 μM	0–6 mM	[32]
Co oxide–GCE	CV	11 μM	0–4 μM	[33]
rGO–PbO/GCE	SWASV	0.01 μM	0.1–0.1 nM	[34]
Gr–nPt/GC	SWASV	1.1 nM	10–100 nM	Present study

Notes: Pt, platinum nanoparticles; Au, gold nanoparticles; Pd, palladium nanoparticles; Co, cobalt; GC, glassy carbon spheres; GCE, glassy carbon electrode; CNT, carbon nanotubes; BPPG, basal plane pyrolitic graphite; rGO, reduced graphene oxide; PbO, lead oxide; Gr–nPt, graphene platinum nanoparticles; CV, cyclic voltammetry; LSV, linear sweep voltammetry; DPASV, differential pulse anodic stripping voltammetry; SWASV, square-wave anodic stripping voltammetry.

Precise and selective measurement of arsenic present in real sample matrices is a challenging task, as the other commonly encountered cations and anions normally present in the real samples along with arsenic may also get deposited and stripped off from the electrode surface under the same optimized conditions used in the present study. The efficiency of the proposed sensor electrode toward arsenic quantification has been examined in the presence of 50 nM of arsenic along with different concentrations of different ions. The tolerance limits of these ions in the present investigation have been listed in Table 2. Above these tolerance limits, the added ions interfere and alter the peak currents produced by the stripping of arsenic. The developed sensor has showed least interference from most of the common ions due to specific and selective interaction of the Gr–nPt composite with As(III) ions. Hence, this sensor can be used for the selective quantification of arsenic from most of the real sample matrices without any significant interference from other ions.

The validation of the proposed electrode using Gr–nPt composite on the surface of GCE toward As(III) measurement has been examined by applying it to the real sample matrices like borewell, polluted water, agricultural soil, spinach and tomato leaf samples. Known quantities of As(III) were added, and its recovery study was carried out. The proposed electrode showed a linear range up to 100 nM, which is well within the limits of As(III) present in the real sample matrices whose compositions are given in Table 3. Hence, the modified electrode of this kind can be used to quantify As(III) from a variety of water samples as well as vegetable samples.

For practical applications, reproducibility of the analytical results is very important as they require very accurate and precise measurement. Therefore, the reproducibility has been examined by using the same electrode for 15 repetitive measurements in 1 M HCl as supporting electrolyte solution containing 50 nM of As(III) under identical conditions.

Table 2. Interference study.

Interferents	Tolerance limit (μM)
Pb^{2+}, Cd^{2+}, Cu^{2+}, Hg^{2+}	100
Ba^{2+}, Ca^{2+}, Zn^{2+}, Ni^{2+}, Co^{2+}, Fe^{2+}	380
Cl^-, F^-, I^-, Br^-	450
Na^+, K^+, Fe^{3+}	620

Table 3.   Determination of As(III) from different environmental sample matrices.

Sample	Originally present arsenic (nM)	Added arsenic (nM)	Total arsenic found (nM)	Recovery (%)
Polluted water	50	10	59.78	99.63
Borewell water	17	10	27.03	100.11
Soil extract	35	10	45.14	100.31
Tomato leaves	12	10	21.83	99.22
Spinach leaves	20	10	30.02	100.06

The relative standard deviation for all these measurements was found to be ±4.6%. Similarly, the reproducibility of the results has been checked using four different electrodes of similar kind. The relative standard deviation for all these measurements was found to be ±5%. These results indicate that the analytical performance of the fabricated electrode is highly reproducible; hence the proposed sensor can be used for several repetitive measurements with good accuracy and precision in its electrocatalytic activity toward arsenic determination.

4. Conclusions

A simple protocol for the chemical modification of graphene with platinum nanoparticles has been described using a mild and environmental friendly reducing agent, i.e., ethylene glycol. The method has resulted in the formation of platinum nanoparticles with an average particle size of 4 nm. The prepared Gr–nPt composite has been used as a sensing platform in the determination of arsenic at nanomolar level from real sample matrices without any severe interference from common cations and anions. The lowest detection limit obtained can be attributed to the superior electrocatalytic activity of platinum nanoparticles present on the graphene layers. The fabricated electrode is quite stable and can be used for repetitive analytical measurements.

Acknowledgments

The authors acknowledge the Department of Science & Technology (DST), New Delhi, India, for the financial support and award of fellowship to G.K. Raghu. The authors are extremely thankful to Prof. K.R. Nagasundara, Department of Chemistry, Bangalore University for providing hexachloroplatinic acid. The authors also express their gratitude to Prof. V. Lakshminarayanan, Raman Research Institution, Bangalore, for useful discussions during the manuscript preparation.

References

[1] F. Arduini, J.Q. Calvo, G. Palleschi, D. Moscone, and A. Amine, *Bismuth-modified electrodes for lead detection*, Trends Anal. Chem. 29 (2010), pp. 1295–1304. doi:10.1016/j.trac.2010.08.003.

[2] A. Cavicchioli, M.A. La Scalea, and I. Gutz, *Analysis and speciation of traces of arsenic in environmental, food and industrial samples by voltammetry: A review*, Electroanalysis. 16 (2004), pp. 697–711. doi:10.1002/elan.200302936.

[3] World Health Organization, *Arsenic in drinking water*, Fact sheet No. 210 Revised May 2001.

[4] M.D. Luh, R.A. Baker, and D.E. Henley, *Arsenic analysis and toxicity – A review*, Sci. Total Environ. 2 (1973), pp. 1–12.

[5] M. Khairy, R.O. Kadara, D.K. Kampouris, and C.E. Banks, *In situ bismuth modified screen printed electrodes for the bio-monitoring of cadmium in oral (saliva) fluid*, Anal. Methods. 2 (2010), pp. 645–649.

[6] R. Feeney and S.P. Kounaves, *On-site analysis of arsenic in groundwater using a micro-fabricated gold ultramicroelectrode array*, Anal. Chem. 72 (2000), pp. 2222–2228. doi:10.1021/ac991185z.

[7] M.C. Daniel and D. Astruc, *Gold nanoparticles: Assembly, supramolecular chemistry, quantum-size-related properties, and applications toward biology, catalysis, and nanotechnology*, Chem. Rev. 104 (2004), pp. 293–346. doi:10.1021/cr030698.

[8] R. Narayanan and M.A. El-Sayed, *Catalysis with transition metal nanoparticles in colloidal solution: Nanoparticle shape dependence and stability*, J. Phys. Chem. B. 109 (2005), pp. 12663–12676. doi:10.1021/jp051066p.

[9] M.M.O. Thotiyl, H. Basit, J.A. Sánchez, C. Goyer, L. Coche-Guerente, P. Dumy, S. Sampath, P. Labbé, and J. Moutet, *Multilayer assemblies of polyelectrolyte-gold nanoparticles for the electrocatalytic oxidation and detection of arsenic(III)*, J. Colloid Interface Sci. 383 (2012), pp. 130–139. doi:10.1016/j.jcis.2012.06.033.

[10] B.K. Jena and C.R. Raj, *Gold nanoelectrode ensembles for the simultaneous electrochemical detection of ultratrace arsenic, mercury, and copper*, Anal. Chem. 80 (2008), pp. 4836–4844. doi:10.1021/ac071064w.

[11] R.S. Dey and C.R. Raj, *Development of an amperometric cholesterol biosensor based on graphene-Pt nanoparticle hybrid material*, J. Phys. Chem. C. 114 (2010), pp. 21427–21433. doi:10.1021/jp105895a.

[12] H. Cui, W. Yang, X. Li, H. Zhao, and Z. Yuan, *An electrochemical sensor based on a magnetic Fe_3O_4 nanoparticles and gold nanoparticles modified electrode for sensitive determination of trace amounts of arsenic(III)*, Anal. Methods. 4 (2012), pp. 4176–4181. doi:10.1039/c2ay25913c.

[13] Q.X. Zhang and L.B. Yin, *Electrochemical performance of heterostructured Au–Pd bimetallic nanoparticles toward As(III) aqueous media*, Electrochem. Commun. 22 (2012), pp. 57–60. doi:10.1016/j.elecom.2012.05.025.

[14] D.A.C. Brownson, D.K. Kampouris, and C.E. Banks, *Graphene electrochemistry: Fundamental concepts through to prominent*, Chem. Soc. Rev. 21 (2012), pp. 6944–6976. doi:10.1039/c2cs35105f.

[15] C. Xu, X. Wang, and J. Zhu, *Graphene − Metal particle nanocomposites*, J. Phys. Chem. C. 112 (2008), pp. 19841–19845. doi:10.1021/jp807989b.

[16] P.V. Kamath, *Virtual graphene*, Analyst. 19 (2010), pp. 100–125.

[17] S. Guo, D. Wen, Y. Zhai, S. Dong, and E. Wang, *Platinum nanoparticle ensemble-on-graphene hybrid nanosheet: One-pot, rapid synthesis, and used as new electrode material for electrochemical sensing*, ACS Nano. 4 (2010), pp. 3959–3968. doi:10.1021/nn100852h.

[18] W.S. Hummers and R. Offeman, *Preparation of graphitic oxide*, J. Am. Chem. Soc. 80 (1958), pp. 1339–1342. doi:10.1021/ja01539a017.

[19] Y. Li, W. Gao, L. Ci, C. Wang, and P.M. Ajayan, *Catalytic performance of Pt nanoparticles on reduced graphene oxide for methanol electro-oxidation*, Carbon. 48 (2010), pp. 1124–1130. doi:10.1016/j.carbon.2009.11.034.

[20] E. Kavlentis, *Spectrophotometry determination of arsenic(III) and antimony(III) by means of isonicotinoyl hydrazones of 4-dimethylaminobezaldehyde (4-DBIH) and 2-hydroxynaphthaldehyde (2-HNIH)*, Anal. Lett. 20 (1987), pp. 2043–2047. doi:10.1080/00032718708078044.

[21] M. Pandurangappa and K. Sureshkumar, *Trace level arsenic quantification through methyl red bromination*, Am. J. Anal. Chem. 3 (2012), pp. 455–461. doi:10.4236/ajac.2012.37060.

[22] S. Stankovich, D.A. Dikin, R.D. Piner, K.A. Kohlhaas, A. Kleinhammes, Y. Jia, Y. Wu, S.T. Nguyen, and R.S. Ruoff, *Synthesis of graphene-based nanosheets via chemical reduction of exfoliated graphite oxide*, Carbon. 45 (2007), pp. 1558–1565. doi:10.1016/j.carbon.2007.02.034.

[23] G.K. Ramesha and S. Sampath, *Exfoliated graphite oxide modified electrode for the selective determination of picomolar concentration of lead*, Electroanalysis. 19 (2007), pp. 2472–2478. doi:10.1002/elan.200704005.

[24] H.-W. Ha, I.Y. Kim, S.-J. Hwang, and R.S. Ruoff, *One-pot synthesis of platinum nanoparticles embedded on reduced graphene oxide for oxygen reduction in methanol fuel cells*, Electrochem. Solid State Lett. 14 (2011), pp. 70–73. doi:10.1149/1.3584092.

[25] X. Dai and R.G. Compton, *Detection of As(III) via oxidation to As(V) using platinum nanoparticle modified glassy carbon electrodes: Arsenic detection without interference from copper*, Analyst. 131 (2006), pp. 516–521. doi:10.1039/b513686e.

[26] X. Dai, O. Nekrassova, M.E. Hyde, and R.G. Compton, *Anodic stripping voltammetry of arsenic (III) using gold nanoparticle-modified electrodes*, Anal. Chem. 76 (2004), pp. 5924–5929. doi:10.1021/ac049232x.

[27] E. Majid, S. Hrapovic, Y. Liu, K.B. Male, and J.H.T. Luong, *Electrochemical determination of arsenite using a gold nanoparticle modified glassy carbon electrode and flow analysis*, Anal. Chem. 78 (2006), pp. 762–769. doi:10.1021/ac0513562.

[28] A. Profumo, M. Fagnoni, D. Merli, E. Quartarone, S. Protti, D. Dondi, and A. Albini, *Multiwalled carbon nanotube chemically modified gold electrode for inorganic as speciation and Bi(III) determination*, Anal. Chem. 78 (2006), pp. 4194–4199. doi:10.1021/ac060455s.

[29] R. Feeney and S. Kounaves, *On-site analysis of arsenic in groundwater using a microfabricated gold ultramicroelectrode array*, Anal. Chem. 72 (2000), pp. 2222–2228. doi:10.1021/ac991185z.

[30] P. Davis, G. Dulude, R. Griffin, W.R. Matson, and E.W. Zink, *Determination of total arsenic at the nanogram level by high-speed anodic stripping voltammetry*, Anal. Chem. 50 (1978), pp. 137–143. doi:10.1021/ac50023a031.

[31] R. Baron, B. Sljukic, C. Salter, A. Crossley, and R.G. Compton, *Electrochemical detection of arsenic on a gold nanoparticle array*, Phys. Chem. Nanoclust. Nanomat. 81 (2007), p. 1443.

[32] J.A. Sánchez, B.L. Rivas, S.A. Pooley, L. Basaez, E. Pereira, I.P. Paintrand, C. Bucher, G. Royal, E.S. Aman, and J.C. Moutet, *Electrocatalytic oxidation of As(III) to As(V) using noble metal–polymer nanocomposites*, Electrochim. Acta. 55 (2010), pp. 4876–4882. doi:10.1016/j.electacta.2010.03.080.

[33] A. Salimi, H. Mamkhezri, R. Hallaj, and S. Soltanian, *Electrochemical detection of trace amount of arsenic(III) at glassy carbon electrode modified with cobalt oxide nanoparticles*, Sens. Actuat. B Chem. 129 (2008), pp. 246–254. doi:10.1016/j.snb.2007.08.017.

[34] G.K. Ramesha and S. Sampath, *In-situ formation of graphene – lead oxide composite and its use in trace arsenic detection*, Sensors Actuat. B Chem. 160 (2011), pp. 306–311. doi:10.1016/j.snb.2011.07.053.

6

Acoustic cloak constructed with thin-plate metamaterials

Pei Li[a], Xuebing Chen[a], Xiaoming Zhou[a]*, Gengkai Hu[a] and Ping Xiang[b]

[a]Key Laboratory of Dynamics and Control of Flight Vehicle, Ministry of Education, and School of Aerospace Engineering, Beijing Institute of Technology, 100081, Beijing, China; [b]Systems Engineering Research Institute, 100036, Beijing, China

We propose a strategy for designing the cylindrical acoustic cloak with thin-plate metamaterials. The inhomogeneous cloaking shell as derived by transformation acoustics is first discretized into a three-layer anisotropic metafluid, and their material parameters are optimized by minimizing the external scatterings. Then these metafluids are practically realized by thin-plate structures according to the metamaterial concept. As an example, an acoustic cloak is designed with nine layers of thin plate and totally 900 plate units. Numerical simulations are performed to assess the cloaking performance of the designed structure.

Keywords: acoustic cloak; thin-plate metamaterial; optimization

1. Introduction

Cloak is a device that keeps objects undetectable to electromagnetic waves or sound waves. This device is becoming true due to the rapid development of the transformation theory and metamaterial technology. Transformation theory is based on the invariance of coordinate transformation of wave system. Pendry et al. [1] first use this theory to design an invisibility cloak that can prevent electromagnetic radiation into the concealed region and cancel the outside scatterings in the meantime. The cloaking materials possess anisotropy and inhomogeneous material parameters. Based on the metamaterial technology, many structured prototypes of the electromagnetic cloak have been proposed and fabricated. Later, the cloaking concept has been extended to the acoustics realm and may open a novel application to acoustic stealth technology. Norris [2] demonstrates that, different from the invariance of Maxwell's equation, the transformation of acoustic wave equation is not unique. It means that acoustic cloaking material can be constructed with either the inertial metafluid with scalar compressibility and inertial mass tensor, or the pentamode material with scalar mass density and anisotropic compressibility. Focusing on the inertial acoustic cloak, Cheng et al. [3] design the cloak with concentric alternating fluid layers based on the effective medium theory. Daniel et al. [4] consider the discrete cloaking layer as being made of sonic crystals containing two types of solid cylinders immersed in fluid, whose elastic parameters should be properly chosen in order to satisfy the acoustic properties under request. Popa et al. [5] realize the ground acoustic cloak by acoustic metamaterials composed of blocks of steel, aluminum foam, and silicon carbide foam [6]. Zhang et al. [7] construct a quasi-two-dimensional inertial cloak with a network

---

*Corresponding author. Email: zhxming@bit.edu.cn

of acoustic circuit elements. Norris et al. [8] discuss the realization of the cloak region with only three acoustic fluids. More recently, the experimental realization of three-dimensional axisymmetric cloak [9] and ground cloak [10] has been reported.

A major issue of inertial acoustic cloak is the narrow frequency bandwidth induced by the dispersion effect of the resonant metafluid. A prospective solution of this issue is the active acoustic cloak with tunable operation frequency. This idea may be realized by active acoustic metamaterials, and their structural units are usually restricted to be the plate or shell equipped with piezoelectric devices. Regarding this issue, the purpose of this work is to develop a strategy for thin-plate acoustic cloak with the potential tunability mechanism. To do so, we first discretize the cloaking material with originally continuous parameter distribution into three effectively homogeneous metafluids. Material parameters of these effective metafluids are determined from the optimization process of minimum background scattering. This process is similar to what has operated for an electromagnetic cloak [11]. This part is presented in Section 2. We then consider thin-plate metamaterials with circular profiles, but still keep straight plates for each unit cells, in order to construct the cloaking layers. The procedure for designing thin-plate acoustic cloak and numerical verification of the cloaking effect are reported in Section 3.

2. Parameter optimization of acoustic cloak made of three-layer anisotropic metafluids

The parameter distribution of the cloaking material given by the transformation theory [12,13] is continuous and inhomogeneous in space. They are not readily realized with the structured metamaterials. To facilitate the structure design, the cloaking material with continuous parameter distribution is discretized into a three-layer anisotropic metafluid. We demonstrate below that the parameters of metafluids can be optimized to produce an imperfect cloak, but with the minor background scattering. The objective function of the optimization algorithm is the scattering cross section of the cloaked object. To this end, the analysis on a plane acoustic wave incident on a rigid cylinder coated with the cloaking material is conducted first.

2.1. Acoustic scattering by a circular cylinder coated with anisotropic metafluids

The cloaking model is shown in Figure 1, where a cylindrical scatterer to be concealed is cloaked by a three-layer metafluid with the scalar compressibility κ_i and anisotropic mass density $\tilde{\rho} = \text{diag}[\rho_{r,i}, \rho_{\theta,i}]$ ($i = 1, 2, 3$). The mass density and compressibility of the background medium (Region 4) are assumed as ρ_0 and κ_0.

The equations of motion in a medium with an anisotropic density can be written as [13]:

$$\nabla p = j\omega \begin{bmatrix} \rho_r & 0 \\ 0 & \rho_\theta \end{bmatrix} \rho_0 v \tag{1}$$

$$j\omega p = \kappa \kappa_0 \nabla \cdot v \tag{2}$$

Combine Equations (1) and (2) to get:

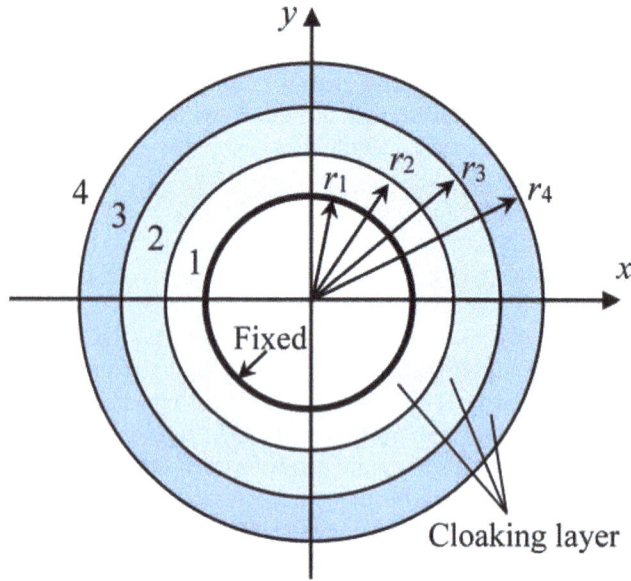

Figure 1. A rigid cylinder covered by three layers of homogeneous metafluid with anisotropic mass density.

$$\kappa \nabla \cdot \left(\begin{bmatrix} 1/\rho_r & 0 \\ 0 & 1/\rho_\theta \end{bmatrix} \nabla p \right) + k_0^2 p = p, \tag{3}$$

where $k_0^2 = \omega^2 \rho_0 / \kappa_0$. In the polar coordinate, Equation (3) can be rewritten as:

$$\frac{1}{r} \frac{\partial}{\partial r} \left(r \frac{\partial p}{\partial r} \right) + \frac{\rho_r}{r^2 \rho_\theta} \frac{\partial^2 p}{\partial \theta^2} + \frac{\rho_r k_0^2 p}{\kappa} = 0. \tag{4}$$

The method of separation of variables can be used to derive the general solution of Equation (4). Regarding the cloaking model under study, the pressure field $p^{(i)}$ in each region of the system can be written as:

$$\begin{cases} p^{(i)}(r,\theta) = \sum\limits_{m=0}^{\infty} \left[A_m^i J_{q_i}\left(k_0 r \sqrt{\frac{\rho_{r,i}}{\kappa_i}} \right) + B_m^i H_{q_i}\left(k_0 r \sqrt{\frac{\rho_{r,i}}{\kappa_i}} \right) \right] e^{jm\theta}, r_i < r < r_{i+1}; i = 1,2,3 \\ p^{(4)}(r,\theta) = \sum\limits_{m=0}^{\infty} \left[A_m^4 J_m(k_0 r) + B_m^4 H_m(k_0 r) \right] e^{jm\theta}, r > r_4 \end{cases} \tag{5}$$

where J_{q_i} and H_{q_i} are Bessel and Hankel functions of the first kind with the order $q_i = m\sqrt{\rho_{r,i}/\rho_{\theta,i}}$. A_m^i (i = 1,2,3) and B_m^i (i = 1,2,3,4) are unknown scattering coefficients. A_m^4 is related to the incident plane wave and is given by:

$$A_m^4 = \begin{cases} 1, m = 0 \\ 2j^m, m > 0 \end{cases}. \tag{6}$$

The radial velocity can be written in terms of the pressure as:

$$v_r^{(i)} = \frac{1}{j\omega\rho_{r,i}\rho_{\theta,i}}\frac{\partial p^{(i)}}{\partial r}.$$ (7)

The continuity conditions of pressure and radial velocity of the system are:

$$p^{(i)}(r_{i+1}) = p^{(i+1)}(r_{i+1}), i = 1, 2, 3$$ (8)

$$v^{(i)}(r_{i+1}) = v^{(i+1)}(r_{i+1}), i = 1, 2, 3.$$ (9)

At the boundary $r = r_1$, the fixed boundary condition is considered, namely

$$v^1(r_1) = 0.$$ (10)

All unknown coefficients A_m^i and B_m^i can be determined according to Equations (8)–(10). The scattering field in the background medium is given by:

$$P_{\text{sc}} = \sum_{m=0}^{\infty} B_m^4 H_m(k_0 r)\cos(m\theta).$$ (11)

The scattering width (SW) of the coated cylinder is defined as:

$$\sigma = 2\pi R|P_{\text{sc}}(\theta, R)/P_{\text{inc}}|^2$$ (12)

where P_{inc} is the incident pressure written as

$$P_{\text{inc}} = J_0(k_0 r) + 2\sum_{m=1}^{\infty} j^m J_m(k_0 r)\cos(m\theta)$$ (13)

2.2. Optimization of the cloaking parameters based on scattering cancellation

The optimization process consists in searching the proper parameters of the cloaking layer for the least scatterings σ in the background. Considering the properties of Hankel function, the scattering field in the far field is inversely proportional to the distance, where scattering width σ becomes almost independent on the location under test. Consider also that the forward scattering ($\theta = 0$) is usually the largest among all directions. The goal of the optimization is defined as the minimum forward scattering in the far field region. The objective function σ is related to only the cloaking parameters, as follows:

$$\sigma(\theta = 0, R) = f(\rho_{ri}, \rho_{\theta i}, \kappa_i)$$ (14)

The optimization algorithm is based on the *fminsearch* function implemented in Matlab. Remind first that the continuous parameter distribution of the perfect cloaking layer [13] is given by:

$$\tilde{\rho}_{ri} = \frac{r}{r - r_1}, \tilde{\rho}_{\theta i} = \frac{r - r_1}{r}, \tilde{\kappa}_i = \left(\frac{r_4 - r_1}{r_4}\right)^2 \frac{r}{r - r_1}, i = 1, 2, 3. \tag{15}$$

The initial value, that is important for reducing the time cost in the optimization, can be taken as the value at the mid-point $(r = (r_i + r_{i+1})/2)$ of the layer and expressed as:

$$\rho_{ri} = \frac{r_i + r_{i+1}}{r_i + r_{i+1} - 2r_1}, \rho_{\theta i} = \frac{r_i + r_{i+1} - 2r_1}{r_i + r_{i+1}}, \kappa_i = \left(\frac{r_4 - r_1}{r_4}\right)^2 \frac{r_i + r_{i+1}}{r_i + r_{i+1} - 2r_1}, i = 1, 2, 3. \tag{16}$$

As an example, three layers with $r_2 = 0.4$ m, $r_3 = 0.5$ m and $r_4 = 0.6$ m are used to cloak a rigid cylinder of radius $r_1 = 0.3$ m at 500 Hz. The density and bulk modulus of the air background are $\rho_0 = 1.25$ kg/m^3 and $\kappa_0 = 0.147$ MPa. The initial and optimized parameters of the cloaking layers relative to the background medium's parameters are listed in Table 1.

The total pressure fields for the uncloaked cylinder and cloaked one with initial and optimized parameters are shown in Figure 2. The numerical simulation is performed based on the commercial software package COMSOL Multiphysics PDE module. Compared with the scattering effect observed for the bare cylinder, the scattering has been efficiently weakened by the cloak having initial parameters (see Figure 2(b)). Furthermore, the optimized material parameters can lead to almost perfect cloaking effect, as shown in Figure 2(c). This example demonstrates that the optimization algorithm is efficient to improve the cloaking effect with only three coating layers.

Table 1.   Initial and optimized material parameters in the cloaking region.

	Initial values			Optimized values		
	ρ_r	ρ_θ	κ	ρ_r	ρ_θ	κ
Layer 1	7	0.29	1.75	5.46	0.15	2.28
Layer 2	3	0.67	0.75	2.79	0.28	0.63
Layer 3	2.2	0.46	0.55	2.07	0.50	0.55

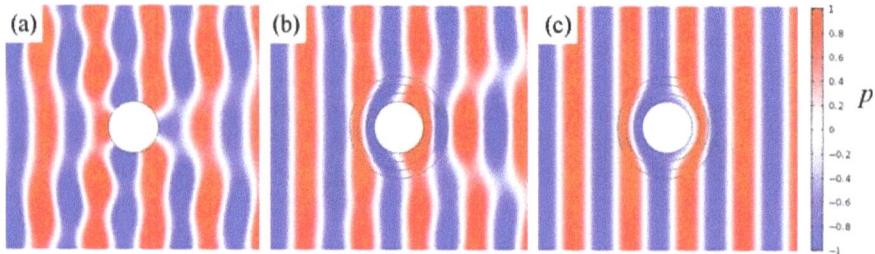

Figure 2.   Total pressure fields at frequency 500Hz: (a) the cylinder without the cloak; (b) the cylinder with the cloak having initial parameters; (c) the cylinder with the cloak having optimized parameters.

3. Designing the cloaking structure with thin-plate metamaterials

3.1. *Acoustic characteristics of thin-plate metamaterials with the circular profile*

Thin-plate acoustic metamaterials are composed of parallel-stacked thin plates, attached periodically with local resonant units. A simple version of thin-plate metamaterials is the simply supported plate (see Figure 3(a)), and its effective acoustic property has been extensively analyzed regarding normally and obliquely incident waves [14]. It is found that the effective density of the simply supported plate is anisotropic, in which the radial density follows the Drude-medium model, while the annular density is almost the background one. It is also found that effective radial density won't vary with the change of the incident angle, when the wavelength of the incident wave is greater than the lattice constant of the plate. These properties make the thin-plate metamaterial a suitable candidate for the cloaking structure. To construct the cloaking layer, we change the straight shape of the plate into the circular profile, while keeping the plates straight in unit cells, as shown in Figure 3(b). The problem arising in the case of circular profile is that the length of straight plate locating in the different radial position will be different. But in order for contrasting an effective metafluid with a multilayered plate, it is convenient to make sure the same effective length of straight plates. The solution is that, the joint between plates at the innermost layer is fixed, and a part of the plate is fixed in other layers, so that the length of the free part of the plate is same to that of the innermost plate.

We assume and will verify later that the effective properties of the circular metamaterials inherit the ones of the straight plates. Consider the Young's modulus E_Y = 0.12 GPa, mass density ρ_p = 1000 kg/m^3, and Poisson's ratio v = 0.33 for the plate and the geometric parameters L = 19.9 mm, d = 33 mm, and the thickness of plate h = 0.5 mm. The background medium is the air with mass density ρ_0 = 1.25 kg/m^3 and sound velocity c_0 = 343 m/s. Figure 4 shows the density of a three-layer plate in the direction vertical to the plate surface, which is also the radial density in the circular model, as a function of frequency for different incident angles θ = 0, $\pi/6$, $\pi/3$. It is seen that effective radial density is almost irrelevant to the incident angles below 500 Hz. We will show below that this angle-independent behavior is necessary for the cloak design.

To verify the effective properties of the circular model, we compute the scatterings of a rigid cylinder covered by the structured plates and their effective medium. For the plate structure considered in Figure 4, the thickness of the coating layer is set as 100 mm and the radius of the cylindrical core is 30 mm. At frequency 500 Hz, the effective parameters are ρ_e = diag[2.702, 1.015]ρ_0 and κ_e = 1.026κ_0 . The scattering width at R = 2 m for the two systems are shown in Figure 5(a). It can be seen that theoretical results based on

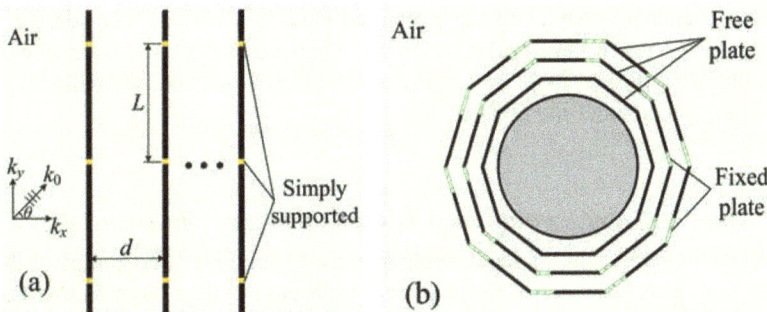

Figure 3.   (a) The model of thin-plate acoustic metamaterials; (b) Schematic diagram of thin-plate metamaterials with the circular profile.

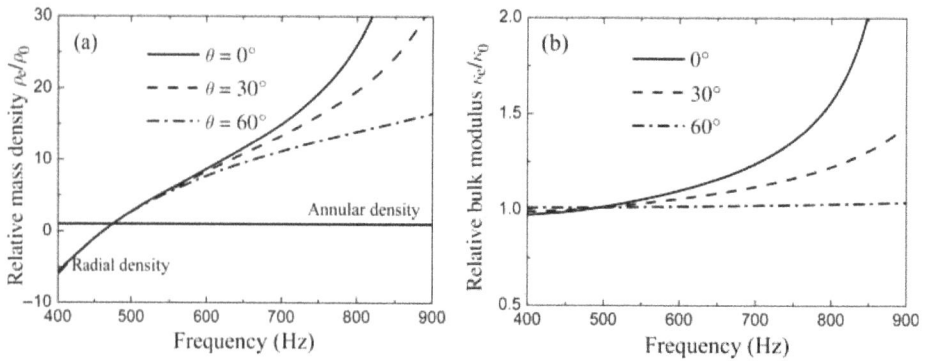

Figure 4.   (a) Effective mass density and (b) effective bulk modulus of thin-plate metamaterials as a function of frequency at three different incident angles, $\theta = 0$, $\pi/6$, $\pi/3$.

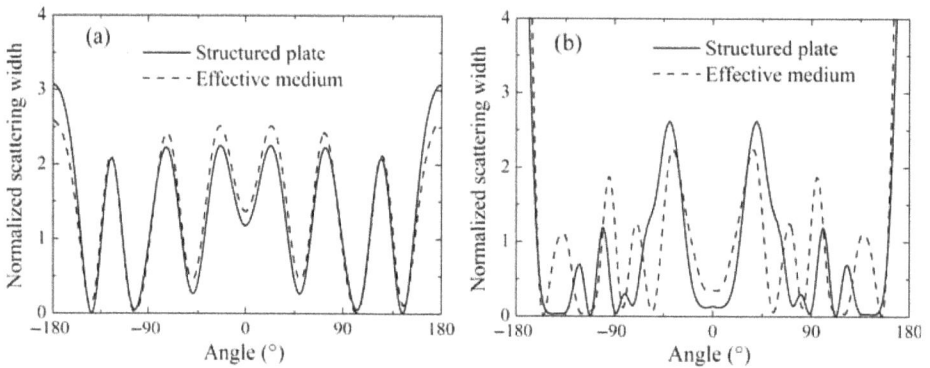

Figure 5.   Normalized scattering width calculated at $R = 2$ m for the structured plate and their effective medium at frequencies (a) 500 Hz and (b) 800Hz.

effective parameters fit very well the scatterings of the actual structure. As a comparison, we show in Figure 5(b) the results of scattering width at frequency 800 Hz, where effective parameters are varied with respect to the incident angle. If effective radial density and modulus in normally incident case are considered, $\rho_e = \mathrm{diag}[26, 1.015]\rho_0$ and $\kappa_e = 1.56\kappa_0$, the theoretical model gives a wrong prediction. Above results mean that the angle-independent behavior of the structured plate is necessary, considering the fact that the straight plates in the circular model must behave in the same manner when they interact with diffracted waves of any direction. In the next section, we will use the thin-plate structure to construct the acoustic cloak.

3.2.   Structure design of acoustic cloak based on thin-plate metamaterials

Figure 6 shows the structure of acoustic cloak constructed with three groups of thin-plate structures; each group comprises three layers of plates to realize the effective fluid with the optimized parameters listed in Table 1. To add a new degree of freedom for tuning the annular density, the region between plates in each group is filled with a specific fluid different from the background medium. The number of plates for each layer is set as 100,

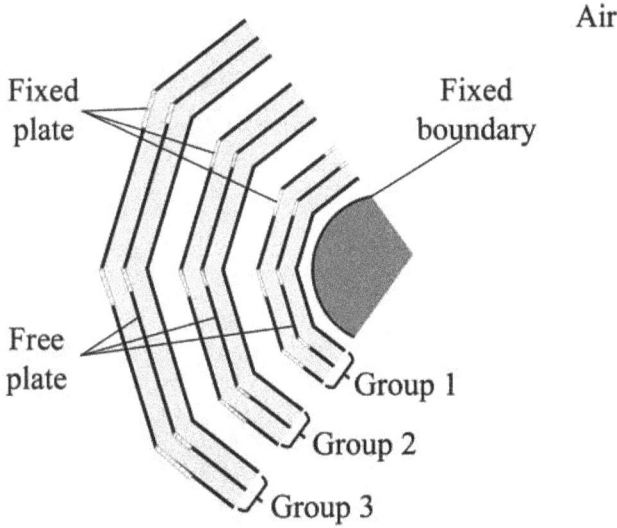

Figure 6.   Schematic of the cloak structure made of thin-plate metamaterials.

then totally 900 thin plates are used to construct the acoustic cloak. For the targeting frequency 500 Hz, the corresponding air wavelength is about 20 times larger than the plate size. This ensures that the angle-dependent behavior of the plate will not appear in the current system.

The fluid filled between plates in each group plays the role of controlling the annular density and bulk modulus of the effective metafluid. Effective annular density of the N-layer plate system with the filling fluid is predicted by:

$$\frac{Nd}{\rho_{\text{eff}}} = \frac{(N-1)(d-h)}{\rho_{\text{filling fluid}}} + \frac{Nh}{\rho_{\text{plate}}} + \frac{d-h}{\rho_{\text{air}}}. \tag{17}$$

The desirable bulk modulus of the metafluid can be achieved by choosing the proper compressibility of the filling fluid. Once the filling fluid is finalized, the effective radial density is controlled by adjusting the thickness of the plate to fit the value required for the cloaking parameter. Note that this thickness variation has little influence on the annular density. For more detail, both effective modulus and radial density are predicted by transfer matrix method [14]. For an effectively homogeneous medium with anisotropic mass density $\tilde{\rho}_{\mathbf{e}} = \text{diag}\left[\rho_x, \rho_y\right]$, the transfer matrix $\mathbf{T}_{\mathbf{e}}$ of the pressure and normal velocity can be calculated analytically. The effective material parameters of thin-plate metamaterials can be retrieved by enforcing $\mathbf{T} = \mathbf{T}_{\mathbf{e}}$, where $\mathbf{T}$ is the transfer matrix of the multilayered plates and can be computed with numerical methods. The geometric and material parameters of the cloaking structure are listed in Table 2, which are designed to fulfill the optimized parameters shown in Table 1.

We perform numerical simulation to verify the cloaking effect of the designed structure. Figure 7(b) shows the pressure distribution of the cloak structure under a plane incident wave at 500 Hz. Compared with the uncloaked cylinder (see Figure 2(a)), the scatterings can be drastically reduced with the cloaking structure. The

Table 2.   A set of geometric and material parameters of acoustic cloak constructed with thin-plate structures.

		Group 1	Group 2	Group 3
Plate	Length (mm)	19.9	26.2	32.5
	Thickness (mm)	0.5	0.5	0.5
	Distance between plates (mm)	33	33	33
	Mass density (10^3 kg/m^3)	1.5	0.5	0.5
	Young's modulus (GPa)	0.16	0.15	0.39
	Poisson's ratio	0.33	0.33	0.33
Filling fluid	Mass density (kg/m^3)	0.126	0.275	0.5
	Bulk modulus (MPa)	0.58	0.091	0.078

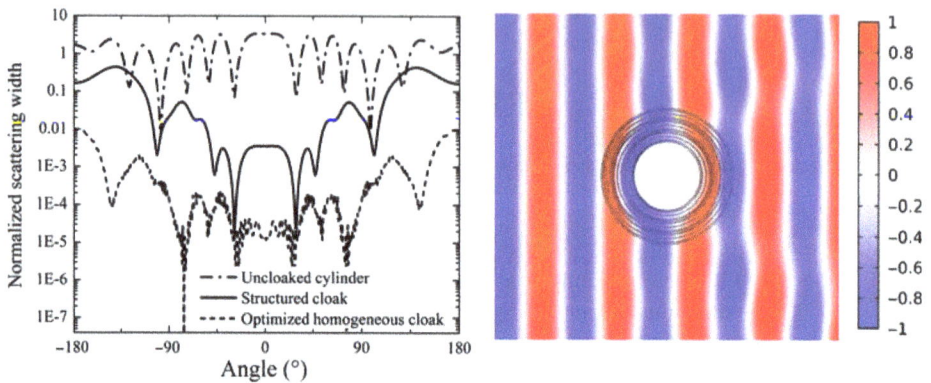

Figure 7.   (a) Normalized scattering width of the uncloaked cylinder, the cylinder with the structured cloak, and that with optimized homogenous cloak; (b) Total pressure fields for the structured cloak incident by a plane acoustic wave.

scattering width calculated in the far field is presented in Figure 7(a) and shows that the scattering level has been reduced to 0.1 times less than the bare cylinder in almost all directions, even though the structured cloak is not as good as the imaginary cloak with optimized parameters.

We also calculate the total and back scattering widths at different frequencies, as shown in Figure 8(a) and (b), respectively. The total scattering can be suppressed in a narrow frequency band and reach a minimum value at the targeting frequency 500 Hz. This cloaking phenomenon is what we want to achieve by the parameter optimization and structure design. Due to the frequency-dependent properties of the plate structure, the narrow-band behavior of the cloak is inevitable, however, may be overcome by active thin-plate metamaterials with the tunable mechanism [15–17]. If the reduction of the wave reflection is only considered, the bandwidth of the back scattering width becomes wider in comparison with the total scattering results.

From the fabrication point of view, we can add constraints in the optimization procedure for pursuing physically realizable materials. However, it is inevitable that the annular density of the cloaking layer must be less than the background one because the concept of the parameter design is based on transformation acoustics. This brings the most difficult part of practical realization, as it is not easy to find a fluid with smaller density

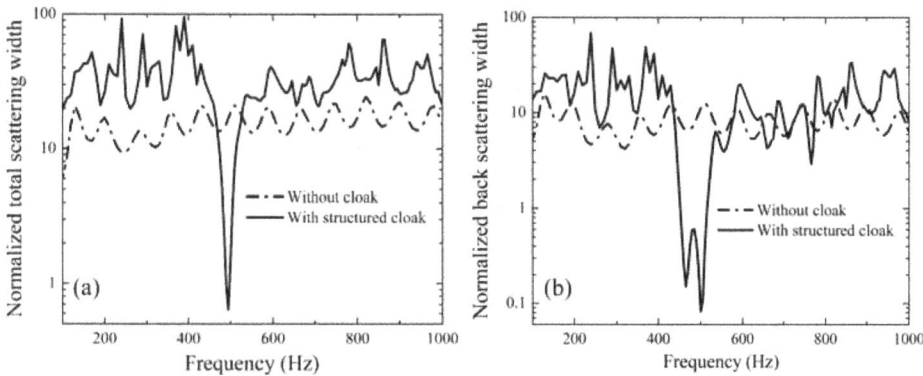

Figure 8.   (a) Total scattering width and (b) back scattering width as a function of frequency in the absence and presence of the structured cloak.

than the air. However, it is worth to stress that the proposed approach for designing acoustic cloak is applicable to any fluid environment. If the background is water, it will be easier than in the air case to find a fluid material with density less than the water one. One example of this fluid can be the water filled with solid spherical shell containing the air.

4. Conclusions

In this work, we propose an idea to realize acoustic cloak with thin-plate metamaterials. The cloaking material is made of three layers of metafluids with anisotropic density. Using the optimization algorithm, the material parameters in each region of metafluids are determined with the goal of the minimized acoustic scatterings. The optimized parameters are anisotropic in density and can be realized by thin-plate metamaterials. As an example, a cloak is designed with nine layers of the plate structure and totally 900 plate units. Numerical simulation is conducted to verify the cloaking effect of the structure near the designed frequency. The proposed cloak based on thin-plate structures can be combined potentially with active piezoelectric devices [16], opening then a new direction for active control of acoustic cloaking in a wide frequency band.

Disclosure statement

No potential conflict of interest was reported by the authors.

Funding

This work was supported by the National Natural Science Foundation of China (grant numbers 10832002, 11172038, 11072031, and 11221202); the National Basic Research Program of China (grant number 2011CB610302); Program for New Century Excellent Talents in University (grant number NCET-11-0794); and Beijing Higher Education Young Elite Teacher Project.

References

[1] J.B. Pendry, D. Schurig, and D.R. Smith, *Controlling electromagnetic fields*, Science. 312 (2006), pp. 1780–1782. doi:10.1126/science.1125907

[2] A.N. Norris, *Acoustic metafluids*, J. Acoust. Soc. Am. 125 (2009), pp. 839–849. doi:10.1121/1.3050288

[3] Y. Cheng, F. Yang, J.Y. Xu, and X.J. Liu, *A multilayer structured acoustic cloak with homogeneous isotropic materials*, Appl. Phys. Lett. 92 (2008), pp. 151913 (3 pp). doi:10.1063/1.2903500

[4] T. Daniel and S.-D. José, *Acoustic cloaking in two dimensions: a feasible approach*, New J. Phys. 10 (2008), pp. 063015 (10 pp). doi:10.1088/1367-2630/10/6/063015

[5] B.-I. Popa and S.A. Cummer, *Homogeneous and compact acoustic ground cloaks*, Phys. Rev. B. 83 (2011), pp. 224304 (6 pp). doi:10.1103/PhysRevB.83.224304

[6] B.-I. Popa, L. Zigoneanu, and S.A. Cummer, *Experimental acoustic ground cloak in air*, Phys. Rev. Lett. 106 (2011), pp. 253901 (4 pp). doi:10.1103/PhysRevLett.106.253901

[7] S. Zhang, C. Xia, and N. Fang, *Broadband acoustic cloak for ultrasound waves*, Phys. Rev. Lett. 106 (2011), pp. 024301 (4 pp). doi:10.1103/PhysRevLett.106.024301

[8] A.N. Norris and A.J. Nagy, *Acoustic metafluids made from three acoustic fluids*, J. Acoust. Soc. Am. 128 (2010), pp. 1606–1616. doi:10.1121/1.3479022

[9] L. Sanchis, V.M. García-Chocano, R. Llopis-Pontiveros, A. Climente, J. Martínez-Pastor, F. Cervera, and J. Sánchez-Dehesa, *Three-dimensional axisymmetric cloak based on the cancellation of acoustic scattering from a sphere*, Phys. Rev. Lett. 110 (2013), pp. 124301 (5 pp). doi:10.1103/PhysRevLett.110.124301

[10] L. Zigoneanu, B.I. Popa, and S.A. Cummer, *Three-dimensional broadband omnidirectional acoustic ground cloak*, Nat. Mater. 13 (2014), pp. 352–355. doi:10.1038/nmat3901

[11] B.-I. Popa and S.A. Cummer, *Cloaking with optimized homogeneous anisotropic layers*, Phys. Rev. A. 79 (2009), pp. 023806 (4 pp). doi:10.1103/PhysRevA.79.023806

[12] H. Chen and C.T. Chan, *Acoustic cloaking in three dimensions using acoustic metamaterials*, Appl. Phys. Lett. 91 (2007), pp. 183518 (3 pp). doi:10.1063/1.2803315

[13] S.A. Cummer and D. Schurig, *One path to acoustic cloaking*, New J. Phys. 9 (2007), pp. 45 (8 pp). doi:10.1088/1367-2630/9/3/045

[14] P. Li, S. Yao, X. Zhou, G. Huang, and G. Hu, *Effective medium theory of thin-plate acoustic metamaterials*, J. Acoust. Soc. Am. 135 (2014), pp. 1844–1852. doi:10.1121/1.4868400

[15] I. Spiousas, D. Torrent, and J. Sánchez-Dehesa, *Experimental realization of broadband tunable resonators based on anisotropic metafluids*, Appl. Phys. Lett. 98 (2011), pp. 244102 (3 pp). doi:10.1063/1.3599849

[16] B.-I. Popa, L. Zigoneanu, and S.A. Cummer, *Tunable active acoustic metamaterials*, Phys. Rev. B. 88 (2013), pp. 024303 (8 pp). doi:10.1103/PhysRevB.88.024303

[17] X. Chen, X. Xu, S. Ai, H. Chen, Y. Pei, and X. Zhou, *Active acoustic metamaterials with tunable effective mass density by gradient magnetic fields*, Appl. Phys. Lett. 105 (2014), pp. 071913 (4 pp). doi:10.1063/1.4893921

7

Improving electromechanical output of IPMC by high surface area Pd-Pt electrodes and tailored ionomer membrane thickness

Viljar Palmre[a,b], Seong Jun Kim[a], David Pugal[a] and Kwang Kim[b]*

[a]Department of Mechanical Engineering, University of Nevada-Reno, Reno, NV, USA; [b]Department of Mechanical Engineering, University of Nevada-Las Vegas, Las Vegas, NV, USA

In this study, we attempt to improve the electromechanical performance of ionic polymer–metal composites (IPMCs) by developing high surface area Pd-Pt electrodes and tailoring the ionomer membrane thickness. With proper electroless plating techniques, a high dispersion of palladium particles is achieved deep in the ionomer membrane, thereby increasing notably the interfacial surface area of electrodes. The membrane thickness is increased using 0.5 and 1 mm thick ionomer films. For comparison, IPMCs with the same ionomer membranes, but conventional Pt electrodes, are also prepared and studied. The electromechanical, mechanoelectrical, electrochemical and mechanical properties of different IPMCs are characterized and discussed. Scanning electron microscopy-energy dispersive X-ray (SEM-EDS) is used to investigate the distribution of deposited electrode metals in the cross section of Pd-Pt IPMCs. Our experiments demonstrate that IPMCs assembled with millimeter thick ionomer membranes and newly developed Pd-Pt electrodes are superior in mechanoelectrical transduction, and show significantly higher blocking force compared to conventional type of IPMCs. The blocking forces of more than 0.3 N were measured at 4V DC input, exceeding the force output of typical Nafion® 117-based Pt IPMCs more than two orders of magnitude. The newly designed Pd-Pt IPMCs can be useful in more demanding applications, e.g., in biomimetic underwater robotics, where high stress and drag forces are encountered.

Keywords: ionic polymer–metal composite; IPMC; electroactive polymer; palladium; electrode

1. Introduction

Ionic polymer–metal composites (IPMCs) are a type of electroactive polymers (EAPs) that have recently become increasingly important as soft actuators and sensors [1]. A typical IPMC material consists of a perfluorinated ion-exchange polymer membrane, such as Nafion®, which is plated on both faces with noble metal (Pt or Au) electrode layers [2]. The membrane also contains water as a working solvent and mobile cations such as Li^+ or Na^+. When a voltage is applied across the metal electrodes, IPMC exhibits a large bending deformation. Conversely, mechanical bending of the material can also induce the voltage [3]. Other notable features inherent to IPMCs include soft and flexible structure, low driving voltage (<5 V), large bending strain (~1%), and ability to operate in aqueous environment. These properties make IPMCs excellent candidates for underwater robotics,

---

*Corresponding author. Email: kwang.kim@unlv.edu

medical and human affinity applications, and biomimetic robotics as the material can emulate the motion of biological muscles [4,5].

The manufacturing of IPMC materials is traditionally based on the electroless plating (aka impregnation-reduction, chemical deposition) of noble metals onto the surface of ionomer membrane [6]. Platinum and gold are commonly used electrode materials due to their high electrochemical stability and excellent electrical conductivity [2,7]. Platinum electrode, however, after prolonged cyclic deformation tends to develop cracks, leading to a higher surface resistance that in turn can adversely affect the actuation performance of IPMC [8]. Gold is more elastic and offers better mechanical durability [9]; however, the low stability of gold complexes in aqueous solutions makes the electroless plating process rather complicated. Also, several alternative electrode materials and their combinations have been explored. For instance, an electrode composed of platinum and copper has been proposed in which the reversible electrochemical reactions upon the actuation (i.e., dissolution and reduction of Cu^{2+} ions at the inner surface of the electrodes) can maintain electrical connections between the cracks of platinum layer [10]. Also, nickel and silver nanopowder have been used as cost-effective electrode materials [11,12]. However, low electrochemical stability of these materials limits the cycle life of IPMC.

Besides electrical conductivity, an interfacial surface area of electrodes is another important parameter that affects the performance of IPMC. It is becoming clear that the large specific surface area of electrodes leads to a higher electromechanical output [13]. This can be directly related to the accumulated charge near the electrodes. Namely, a study by Wallmersberger et al. [14] shows that in the calculations, large surface area of the electrodes corresponds to large effective dielectric permittivity value in the governing equations. This is illustrated in Figure 1, where the accumulated charge (normalized) is calculated for various dielectric permittivity values. It can be seen that the charge increases with the increase of the dielectric permittivity, or the area of the electrodes. In this regard, some authors have incorporated porous conductive powder materials into the electrodes to create a larger interface for the double-layer charging. These examples

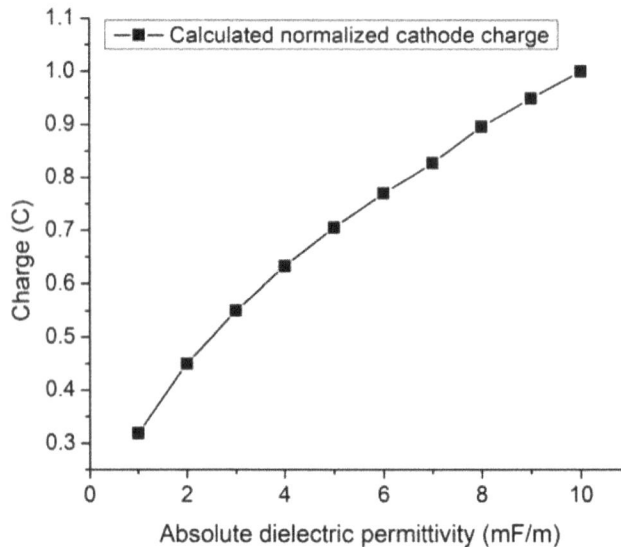

Figure 1. Calculated normalized accumulated charge in the vicinity of cathode for different absolute dielectric permittivity values.

include use of ruthenium dioxide powder [15], carbon nanotubes [16], highly porous carbide-derived carbons [17], and carbon aerogels [18] in IPMC electrodes. However, powder materials require physical assembly by heat pressing and therefore are mainly suited for fabricating the so-called 'dry type' actuators based non-volatile electrolytes such as ionic liquids.

Recently, we reported palladium buffer-layered IPMCs [19] in which the palladium was used as a supporting layer on the surface of Nafion® 117 membrane (thickness 0.178 mm) underneath the platinum top layer. The Pd-Pt electrodes showed noticeable improvement in the actuation range and better mechanical stability compared to the conventional Pt electrodes. In this study, we attempted to increase the electromechanical performance of IPMC by developing high surface area Pd-Pt electrodes and tailoring the ionomer membrane thickness. First, IPMCs were fabricated on 0.5 and 1 mm thick perfluorinated sulfonic acid membranes. Thicker membranes were used in order to increase the maximum force output. For the two actuators both having the same length L, width b, but different thicknesses t and $2t$, the ratio of the stresses for the same tip displacement can be expressed as [20]:

$$\frac{\sigma_{2t}}{\sigma_t} \approx 2 = \frac{F_{2t}}{2^2 F_t},$$ (1)

where

$$\sigma_t = \frac{6F_t L}{bt^2}.$$ (2)

Therefore, the maximum blocking force for the thicker actuator is expected to be $F_{2t} = 8F_t$, i.e., eight times larger than that of the actuator with thickness of t. Second, we have modified the conditions of palladium electroless plating process and created highly dispersed Pd particles not only at the ionomer surface, but deep in the polymer membrane, thereby increasing notably the interfacial surface area of the electrodes. The latter has given a significant increase to the blocking force output of IPMC. This paper reports and discusses the electromechanical, mechanoelectrical, electrochemical and mechanical properties of the newly designed 0.5 and 1 mm thick Pd-Pt IPMCs. Also, using the same ionomer membranes, 0.5 and 1 mm thick IPMCs with conventional Pt electrodes were fabricated and comparatively studied.

2. Experimental details

2.1. Chemicals and materials used

Perfluorinated sulfonic acid membranes with thicknesses of 0.5 and 1 mm were purchased from GEFC, Co., Ltd (Beijing, China) and were conditioned as described further in Section 2.2. The following reagents were of analytical grade and were used as received: tetraammineplatinum (II) chloride monohydrate ($Pt(NH_3)_4Cl_2 \cdot H_2O$, 98%, Sigma), tetra-amminepalladium (II) chloride monohydrate ($Pd(NH_3)_4Cl_2 \cdot H_2O$, 98%, Sigma), sodium borohydrate ($NaBH_4$, 98%, Sigma), hydroxylamine hydrochloride ($H_2NOHHCl$, 98%, Sigma), hydrazine monohydrate ($H_2NNH_2 \cdot H_2O$, 98%, Sigma), lithium chloride (LiCl, 99%, Sigma), hydrogen peroxide (H_2O_2, 50%, Sigma), sulfuric acid (98%, Pharmaco-AAPER), ammonium hydroxide (35%, EM Industries, Inc.).

2.2. Fabrication of IPMC materials

First, the ionomer membranes were pretreated by roughening both surfaces with sandpaper (Grit #600) in order to enhance the physical bonding between the polymer and metal electrode. Both the surfaces were polished until the membrane was uniformly flat and non-transparent, after which it was thoroughly rinsed with deionized water. Then, the ionomer membrane was cleaned in 3 wt% hydrogen peroxide (H_2O_2) solution at 80°C for 45 min and then in 1 M sulfuric acid (H_2SO_4) solution at the same temperature and time. Finally, the membrane was boiled in deionized water for 30 min (twice) to remove acid residues.

The procedure for fabricating Pd-Pt electrodes was as follows: First, an impregnation-reduction process was carried out to incorporate palladium particles into the inner surface of the membrane. The pretreated membrane was soaked in palladium salt ($Pd(NH_3)_4Cl_2·H_2O$) solution for 1.5 h to impregnate the polymer with palladium complex ions. After rinsing thoroughly with deionized water, the membrane was immersed for 4 h in aqueous solution containing reducing agent such as sodium borohydride ($NaBH_4$) and ammonium hydroxide (NH_4OH) for pH adjustment. During this step, the palladium salt contained in the ionomer was chemically reduced to palladium metal ($Pd°$) at the inner surface of the membrane. The obtained composite was then cleaned in 1 M sulfuric acid solution for 30 min at 70°C and then in deionized water at the same temperature and time. Next, the platinum particles were deposited at the outer surface of the ionic polymer membrane using a chemical deposition method. The membrane was immersed in an aqueous platinum complex ($Pt(NH_3)_4Cl_2·H_2O$) solution and while moderately stirring, a mild reducing agents such as hydrazine monohydrate and hydroxylamine hydrochloride were added in every 30 min for 4 h to deposit the platinum on the palladium surface. After typical cleaning procedure in sulfuric acid and deionized water, the chemical deposition step was repeated 1–2 times, depending on the electrode surface conductivity.

IPMCs with Pt electrodes were fabricated using a traditional impregnation-reduction method with platinum complex ($Pt(NH_3)_4Cl_2·H_2O$) and sodium borohydride ($NaBH_4$) [6]. The impregnation-reduction cycle was performed twice in order to grow the platinum layer deeper into the polymer surface, thereby improving the reliability of electrodes against mechanical deformations. Additionally, the chemical deposition was carried out to deposit extra platinum particles onto the outer surface of the composite to further increase the electrical conductivity.

Using the aforementioned steps, four different sets of IPMC samples were fabricated – 0.5 and 1 mm thick Pd-Pt-IPMCs, and 0.5 and 1 mm thick Pt-IPMCs. All samples were cut into 10 mm × 50 mm rectangular shape and ion-exchanged to Li^+-form by soaking in 1 M lithium chloride (LiCl) solution for overnight. Li^+ ion due to its relatively small ionic radius and strong solvation in water provides optimal deformation performance and actuation speed, and is therefore commonly used for counter ion in IPMC studies.

2.3. Electromechanical and electrochemical characterization

The electromechanical responses such as displacement and blocking force of the prepared IPMCs were measured by a laser displacement sensor (optoNCDT-1401, Micro-Epsilon, Ortenburg, Germany) and a load cell (GSO-30, Transducer techniques) with a sample size of 50 mm × 10 mm in the test setup composed of a signal generator (FG-7002C, EZ digital, Bucheon, South-Korea), a power amplifier (LVC-608, AE Techron, Elkhart, IN, USA), a DC

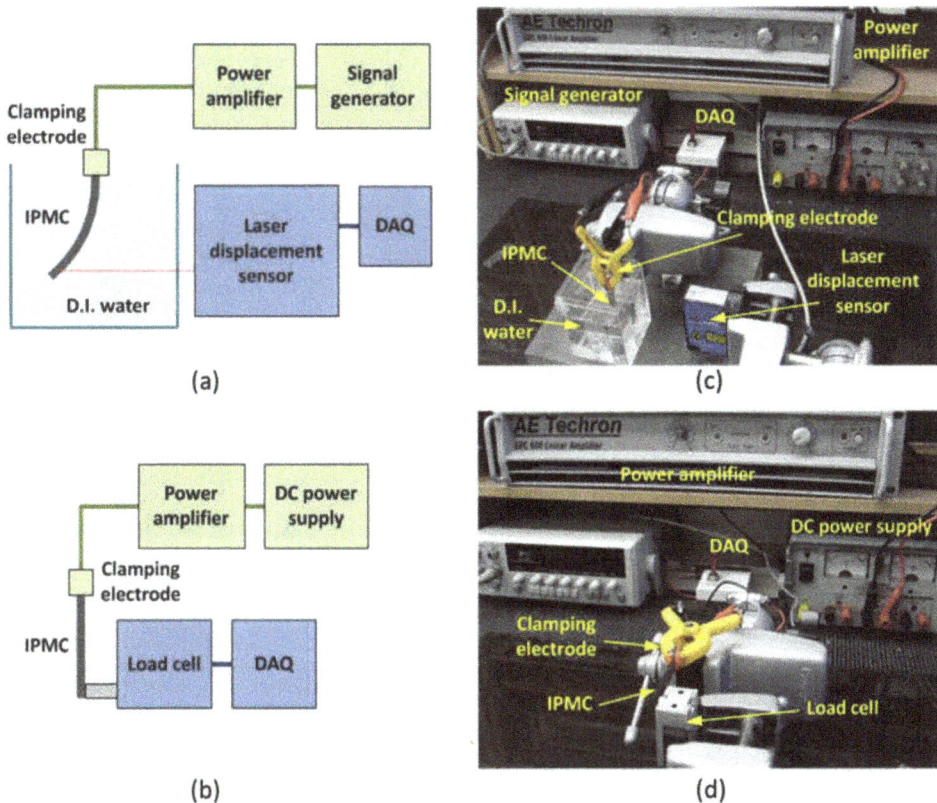

Figure 2.   Schematic diagrams of experimental setups used for displacement (a) and blocking force (b) characterization and corresponding photographs (c and d).

power supply (Tektronix), and a DAQ (SCB-68, National Instruments, Austin, TX, USA) as illustrated in Figure 2. An IPMC was clamped in a cantilever configuration with a free beam length of 40 mm. The blocking force measurements were performed in air condition at room temperature (22°C) and normal humidity (~30%). The experiments were repeated with three samples from each type and the standard deviation percentages were within 16% from average, which is considered reasonable margin for soft polymeric actuators.

The mechanoelectrical transduction response was measured using an oscilloscope (PicoScope 2203, Pico technology, St. Neots, UK) and a vibration test system composed of a vibration controller (VR-8500, Vibration Research Corp., Jenison, MI, USA), a power amplifier (BAA-120, TIRA, Schalkau, Germany), and an electrodynamic shaker (S-521, TIRA) as shown in Figure 3. An IPMC sample with a size of 50 mm × 10 mm was clamped from both ends with a free beam length of 40 mm. The voltage output was measured at the clamp contacts.

Cyclic voltammetry (CV) and electrochemical impedance spectroscopy (EIS) measurements were performed at ambient temperatures using a Radiometer Analytical PGZ-402 potentiostat (Radiometer Analytical, Lyon, France) in a three-electrode configuration. The two IPMC electrodes served as a working and counter electrode, and saturated calomel electrode (SCE) was used as a reference electrode. All measurements were performed in 0.5 M H_2SO_4 (aq) solution. Cyclic voltammograms of IPMC samples were recorded in a potential range of −0.25 to 1.2 V at a scan rate of 20 mV/s. The

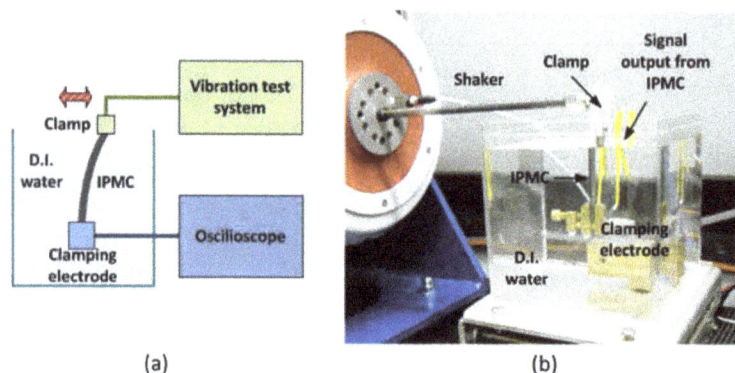

(a)   (b)

Figure 3.   Schematic of experimental setup used for mechanoelectrical characterization of IPMC (a) and corresponding photograph (b).

capacitance versus frequency plots were obtained from EIS measurements at a frequency range of 1 kHz to 0.1 Hz using an AC perturbation of 10 mV and DC potential bias of 0.1 V versus SCE.

2.4.   Scanning electron microscopy (SEM), energy dispersive X-ray (EDS) analysis, and flexural modulus characterizations

Scanning electron micrographs of IPMC cross sections were obtained using an Hitachi S-4200 microscope (Hitachi, Japan) in secondary electron image mode with a 20 kV accelerating voltage. EDS detector was used to perform chemical line-scan analysis in order to investigate the distribution of deposited electrode metals in the cross section of Pd-Pt IPMCs. The flexural (bending) modulus of the samples was determined from stress–strain measurements in a three-point bending mode using an Instron universal testing machine (5565, Norwood, MA, USA) (Figure 4). The three-point bending tests were performed in air condition at room temperature (22°C) and normal humidity (~30%).

Figure 4.   The three-point bending test setup used for determining the flexural modulus of IPMC.

3. Results and discussion

The cross sections of Pd-Pt IPMCs were examined using SEM-EDS to investigate the distribution of deposited electrode metals in the composite (Figure 5). The EDS line-scan analysis shows clearly that the palladium particles have penetrated deep into the polymer membrane, ranging more than 100 microns from the surface of 0.5 mm thick sample and more than 200 microns from the surface of 1 mm thick IPMC sample.

The widely dispersed palladium particles near the polymer surface provide a large interfacial area and the platinum layer at the outer surface of the composite serves as a highly conductive current collector. A higher content of palladium in case of 1 mm thick sample is probably due to the fact that the thicker membrane is able to absorb more palladium complex during the ion-exchange process, thus resulting in more Pd particles in the polymer after the reduction step. As can be seen, the middle region of the cross section is relatively void of Pd particles, indicating that the electrodes are still electrically insulated from each other. The low intensity for Pt or Pd in this region is mainly due to the noise signal. Electrolessly plated polymer membrane can have trace amounts of metal particles in the polymer, which can be also a reason for low-level detection of Pt or Pd through cross section. The thickness of the platinum layer at the membrane surface is around 7–10 microns, which is typical for Pt-electroded IPMCs and also for the Pt IPMCs fabricated in this work.

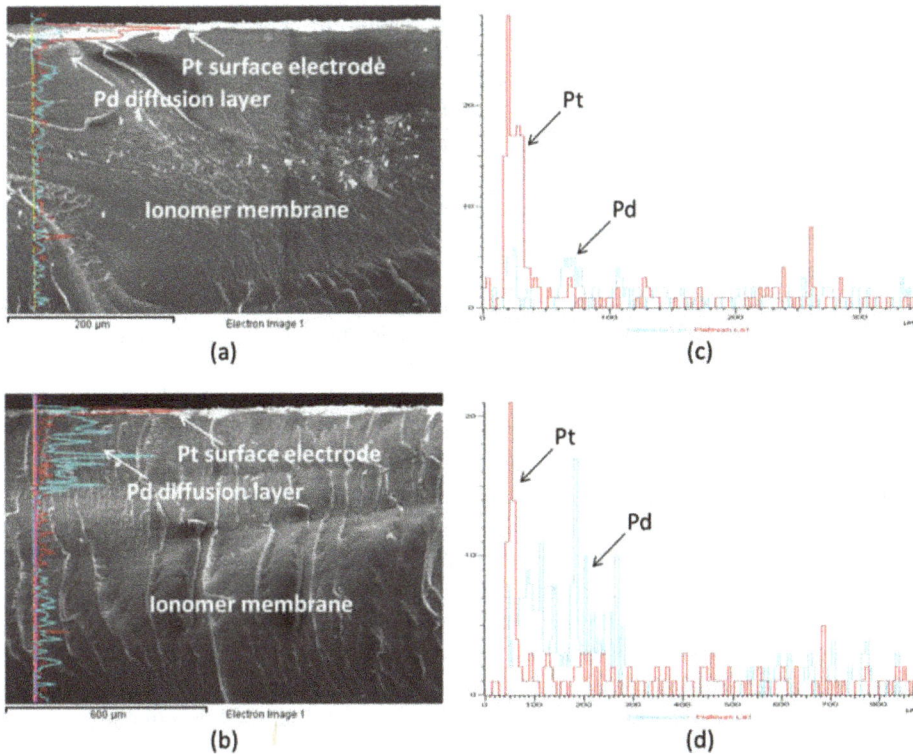

Figure 5. SEM micrographs of the cross sections of 0.5 mm (a) and 1 mm (b) thick Pd-Pt IPMCs, and corresponding EDS line-scan profiles for elements of Pt and Pd (c and d).

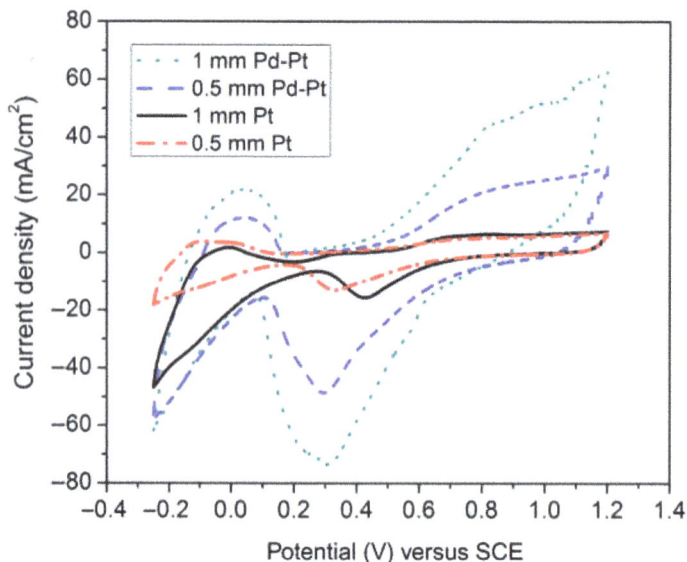

Figure 6. Cyclic voltammograms of Pt and Pd-Pt-IPMCs measured in 0.5 M H_2SO_4 (aq) solution at 20 mV/s.

The CV was employed to investigate the electrochemical properties of the Pt and Pd-Pt IPMCs. Figure 6 shows the cyclic voltammograms measured in 0.5 M H_2SO_4 (aq) solution in a potential range of -0.25 to 1.2 V at a scan rate of 20 mV/s. An approximately exponential increase in the current density corresponds to the cathodic hydrogen evolution process and the flat wide maximums characterize the ionization of adsorbed hydrogen at the Pt and porous Pd-Pt electrodes. A large potential difference between the overpotential of cathodic evolution and ionization process refers to a slow mass transport in the porous Pd-Pt system. The region between 0.2 and 0.5 V corresponds to ideal polarizability, where the electrical double-layer formation occurs. It can be seen that the Pd-Pt samples have the highest peaks in that potential range, especially the 1 mm thick sample. This is in agreement with SEM-EDS analysis (Figure 5) that also shows higher content of palladium for 1 mm Pd-Pt IPMC. Starting from 0.6 V, the adsorption of oxygen occurs at the Pt electrodes and more intensely at the Pd-Pt electrodes. The wide peaks at the reduction curve at 0.3 V characterize the electroreduction of adsorbed oxygen at the Pd-Pt and Pt electrode surface. A large difference between the peak potentials in oxidation and reduction curves indicates the irreversible nature of oxygen adsorption. All these processes occur more intensely at the Pd-Pt electrodes, indicating their large interface.

Figure 7 presents the capacitance of different IPMC samples as a function of frequency, determined from EIS measurements. The data indicates that the widely dispersed palladium in the electrodes increases significantly the capacitance of IPMC at the lower frequencies (<10 Hz). The capacitances at 0.1 Hz are up to 110 and 70 mF/cm^2 for 1 and 0.5 mm Pd-Pt IPMCs, and 43 and 10 mF/cm^2 for 1 and 0.5 mm Pt IPMCs. These results are consistent with the CV measurements that also indicate higher capacitance for the Pd-Pt samples (Figure 6). It should be noted that while the CV results are in good agreement with our previous paper, it is difficult to compare the EIS data due to very different measurement conditions used [19]. Namely, the EIS measurements in our previous study were performed at very high DC bias potentials – at 2 V (as opposed to

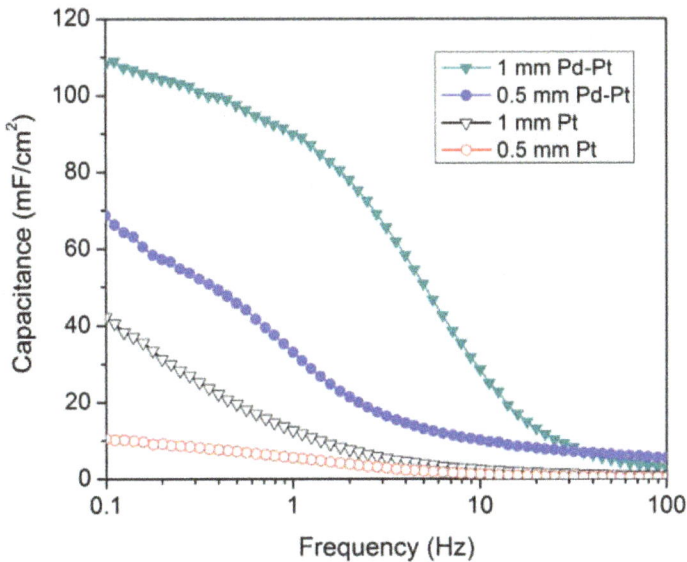

Figure 7.   EIS capacitance of Pt and Pd-Pt IPMCs as a function of frequency at 0.1 V versus SCE.

0.1 V used in this work). AC voltage perturbation at high bias potentials can prevent electrolyte ions from adsorbing effectively at the electrodes with porous interface. The faradic processes (i.e., water electrolysis) occurring at this voltage can cause a blocking effect of the pores by adsorbed gaseous products (H_2 and O_2), leading to lower value of capacitance [21]. This explains the lower EIS capacitance observed for Pd-Pt electrodes in our previous study compared to Pt electrodes, while the CV measurements (at 20 mV/s) clearly indicated the opposite and are consistent with this study.

Figure 8 shows the flexural (bending) moduli of Pt and Pd-Pt IPMCs with different thicknesses determined from the three-point bending test. It should be noted that the IPMC's flexural rigidity can have a great impact on its electromechanical performance. Generally, a stiffer sample exhibits less displacement, but on the other hand, is capable of

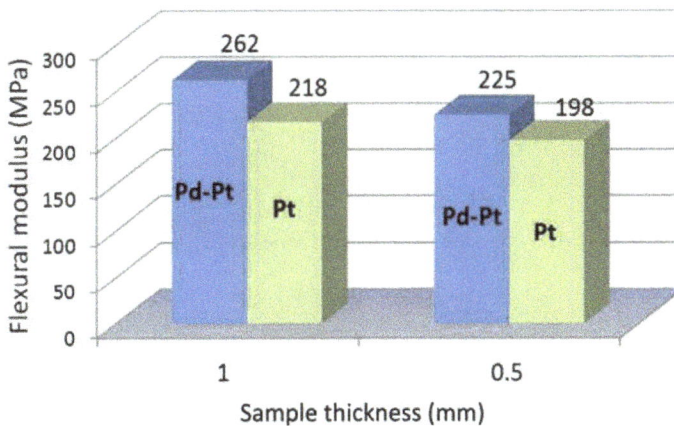

Figure 8.   Flexural moduli of Pt and Pd-Pt IPMCs measured in a three-point bending mode.

generating higher blocking force, since the rigid structure of material is able to sustain more stress. The measurements show that IPMCs with Pd-Pt electrodes have noticeably higher stiffness compared to Pt IPMCs. The respective flexural modulus values for Pd-Pt samples are 225 MPa (0.5 mm) and 262 MPa (1 mm), whereas the flexural modulus for Pt samples are 198 MPa (0.5 mm) and 218 MPa (1 mm). This result can be expected since the ionomer membranes in case of Pd-Pt IPMCs contain palladium that gives rise to the stiffness.

The displacement measurements were carried out at different actuation frequencies from 0.1 to 1 Hz at ±3 V square-wave input. Figure 9 shows the measured actuation in terms of displacement (mm) and Figure 10 in terms of bending strain (%). The bending strain (ε) was obtained from the measured displacement (δ) using the following relation [2]:

$$\varepsilon = \frac{2\delta t}{L^2 + \delta^2},$$ (3)

where L is the free beam length and t is the cross-sectional thickness of IPMC. Due to a large variation of thicknesses of the samples, the data in terms of displacement provides better perspective of the actuation range. The results show that the displacements of the samples are relatively minor at 1 Hz and increase noticeably as the actuation frequency is decreased to 0.1 Hz. It should be noted that these IPMCs have slower actuation response compared to typical Nafion 117 membrane-based Pt IPMCs that are capable of generating adequate displacement also at 1 Hz input signal [3]. This can be related to the much larger thickness of the samples, requiring longer time for electrolyte migration and diffusion between the electrodes. As observed in Figure 9, the 0.5 mm thick samples show noticeably larger deflection at lower frequency (0.1 Hz) compared to 1 mm samples.

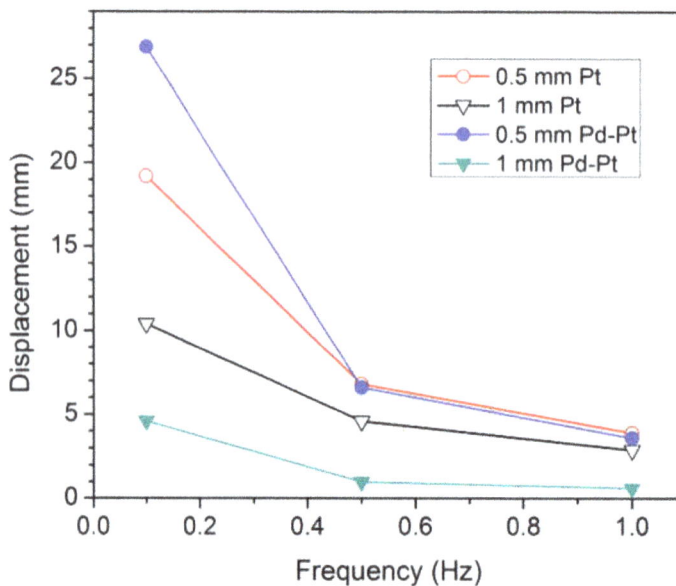

Figure 9.   Displacement performance of different IPMCs versus frequency at ±3 V square-wave input.

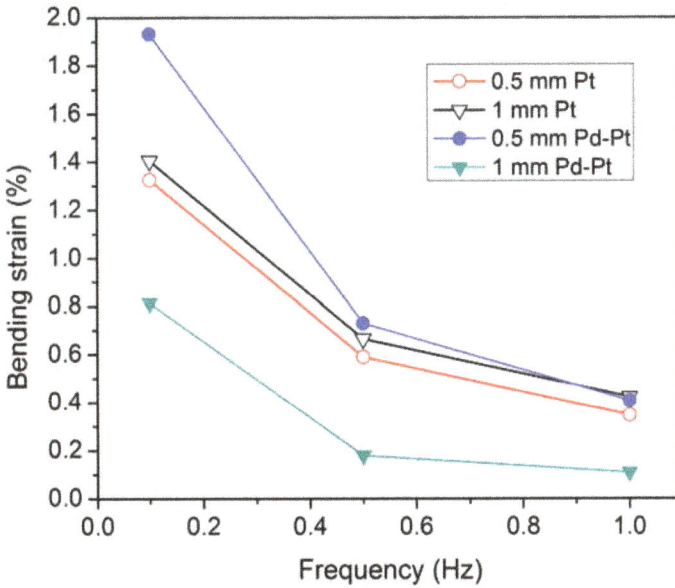

Figure 10.   Maximum bending strain as a function of frequency at 0.1 Hz, ±3 V square-wave input.

Figure 11 shows the strain rates (actuation speed) of different samples at 0.1 Hz square-wave input, calculated by taking the time derivative of the bending strain. It can be seen that 0.5 mm thick samples are capable of faster actuations with much higher peak strain rates compared to 1 mm thick IPMCs. The 1 mm thick Pd-Pt IPMC shows the lowest strain rate among the samples.

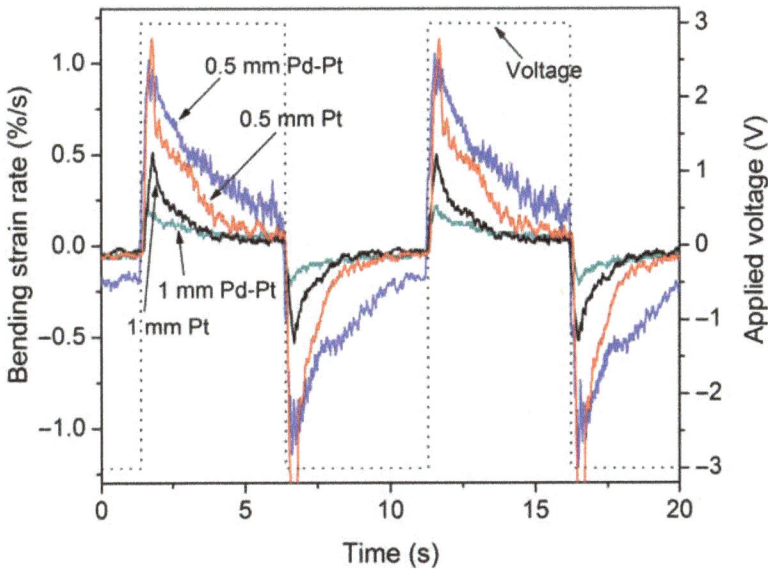

Figure 11.   Strain rates (%/s) of Pt and Pd-Pt IPMCs actuation frequency at ±3 V square-wave input.

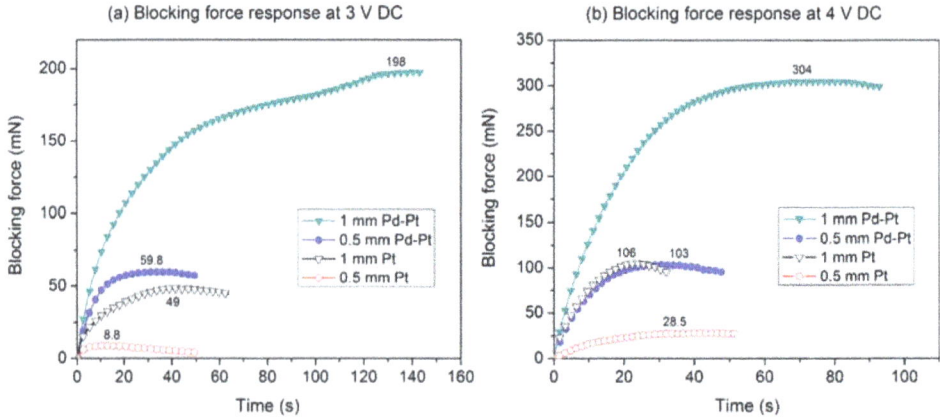

Figure 12.   Blocking force responses in time for Pt and Pd-Pt IPMCs measured at 3 and 4 V DC input.

The electromechanical performance of Pt and Pd-Pt IPMCs was also evaluated in terms of blocking force. Figure 12 presents the measured blocking force response in time at 3 and 4 V DC input. It can be seen that IPMCs with Pd-Pt electrodes show significantly higher blocking force output compared to Pt IPMCs with respective thicknesses. At 3 V DC input, the peak blocking forces for 1 and 0.5 mm Pd-Pt IPMCs are nearly 200 and 60 mN, whereas the forces for 1 and 0.5 mm Pt samples are 50 and 9 mN. Moreover, at 4 V DC input, the blocking forces more than 0.3 and 0.1 N were recorded for 1 and 0.5 mm Pd-Pt samples. The peak forces for Pt samples were 0.1 and 0.03 N. The higher performance of Pd-Pt electrodes is related to highly capacitive interface of dispersed Pd particles near the Pt surface, leading to higher electrolyte transport and adsorption at the electrodes, as observed from EIS capacitance measurements (Figure 7). A higher performance of 1 mm samples over their 0.5 mm thick versions can be due to a combination of several effects: twice larger thickness, higher interfacial surface area of electrodes, and higher stiffness. It should be noted that the 1 mm Pd-Pt IPMC has noticeably higher capacitance and stiffness compared to other samples, which can contribute to a higher force output. However, the palladium in the polymer matrix can inhibit the electrolyte diffusion, lowering the rate of double-layer charging, which is reflected in the slower force response in the case of 1 mm Pd-Pt IPMC. For comparison, the blocking forces of the previously reported Pt and Pd-Pt IPMCs made on conventional Nafion 117 membranes were 0.3 and 1.1 mN, respectively [19]. Thus, by incorporating thicker ionomer membranes and high surface area Pd-Pt electrodes, the blocking force performance of IPMC was significantly increased.

Besides functioning as an electromechanical actuator, IPMC can also operate in a reverse mode, as a mechanoelectrical transducer, i.e., generate voltage upon mechanical deformation. The mechanoelectrical transduction of Pt and Pd-Pt IPMCs was monitored using an electrodynamic shaker assembly, which was programmed to apply a constant sinusoidal displacement to the IPMC tip in a sweeping frequency mode. Since the shaker had limits regarding the displacement range at higher frequencies, the measurements were performed at two different displacement and frequency settings: (a) at 4–40 Hz with 5 mm displacement, and (b) at 4–20 Hz with 10 mm displacement. Figure 13 shows the induced voltage amplitude as a function of imposed actuation frequency at these testing regimes. It

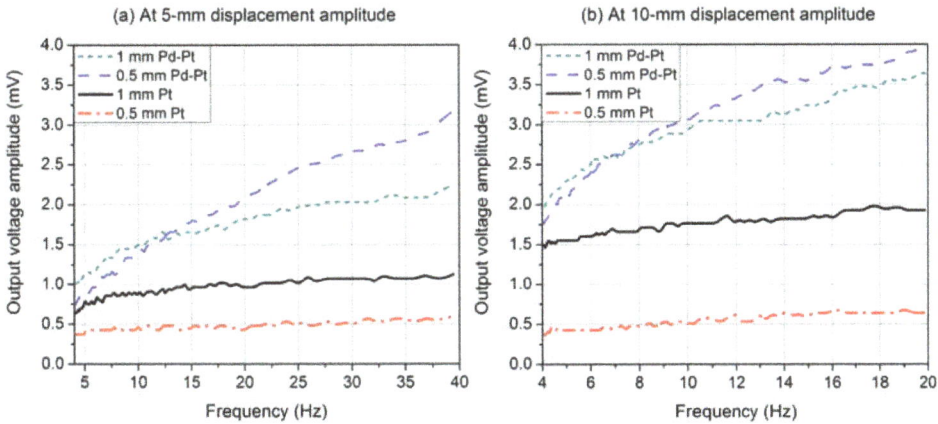

Figure 13.   Induced voltage amplitude of Pt and Pd-Pt IPMCs in response to imposed sinusoidal displacement with 5 mm (a) and 10 mm (b) amplitude.

can be seen that IPMCs with Pd-Pt electrodes generate considerably higher output voltage compared to Pt IPMCs, especially at the higher frequencies. This can be related to the larger interface of Pd-Pt electrodes, allowing more effective surface charging. It is interesting to note that the output voltage generated from Pd-Pt electrodes increases drastically with the increase of actuation frequency, while in case of Pt electrodes the overall change in the signal amplitude is rather minor. It can be also noticed that the voltage produced at 10-mm displacement is almost double the voltage amplitude at 5-mm displacement, indicating that the mechanoelectrical output of IPMC is strongly dependent on the amount of applied deformation.

4.   Conclusions

In this work, IPMCs with high surface area Pd-Pt electrodes and thick ionomer membranes (0.5 and 1 mm) were developed. Also, using the same ionomer membranes, IPMCs with conventional Pt electrodes were fabricated and comparatively investigated. The electromechanical, mechanoelectrical, electrochemical and mechanical properties of different samples were evaluated and analyzed. The results demonstrate that the newly designed Pd-Pt IPMCs are superior in mechanoelectrical transduction and exhibit significantly higher blocking force compared to the IPMCs made with conventional Pt electrodes. The peak blocking forces more than 0.3 N were measured for Pd-Pt IPMCs and more than 0.1 N for the Pt-IPMCs at 4 V DC input, exceeding the force output of typical Nafion 117-based Pt IPMCs more than two orders of magnitude. The higher performance of Pd-Pt electrodes is related to highly capacitive interface of dispersed Pd particles near the Pt surface, leading to higher electrolyte transport and adsorption at the electrodes – responsible for the actuation process of IPMC. Both types of IPMCs showed adequate displacement performance, however, the actuation response was slower compared to typical Nafion 117-based Pt IPMCs. The SEM-EDS cross-sectional analysis indicated that the palladium particles were widely dispersed in the ionomer membrane, thereby increasing the interfacial surface area of electrodes and the capacitance of IPMC. The measured capacitances were in good correlation with the blocking force data of the samples. Thus, by incorporating millimeter thick ionic polymer membranes and high surface area Pd-Pt electrodes, the IPMC performance was successfully improved.

Acknowledgments

This work was supported in part by the Office of Naval Research under Grant N000140910218 and N000141310274.

References

[1] C. Jo, D. Pugal, I.-K. Oh, K.J. Kim, and K. Asaka, *Recent advances in ionic polymer–metal composite actuators and their modeling and applications*, Prog. Polym. Sci. 38 (2013), pp. 1037–1066. doi:10.1016/j.progpolymsci.2013.04.003.

[2] M. Shahinpoor and K.J. Kim, *Ionic polymer-metal composites: I. Fundamentals*, Smart Mater. Struct. 10 (2001), pp. 819–833. doi:10.1088/0964-1726/10/4/327.

[3] M. Shahinpoor, Y. Bar-Cohen, J.O. Simpson, and J. Smith, *Ionic polymer-metal composites (IPMCs) as biomimetic sensors, actuators and artificial muscles – a review*, Smart Mater. Struct. 7 (1998), pp. R15–R30. doi:10.1088/0964-1726/7/6/001.

[4] Y. Bar-Cohen, *Biomimetics: Biologically Inspired Technologies*, CRC Press, Boca Raton, FL, 2006.

[5] M. Shahinpoor and K.J. Kim, *Ionic polymer-metal composites: IV. Industrial and medical applications*, Smart Mater. Struct. 14 (2005), pp. 197–214. doi:10.1088/0964-1726/14/1/020.

[6] K.J. Kim and M. Shahinpoor, *Ionic polymer-metal composites: II. Manufacturing techniques*, Smart Mater. Struct. 12 (2003), pp. 65–79. doi:10.1088/0964-1726/12/1/308.

[7] K. Onishi, S. Sewa, K. Asaka, N. Fujiwara, and K. Oguro, *Morphology of electrodes and bending response of the polymer electrolyte actuator*, Electrochim. Acta. 46 (2001), pp. 737–743. doi:10.1016/S0013-4686(00)00656-3.

[8] A. Punning, M. Kruusmaa, and A. Aabloo, *Surface resistance experiments with IPMC sensors and actuators*, Sens. Actuators A Phys. 133 (2007), pp. 200–209. doi:10.1016/j.sna.2006.03.010.

[9] N. Fujiwara, K. Asaka, Y. Nishimura, K. Oguro, and E. Torikai, *Preparation of gold–solid polymer electrolyte composites as electric stimuli-responsive materials*, Chem. Mater. 12 (2000), pp. 1750–1754. doi:10.1021/cm9907357.

[10] U. Johanson, U. Mäeorg, V. Sammelselg, D. Brandell, A. Punning, M. Kruusmaa, and A. Aabloo, *Electrode reactions in Cu-Pt coated ionic polymer actuators*, Sens. Actuators B Chem. 131 (2008), pp. 340–346. doi:10.1016/j.snb.2007.11.044.

[11] M. Siripong, S. Fredholm, Q.A. Nguyen, B. Shih, J. Itescu, and J. Stolk, *A Cost-Effective Fabrication Method for Ionic Polymer-Metal Composites*, Materials Research Society Symposia Proceedings, Materials Research Society, Boston, MA, 2005, p. 889.

[12] C.K. Chung, P.K. Fung, Y.Z. Hong, M.S. Ju, C.C.K. Lin, and T.C. Wu, *A novel fabrication of ionic polymer-metal composites (IPMC) actuator with silver nano-powders*, Sens. Actuators B Chem. 117 (2006), pp. 367–375. doi:10.1016/j.snb.2005.11.021.

[13] B. Akle, S. Nawshin, and D. Leo, *Reliability of high strain ionomeric polymer transducers fabricated using the direct assembly process*, Smart Mater. Struct. 16 (2007), pp. S256–S261. doi:10.1088/0964-1726/16/2/S09.

[14] T. Wallmersperger, B.J. Akle, D.J. Leo, and B. Kroplin, *Electrochemical response in ionic polymer transducers: An experimental and theoretical study*, Compos. Sci. Technol. 68 (2008), pp. 1173–1180. doi:10.1016/j.compscitech.2007.06.001.

[15] B.J. Akle, M.D. Bennett, D.J. Leo, K.B. Wiles, and J.E. McGrath, *Direct assembly process: A novel fabrication technique for large strain ionic polymer transducers*, J. Mater. Sci. 42 (2007), pp. 7031–7041. doi:10.1007/s10853-006-0632-4.

[16] D.Y. Lee, I.-S. Park, M.-H. Lee, K.J. Kim, and S. Heo, *Ionic polymer-metal composite bending actuator loaded with multi-walled carbon nanotubes*, Sens. Actuators A Phys. 133 (2007), pp. 117–127. doi:10.1016/j.sna.2006.04.005.

[17] V. Palmre, D. Brandell, U. Mäeorg, J. Torop, O. Volobujeva, A. Punning, U. Johanson, M. Kruusmaa, and A. Aabloo, *Nanoporous carbon-based electrodes for high strain ionomeric*

bending actuators, Smart Mater. Struct. 18 (2009), pp. 095028. doi:10.1088/0964-1726/18/9/095028.

[18] V. Palmre, E. Lust, A. Jänes, M. Koel, A.-L. Peikolainen, J. Torop, U. Johanson, and A. Aabloo, *Electroactive polymer actuators with carbon aerogel electrodes*, J. Mater. Chem. 21 (2011), pp. 2577–2583. doi:10.1039/c0jm01729a.

[19] S.M. Kim and K.J. Kim, *Palladium buffer-layered high performance ionic polymer-metal composites*, Smart Mater. Struct. 17 (2008), pp. 035011. doi:10.1088/0964-1726/17/3/035011.

[20] K.J. Kim and M. Shahinpoor, *A novel method of manufacturing three-dimensional ionic polymer-metal composites (IPMCs) biomimetic sensors, actuators and artificial muscles*, Polymer. 43 (2002), pp. 797–802. doi:10.1016/S0032-3861(01)00648-6.

[21] H. Kurig, A. Jänes, and E. Lust, *Electrochemical characteristics of carbide-derived carbon / 1 -ethyl-3-methylimidazolium tetrafluoroborate supercapacitor cells*, J. Electrochem. Soc. 157 (2010), pp. A272–A279. doi:10.1149/1.3274208.

8

Functionalization of carbon nanotube yarn by acid treatment

H.E. Misak[a], R. Asmatulu[b], M. O'Malley[c], E. Jurak[b] and S. Mall[a]*

[a]Department of Aeronautics and Astronautics, Air Force Institute of Technology, 2950 Hobson Way, Wright-Patterson AFB, OH 45433-7765, USA; [b]Department of Mechanical Engineering, Wichita State University, 1845 Fairmount, Wichita, KS 67260-0133, USA; [c]Air Force Research Laboratory, Materials and Manufacturing Directorate, 2941 Hobson Way, Wright-Patterson AFB, OH 45433-7750, USA

Carbon nanotube (CNT) yarn was functionalized using sulfuric and nitric acid solutions in 3:1 volumetric ratio. Successful functionalization of CNT yarn with carboxyl and hydroxyl groups (e.g., COOH, COO–, OH, etc.) was confirmed by attenuated total reflectance spectroscopy. X-ray diffraction revealed no significant change to the atomic in-plane alignment in the CNTs; however, the coherent length along the diameter was significantly reduced during functionalization. A morphology change of wavy extensions protruding from the surface was observed after the functionalization treatment. The force required to fracture the yarn remained the same after the functionalization process; however, the linear density was increased (310%). The increase in linear density after functionalization reduced the tenacity. However, the resistivity density product of the CNT yarn was reduced significantly (234%) after functionalization.

Keywords: carbon nanotube yarns; functionalization; textile; conductivity

1. Introduction

There is a need in commercial applications for textiles that have high electrical conductivity and low density. These fabrics can be used in many applications such as static charge dissipation, conducting wire, shielding against electromagnetic interference, wearable electronics, etc. [1]. In order to make common textiles conductive, fabrics are coated with polypyrrole that has a resistivity of $\sim 5.56E^{-3}$ Ohms cm and a density of ~ 1.44 g/cm^3, which translates to a resistivity density product of $\sim 80{,}000$ nOhms m g/cm^3 [2]. This property is considerably inferior when compared to metals. Iron has a resistivity density product of 757 nOhms m g/cm^3, and the more conductive copper has 150 nOhms m g/cm^3. However, metals are not conducive for use in textile and textile-type applications owing to weight, stiffness, and safety. Thus, carbon nanotube (CNT)-based structures

---

*Corresponding author. Email: Shankar.Mall@afit.edu
The views expressed in this article are those of the authors and do not reflect the official policy or position of the US Air Force, Department of Defense, or the US Government.

have potential that can bridge this gap when low electrical resistivity and low density in these cases are desirable [3].

A great deal of research has been conducted to create long CNT structures for various technological applications such as conductive textiles. One such method is drawing multitudes of short CNTs into a long yarn [4]. Unfortunately, the mechanical properties of the CNT yarn are not as impressive as an individual CNT [5]. This is due to the fact that CNT yarn is made up of multitudes of short CNTs, so the mechanical properties are constrained by the forces required to move one CNT relative to other. Thus, weaker mechanical interlocking and van der Waals forces between CNTs play an important role in fracture resistance of CNT yarn, and not strong covalent bonding as is seen in an individual CNT. Also, the yarn density and the large difference between Young's modulus in tension and Young's modulus in compression play an important role in the mechanical properties of CNT yarn [6–8]. Much success in improving mechanical properties has been linked to the optimization of the CNT drawing process parameters [9,10]. Zhong et al. have used transmission electron microscopy and X-ray diffraction (XRD) studies to characterize how CNTs collapse, thus increasing the density and potential benefits of mechanical interlocking [11]. Gui et al. have studied the control of macro- and microstructure of CNT-based structures for applications including mechanical energy absorption [12].

Others have tried to improve the properties of CNT yarns by secondary treatments after the drawing process. Xiao et al. functionalized CNTs in the yarn by drawing the yarn through barium nitrate. The barium-functionalized yarn performed well as cathodes at high temperatures (1317 K) [13]. Trakakis et al. functionalized buckypaper by oxidation and epoxidation reactions [14]. The oxidation created a denser buckypaper, while the epoxidation created a more foam-like structure. Also, they found that the mechanical and electrical properties increased with higher density. Wei et al. made a CNT/polyvinyl alcohol (PVA) composite yarn with a reported 2.2 GPa tensile strength by utilizing oxygen plasma process to functionalize the CNT ribbon and then dipping in a PVA solution [15]. Malik et al. also made CNT/PVA composite but in the form of a sheet, and the manufacturing process is discussed in detail in [16]. Carretero-González et al. modified CNT yarn by chemically unzipping the CNTs to form nanoribbons of graphene. The resulting nanoribbon yarn has very high specific capacitance properties [17]. Jang et al. also produced nanoribbons by unzipping CNTs to use as point emitters for high current field emission [18].

In this study, CNT yarn was functionalized via sulfuric acid and nitric acid treatment in 3:1 volumetric ratio. The treated and non-treated CNTs were characterized by attenuated total reflectance (ATR) module connected to a Fourier transform infrared spectroscopy, scanning electron microscopy (SEM), and XRD. The resistance, tensile strength, and mass of the prepared samples were measured for the treated and non-treated CNT yarns.

2. Experiments

2.1. Materials

CNT yarn was procured from the Nanocomp Technologies, Inc., Concord, New Hampshire, USA. The CNT 1 yarn has a linear density of 0.98 ± 0.10 tex. Tex is the linear density commonly used in textile applications where the density of the fiber can change depending on the packing factor and has the unit of g/km. Acid treatment of the as-received CNT yarn was performed by submerging 100 mm yarn segments into a 3:1 volumetric solution of sulfuric acid and nitric acid for 35 min under slight tension. Slight tension was applied to hinder the yarn from unraveling, however this still occurred to a little extent. The 35 min soak time was chosen based on a preliminary experiment where the resistance of the wire in the bath was measured and was found to increase significantly after 35 min. Following acid treatment, the yarn was dipped in deionized water multiple times to remove the residue acid.

2.2. Experiments

2.2.1. Linear density

A Mettler Toledo XP26 microbalance (Schaffhausen, Switzerland) was used to determine the mass of the CNT yarn specimens. The measured mass from 100 mm length was used to find the linear density (g/km or tex) for each specimen. An accurate cross-sectional area of the yarns was difficult to determine due to the varying degrees of CNT packing factors, micro- and nanoscale gaps/voids, and properties that are dependent on processing parameters. For example, since CNTs are tubular structures, there is a gap inside. A yarn has ribbon–ribbon boundaries that would also produce micro gaps [19]. Thus, presentation of the strength properties in terms of tenacity (N/tex) is more appropriate and allows the yarns to be readily compared to other studies and fabrication methods. Tenacity, linear density, and density will be mentioned often henceforth. For clarification, tenacity is force divided by linear density (N/tex), linear density is weight divided by length (g/km), and density is weight divided by volume (g/cm^3).

2.2.2. Tensile test

The CNT yarn specimens were tested until they fractured under a monotonic tensile loading condition using an Agilent T150 UTM test machine with a 500 mN load cell. CNT yarn specimens with a gauge length of 100 mm were glued between two thin pieces of plastic and inserted into the grips. On both ends, the CNT yarn was glued in between 2 thin 20 mm long plastic pieces. The gauge length was 100 mm. The specimens were pulled at a strain rate of 0.001 mm/(mm sec)).

2.3. Resistance

Resistance was measured before and after the process of functionalization. A four-point probe technique was used to measure the electrical resistivity using a Keithley 2400 SourceMeter (Cleveland, OH, USA) with a current of 0.1 A. A jig was used to ensure that the distance between each probe was 1 cm when measuring resistivity.

As stated previously, it is difficult to accurately determine the cross-sectional area of CNT yarn due to nanoscale gaps and varying packing density of CNTs. Resistivity density product ($\rho \times d$) is independent of area as seen below:

$$\rho \times d = R\frac{A_1}{\ell} \times \frac{m}{LA_1} \tag{1}$$

$$\rho \times d = R\frac{m}{\ell L} \tag{2}$$

where ρ is resistivity, d is density, A_1 is cross-sectional area of the yarn, R is resistance along the length (ℓ) of the yarn, and m is mass of yarn of length L of the CNT yarn. Furthermore, presenting the resistivity density product provided a convenient and consistent comparison among different materials.

2.4. Characterization tests

The CNT yarn was characterized via three different methods: ATR, SEM, and XRD. Thermo Nicolet Avatar 360 (Madison, WI, USA) with SmartPerformer ATR accessory was used for ATR. A Quanta 450 (Brno-Královo Pole, Czech Republic) SEM set at 2 kV acceleration voltage with a spot size of 2.5 was used to image the CNT yarn. XRD tests were performed using a Bruker D2 Phaser (Fitchburg, WI, USA; 30 kV, 10 mA) equipped with sealed Cu X-ray tube ($\lambda = 1.54178$ Å) and a Lynxeye linear position sensitive detector. The XRD data were collected with a 0.05° step size at 1 sec dwell time. For these measurements, CNT yarn was cut into 1 cm length samples which were spread out on zero background specimen holders, off-axis silicon. A split Pearson VII function was used to calculate the peak positions and full width at half maxima.

3. Results and discussion

3.1. ATR tests

The CNT yarn was tested using ATR spectroscopy in air with a slight mechanical pressure placed above the yarn to flatten the yarn against the ATR crystal. The results obtained from the ATR spectroscopy test can be seen in Figure 1. Figure 1(a) shows the results for the as-received yarn. As expected, the pure CNTs made up of carbon show very little signal. Carbon strongly absorbs infrared light in a broad range of frequencies making it difficult to resolve the details. Also note that the spectrum is flat from 2500 to 3600 cm^{-1}, which shows there are no hydroxyl groups in the as-received specimen.

Figure 1(b) shows the results for the functionalized CNT yarn. Two peaks are detected at 1140 and 1030 cm^{-1}, which is typical of C–O and C–O–C stretching that were due to the inherent impurities in the as-received CNTs. A strong peak at 1700 cm^{-1} confirms the existence of the C=O bond and the broad peak around 3400 cm^{-1} confirms the existence of the OH bond stretching. Based on the test results, it can be concluded that the CNT yarns were successfully functionalized with carboxyl and hydroxyl groups (e.g., COOH, COO–, OH, etc.), and the results are consistent with other studies conducted on CNT powders [20,21]. The CNTs reacted with nitric acid solution to form carboxyl functional groups on the CNT yarn. This reaction is as follows:

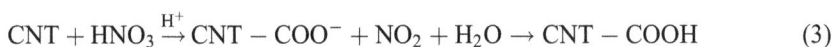

$$CNT + HNO_3 \xrightarrow{H^+} CNT - COO^- + NO_2 + H_2O \rightarrow CNT - COOH \qquad (3)$$

3.2. Morphology analysis

There was a significant morphology change due to the functionalization of the CNT yarn. The morphology of the as-received CNT yarn can be seen in Figure 2(a). The yarn has a smooth

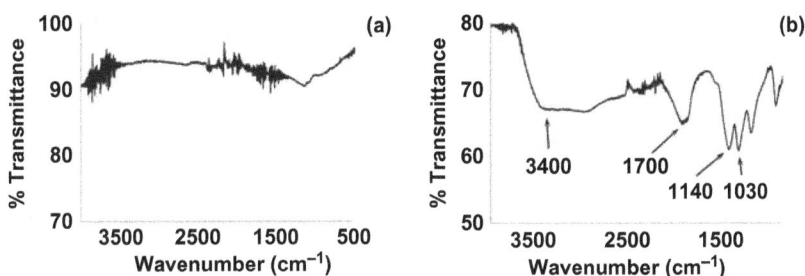

Figure 1.   ATR results of (a) as-received and (b) treated CNT yarn.

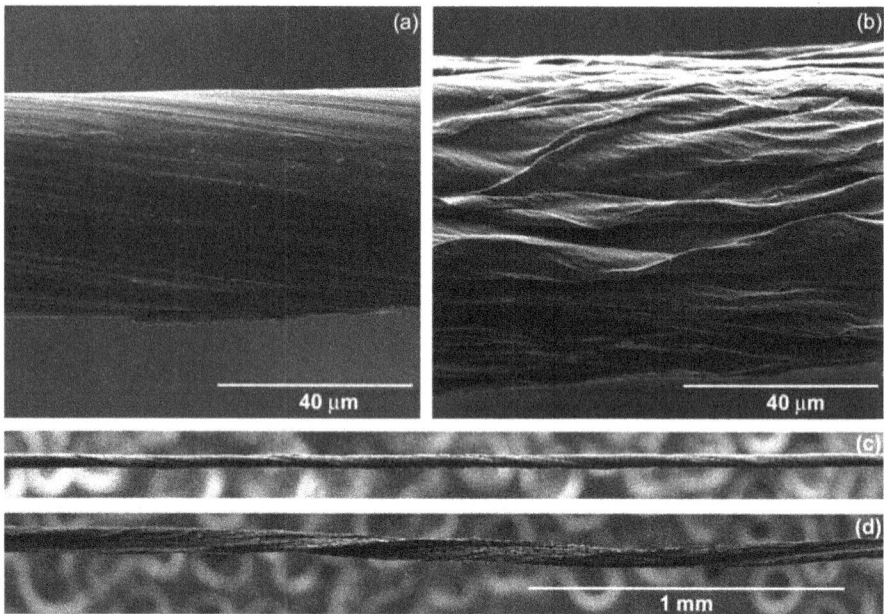

Figure 2.   SEM image of (a, c) as-received and (b, d) functionalized CNT yarn.

surface and a slight twist along the longitudinal axis. After functionalization (Figure 2(b)), the surface still has a smooth appearance, but new wavy extensions protrude out of the surface. Furthermore, the functionalized CNT yarn (Figure 2(c)) reverts back to the ribbons of CNTs, as can be seen in Figure 2(d). This morphology change can be explained by hydrophobic interactions, van der Waals and capillary force interactions between the CNTs, and functional group repulsions (or charge repulsions) [22]. There are many voids/porosity through which the solution can seep into the yarn by capillary force and can rearrange the CNT structure.

3.3.   XRD tests

3.3.1.   (001) Peak

XRD was used to analyze the crystalline nature of individual CNTs in the CNT yarn before and after the functionalization process as shown in Figure 3. The first peak at 25.9°

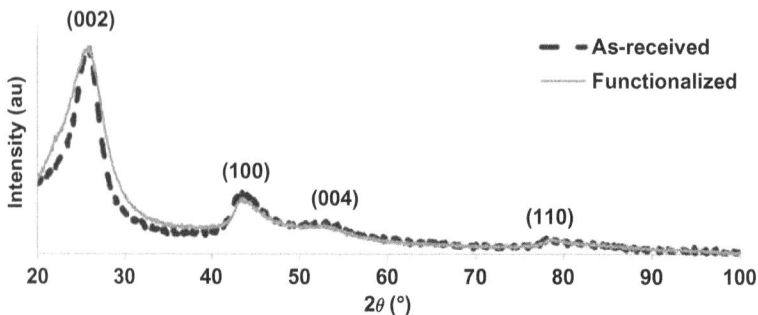

Figure 3.   XRD pattern of as-received and functionalized CNT yarn.

is identified as the (002) peak that is commonly seen due to the spacing between graphitic layers in carbon-based materials. Utilizing Bragg's law, the distance between these layers, or more specifically the distance between individual CNTs in the MWCNT ($d_{(002)}$-spacing) can be determined. The (002) d-spacing was found to be 3.42 Å, which is congruent with the accepted value [23] for CNTs.

From Figure 3, it can be seen that the functionalized CNT yarn (002) peak is considerably broader than the counterpart from the as-received yarn. The increase in broadening likely occurs due to an increased strain between graphic layers as well as partial reduction in the number of layers (walls) as a result of the functionalization process. Thus, it can be concluded that the 35min exposure to the sulfuric and nitric acid solutions in 3:1 volumetric ratio created significant defects disrupting the $d_{(002)}$-spacing and resulting in peak broadening.

3.3.2. (100) Peak

The (100) peak represents the curved graphene sheets that make up the structure of MWCNTs [23]. The (100) peak intensity is affected by diffractions from in-plane regularity. Based on the data in Figure 3, it can be seen that once the (002) peaks are normalized to each other, there is a limited decrease in intensity from the (100) peak after the functionalization process. This shows that functionalization did not catastrophically affect the in-plane regularity due to translation, twisting, or altering the helicity between tubes in a given CNT. Also, the shape of the (100) peak broadening was comparable for both, which suggests no appreciable reduction in the size of the nanotubes or increased strain along the length of the nanotubes due to functionalization. Further, the fact that (100) peak did not decrease in intensity proves that the CNTs did not chemically unzip into nanoribbons.

3.4. Functionalization

The electrical and mechanical properties were tested for the as-received and the functionalized CNT yarns. As described earlier, the area of a CNT yarn is difficult to calculate due to voids and surface irregularities. Thus, the data are reported in the form of resistivity density product. Figure 4 shows the resistivity density product for the as-received and functionalized CNT yarn. There is a significant decrease in the resistivity density product, i.e., from 2860 to 1224 nOhms m g/cm^3, after the functionalization process. This

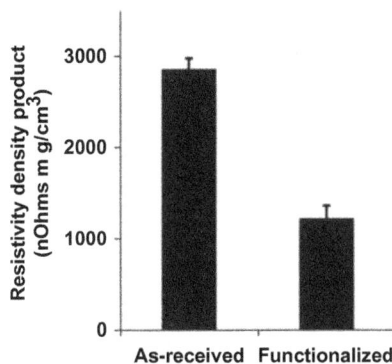

Figure 4.   Resistivity density product of as-received and functionalized CNT yarns.

corresponds to a 234% decrease in resistivity density product. Further, the functionalized CNT yarn is highly conductive among the textile/woven fibers. For example, fabrics, which are commonly coated with polypyrrole, have resistivity density product of ~80,000 nOhms m g/cm^3 [2]. Thus, the functionalized CNT yarn is a better conductive material compared to the commonly used textile materials coated with polypyrrole. On the other hand, the resistivity density product of the functionalized CNT yarn is slightly higher than that of metals, e.g., resistivity density product of iron and copper are 757 and 150 nOhms m g/cm^3, respectively.

A linear density is commonly used in textiles to assess their strength, i.e., strength is expressed in terms of tenacity (N/tex or N km/g). This is because the packing factor of fibers in the yarn can vary depending on the manufacturing process and parameters. The packing factor can significantly alter the physical and mechanical properties. Figure 5 shows the linear density (tex = g/km) of CNT yarn. From the chart, the linear density of the as-received and functionalized CNT yarn is 0.98 ± 0.10 and 3.04 ± 0.15 tex, respectively. This 310% increase in linear weight is not only due to the addition of oxygen and hydrogen from the functionalization process but also due to the contraction of the yarn. Upon soaking in the acid bath, the yarns untwist to form a ribbon-type shape from a cylinder, thus changing its geometry along with contraction in length.

Unlike resistivity density product, there is a considerable decrease in tenacity after the functionalization process. As shown in Figure 6, the tex of the as-received CNT yarn

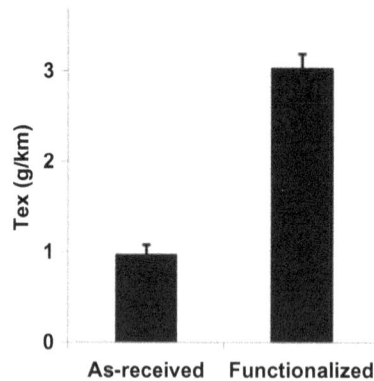

Figure 5.   Tex of as-received and functionalized CNT yarns.

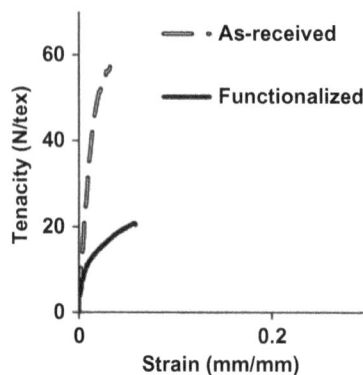

Figure 6.   Tenacity versus strain of as-received and functionalized CNT yarns.

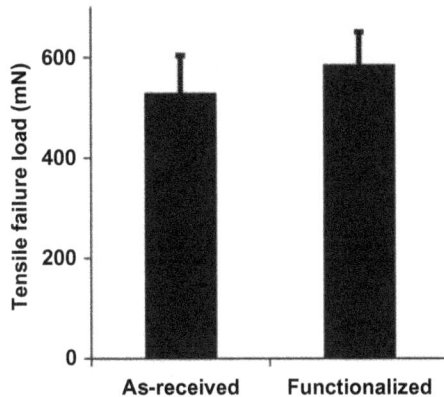

Figure 7. Tensile force of as-received and functionalized CNT yarns.

changed from 55 ± 5.2 to 21 ± 2.1 N/tex after functionalization of the CNT yarn. However, the actual failure loads of the as-received and functionalized CNT yarns were almost comparable (Figure 7). This indicates that the functionalization process had practically no effect on the tensile failure load directly, but the tenacity value was significantly reduced due to an increase in the weight. As a CNT yarn is made up of multitudes of CNTs, the limiting factor involved in tensile strength is the force required to slide one CNT past the other, and not the tensile strength of the individual CNT. Therefore, functional groups did not appear to increase mechanical interlocking within the yarn that would prevent sliding between CNTs and increase strength; however, the conductivity increased (resistivity density products decreased) due to the functional groups which increased the mobility of electrons within the CNT yarn.

4. Conclusion

CNT yarns were successfully functionalized with carboxyl and hydroxyl groups (e.g., COOH, COO–, OH, etc.) by submerging the CNTs into a 3:1 volumetric bath of sulfuric and nitric acid solutions for 35 min. During the submersion of the CNT yarn into the acid bath, the yarn's morphology changed from a smooth surface to a smooth surface with wavy extensions protruding from the surface. Also, the twisted yarn partially untwisted into a yarn ribbon. The functionalization of the individual CNT did not significantly change the in-plane regularity, however, after 35 min of exposure to the acid bath created significant atomic defects that disrupted the atomic pattern along the length of the diameter. The functionalization process increased the linear density by 310% and resistivity density product by 234%. The functionalized CNT yarn has superior resistivity density product (1224 nOhms m g/cm^3) when compared to conventional conductive coatings of polypyrrole (80,000 nOhms m g/cm^3). CNT yarn is a promising conductive fabric material.

References

[1] L. Hu, M. Pasta, F.L. Mantia, L. Cui, S. Jeong, H.D. Deshazer, J.W. Choi, S.M. Han, and Y. Cui, *Stretchable, porous, and conductive energy textiles*, Nano Letters. 10 (2010), pp. 708–714. doi:10.1021/nl903949m

[2] R. Technology, and R.T.L. Smithers Rapra Ltd, *Polymers in Electronics 2007*, iSmithers Rapra Publishing, Shawbury, 2007.

[3] M.F.L. De Volder, S.H. Tawfick, R.H. Baughman, and A.J. Hart, *Carbon nanotubes: Present and future commercial applications*, Science 339 (2013), pp. 535–539. doi:10.1126/science.1222453

[4] K. Jiang, Q. Li, and S. Fan, *Nanotechnology: Spinning continuous carbon nanotube yarns*, Nature 419 (2002), p. 801. doi:10.1038/419801a

[5] V. Sabelkin, H.E. Misak, S. Mall, R. Asmatulu, and P.E. Kladitis, *Tensile loading behavior of carbon nanotube wires*, Carbon 50 (2012), pp. 2530–2538. doi:10.1016/j.carbon.2012.01.077

[6] H.E. Misak, V. Sabelkin, S. Mall, R. Asmatulu, and P.E. Kladitis, *Failure analysis of carbon nanotube wires*, Carbon 50 (2012), pp. 4871–4879. doi:10.1016/j.carbon.2012.06.015

[7] H.E. Misak, R. Asmatulu, V. Sabelkin, S. Mall, and P.E. Kladitis, *Tension–tension fatigue behavior of carbon nanotube wires*, Carbon 52 (2013), pp. 225–231. doi:10.1016/j.carbon.2012.09.024

[8] H.E. Misak, V. Sabelkin, S. Mall, and P.E. Kladitis, *Thermal fatigue and hypothermal atomic oxygen exposure behavior of carbon nanotube wire*, Carbon 57 (2013), pp. 42–49. doi:10.1016/j.carbon.2013.01.028

[9] X. Zhang, K. Jiang, C. Feng, P. Liu, L. Zhang, J. Kong, T. Zhang, Q. Li, and S. Fan, *Spinning and processing continuous yarns from 4-inch wafer scale super-aligned carbon nanotube arrays*, Adv. Mater. 18 (2006), pp. 1505–1510. doi:10.1002/adma.200502528

[10] C.D. Tran, W. Humphries, S.M. Smith, C. Huynh, and S. Lucas, *Improving the tensile strength of carbon nanotube spun yarns using a modified spinning process*, Carbon 47 (2009), pp. 2662–2670. doi:10.1016/j.carbon.2009.05.020

[11] X.H. Zhong, R. Wang, L.B. Liu, M. Kang, Y.Y. Wen, F. Hou, J.M. Feng, and Y.L. Li, *Structures and characterizations of bundles of collapsed double-walled carbon nanotubes*, Nanotechnology 23 (2012), p. 505712. doi:10.1088/0957-4484/23/50/505712

[12] X. Gui, Z. Lin, Z. Zeng, K. Wang, D. Wu, and Z. Tang, *Controllable synthesis of spongy carbon nanotube blocks with tunable macro- and microstructures*, Nanotechnology 24 (2013), p. 085705. doi:10.1088/0957-4484/24/8/085705

[13] L. Xiao, P. Liu, L. Liu, K. Jiang, X. Feng, Y. Wei, L. Qian, S. Fan, and T. Zhang, *Barium-functionalized multiwalled carbon nanotube yarns as low-work-function thermionic cathodes*, Appl. Phys. Lett. 92 (2008), p. 153108. doi:10.1063/1.2909593

[14] G. Trakakis, D. Tasis, J. Parthenios, C. Galiotis, and K. Papagelis, *Structural properties of chemically functionalized carbon nanotube thin films*, Materials 6 (2013), pp. 2360–2371. doi:10.3390/ma6062360

[15] H. Wei, Y. Wei, Y. Wu, L. Liu, S. Fan, and K. Jiang, *High-strength composite yarns derived from oxygen plasma modified super-aligned carbon nanotube arrays*, Nano Res. 6 (2013), pp. 208–215. doi:10.1007/s12274-013-0297-7

[16] R. Malik, Y. Song, N. Alvarez, B. Ruff, M. Haase, B. Suberu, A. Gilpin, M. Schulz, and V. Shanov, *Atmospheric pressure plasma functionalization of dry-spun multi-walled carbon nanotubes sheets and its application in CNT–polyvinyl alcohol (PVA) composites*, MRS Online Proceedings Library 1574 (2013). Available at http://iopscience.iop.org/0957-4484/24/8/085705.

[17] J. Carretero-González, E. Castillo-Martínez, M. Dias-Lima, M. Acik, D.M. Rogers, J. Sovich, C.S. Haines, X. Lepró, M. Kozlov, A. Zhakidov, Y. Chabal, and R.H. Baughman, *Oriented graphene nanoribbon yarn and sheet from aligned multi-walled carbon nanotube sheets*, Adv. Mater. 24 (2012), pp. 5695–5701. doi:10.1002/adma.201201602

[18] E.Y. Jang, J. Carretero-González, A. Choi, W.J. Kim, M.E. Kozlov, T. Kim, T.J. Kang, S.J. Baek, D.W. Kim, Y.W. Park, R.H. Baughman, and Y.H. Kim, *Fibers of reduced graphene oxide nanoribbons*, Nanotechnology 23 (2012), p. 235601. doi:10.1088/0957-4484/23/23/235601

[19] H.E. Misak and S. Mall, *Investigation into microstructure of carbon nanotube multi-yarn*, Carbon (2014, in press). Available at http://dx.doi.org/10.1016/j.carbon.2014.02.012. doi:10.1016/j.carbon.2014.02.012.

[20] Y.-S. Kim, J.-H. Cho, S. Ansari, H.-I. Kim, M. Dar, H.-K. Seo, G.-S. Kim, D.-S. Lee, G. Khang, and H.-S. Shin, *Immobilization of avidin on the functionalized carbon nanotubes*, Synth. Met. 156 (2006), pp. 938–943. doi:10.1016/j.synthmet.2006.06.003

[21] T. Ramanathan, F.T. Fisher, R.S. Ruoff, and L. Catherine Brinson, *Apparent enhanced solubility of single-wall carbon nanotubes in a deuterated acid mixture*, J. Nanotechnol. 2008 (2008), 4p. Article ID 296928.

[22] R. Asmatulu and R.H. Yoon, *Effects of surface forces on dewatering of fine particles*, in *Separation Technologies for Minerals, Coal and Earth Resources*, C. Young and G.H. Luttrell., eds., Society for Mining Metallurgy, and Explorations (SME), Englewood, CO, 2012, pp. 95–102.

[23] P.J.F. Harris, *Carbon Nanotube Science: Synthesis, Properties and Applications*, Cambridge University Press, Cambridge, 2009.

9

Improving the performance of structure-embedded acoustic lenses via gradient-index local inhomogeneities

H. Zhu and F. Semperlotti*

Department of Aerospace and Mechanical Engineering, University of Notre Dame, Notre Dame, Indiana 46556, USA

We investigate the use of graded inhomogeneities in order to enhance the focusing and collimation performance of structure-embedded acoustic metamaterial lenses. The type of inhomogeneity exploited in this study consists in axial symmetric exponential-like gradients of either material or geometric properties that create gradient-index inclusions able to bend and redirect propagating waves. In particular, we exploit the concept of gradient index inclusions to achieve focusing and collimation of ultrasonic beams created by embedded drop-channel lenses in both bulk and thin-walled structures. In the latter, the implementation is possible thanks to geometric exponential tapers known as Acoustic Black Holes (ABH). ABH tapers allow accurate control of the characteristics of the acoustic beam emanating from the lens channel which in the conventional design is severely affected by diffraction. The concept of beam control via graded inclusions is numerically illustrated and validated by using a combination of methodologies including geometric acoustics, finite difference time domain, and finite element methods.

Keywords: acoustic lens; acoustic black hole; gradient index inhomogeneity; drop-channel; acoustic metamaterials

1. Introduction

The ability to generate highly directional and focused ultrasonic excitation has become a major area of interest for the structural health monitoring (SHM) field since it enables selective damage interrogation and leads to improved accuracy and resolution. The possibility to send ultrasonic energy in a preferential direction can also help counteracting the highly directional character of certain material systems, such as orthotropic layered composites [1] where the direction of energy propagation can be largely different from the initial direction of the emitted wave. In recent years, phased array [2] (PA) technology has emerged as one of the most promising approaches to the generation of directional and focused ultrasonic excitation. PA exploits a network of transducers that emit mostly omnidirectional waves with prescribed phase differences. Although a robust and efficient technology, PA has two major limitations that prevent its extensive use in practical applications: (1) it requires a large number of transducers, and (2) it cannot produce collimated beams. Both limitations can be traced back to the same cause that is PA operates based on principles of constructive/destructive interference of the multiple

---

*Corresponding author. Email: fsemperl@nd.edu

omnidirectional wavefronts generated by the array of transducers. An effective PA system typically requires dozens of transducers, which is a major downside from the SHM perspective because it leads to higher system complexity and to increased probability of malfunctions and false alarms.

In an effort to develop novel technologies able to address these important limitations of PA, previous studies have explored the design of embedded metamaterial lenses [3,4]. This approach was proposed as a possible alternative to achieve selective beam-forming and beam-steering via a single ultrasonic transducer. The embedded lens design exploited the host structure in order to mold the wavefronts generated by a single ultrasonic transducer into a highly directional beam; that is the structure itself was used as a mechanical filter to produce selective excitation. The lens was initially developed according to a locally resonant [4] design approach. It was shown that a proper design of the equi-frequency-contour characteristics of the lens would result in beaming, steering, focusing, and collimation properties, all achievable by simply tuning the excitation frequency of a single acoustic source. The locally resonant design, however, presented some fabrication complexities that made it not suitable for certain applications, particularly those involving the design of structural materials (that is materials exhibiting load-bearing capabilities). For these reasons, a nonresonant design [3] was proposed in an effort to retain the advantages of the embedded lens concept while reducing the fabrication complexity. The working mechanism was based on the well-known concept of drop-channel [5–7], which allows creating preferential paths for wave propagation inside a periodic material. In this approach, the lens was created by embedding periodic homogeneous inclusions in the host structure while a network of carefully engineered drop-channels was used to produce and deliver the acoustic signal in prescribed directions. The individual channels were activated by simply tuning the frequency of excitation of a single transducer thanks to a coupling mechanism based on locally resonant defects. While this design allowed achieving very effective beam-forming and beam-steering characteristics, the overall performance was more limited than in the locally resonant case. In particular, the ultrasonic beam was subjected to diffraction at the exit of the acoustic channel that ultimately degraded the performance by preventing the generation of focused or collimated beams. In fact, the diffraction process occurring at the exit of the channel resulted in large angle of apertures that produced a diffused ultrasonic beam not unlike that generated by a plane wave source through a slit.

In this paper, we investigate the possibility to achieve focusing and collimation of the ultrasonic beam emitted by the embedded drop-channel lens by using gradient index (GRIN) inclusions implemented via local inhomogeneity. GRIN [8,9] materials exploit a spatially variable distribution of the physical parameters to produce a position-dependent phase velocity. The existence of phase velocity gradients results in acoustic rays having curved trajectories and, ultimately, in wavefronts that are largely distorted upon propagation. In recent years, phononic crystals and/or acoustic metamaterials have been extensively used to tailor the physical parameters of the host material in order to control the propagation of acoustic and elastic waves. In the long wavelength limit approximation, effective parameters [10] can be used to characterize the behavior of these composites and to design refraction index and phase velocity profiles [11]. Note that the conventional metamaterial design of GRIN, which relies on multiple phase inclusions, cannot provide a smooth and continuous variation of the refraction index but only an approximation via a step-like distribution of the material properties. At high frequency, this characteristic yields limited performances and very restrictive requirements on fabrication precision. An alternative approach to tailoring the phase velocity profile is based on graded elastic properties of the host material (e.g. density, modulus, etc.), the so-called functionally graded materials [12,13] (FGM). Although these materials could in principle achieve a continuous variation of the elastic parameters, such profiles are very challenging to obtain in practice. An alternative approach, conceptually analogous to FGM but limited to applications on structural waveguides, is based on tailoring the geometric properties. In structural waveguides, spatial gradients of the flexural phase velocity can be enforced by tailoring the geometric parameters of the guide such as, for example, the thickness. In previous studies, this approach has been applied to achieve extreme damping for flexural waves by exploiting the concept of Acoustic Black Hole (ABH) [14]. The fundamental physical principle exploited in the ABH tapers was first observed by Pekeris [15] for waves propagating in stratified fluids and later extended to acoustics in solids by Mironov [16]. Afterwards, Krylov [13,17] applied this concept to achieve passive vibration control of structural elements. The intrinsic wave-focusing characteristic of the ABH tapers was also exploited in the design of high-performance vibration-based energy harvesting systems [18].

The ABH consists of a circular taper with exponentially variable thickness that is embedded into the supporting structural element and able to produce smooth gradients of both the phase and group velocities. The existence of these gradients has two main effects: (1) it steers flexural waves towards the center of the ABH and (2) it gradually reduces the phase and group velocities that, in the ideal case, will gradually approach a zero value at the ABH center. A typical ABH thickness profile is described by the relation $h(x) = h_r + \varepsilon x^m$, where h_r is the residual thickness, $m \geq 2$ is the taper coefficient and ε is a real constant. Both m and ε must be chosen to satisfy the smoothness condition [15,19]. Examples of applications of geometric tailoring to thin plate structures were recently reported by Dehesa [20] and Zhu [21] who designed, respectively, refractive lenses and phononic plates based on embedded thickness profiles.

The paper is structured as follows: we first briefly review the concept of drop-channel acoustic lens, and then we illustrate via numerical computations the concept of beam focusing and collimation achievable via a graded elastic modulus in bulk structures. Successively, we extend the design to thin plate structures where the control on the beam properties is achieved via embedded ABH tapers. The performance of the graded design is explored using a combination of geometric acoustics, finite difference time domain (FDTD), and finite element methods. It will be shown that, compared with the initial design studied by Zhu [3] without tapered inclusions, the proposed approach provides a practical methodology to control the characteristics of the ultrasonic excitation especially in terms of beam collimation (i.e. angle of aperture), reduction of side-lobes, and adjustment of the beam width.

2. Drop-channel lens design and numerical model

In the following, we briefly review the main design and modeling approach for the drop-channel-based acoustic lenses [3]. The lens design relies on the use of periodic distributions of inclusions embedded in the surrounding structure. These inclusions effectively create a metamaterial structure that is designed to provide full bandgaps [22,23] in the frequency range of interest for the operation of the lens. In this frequency range, the metamaterial acts as a stopband mechanical filter therefore preventing energy from propagating through. Figure 1(a) shows the conceptual

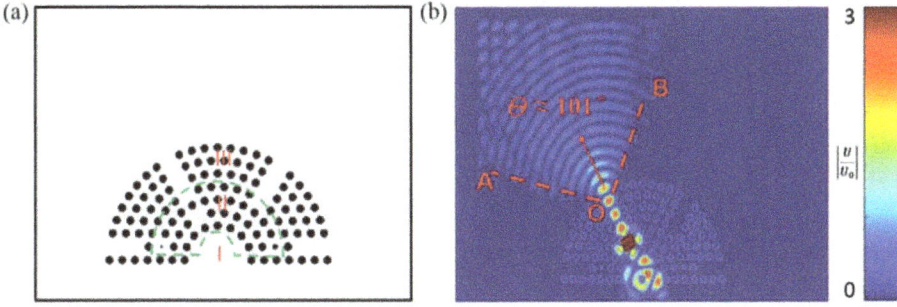

Figure 1. (a) Conceptual schematic of the drop-channel based lens with four embedded channels. The lens is divided in three main sections: (I) an inner section without cylindrical inclusions, (II) an intermediate section with embedded point defects, and (III) an outer section containing a set of radial waveguides at prescribed azimuthal locations. (b) Shows the magnitude of the nondimensional displacement field generated when the lens is excited by a single-tone harmonic input at $f_z = 24.95$ kHz. U_0 indicates the maximum displacement at the excitation point.

schematic and the geometry of such a lens. The lens is mostly divided in three sections: (I) an inner section without inclusions, (II) an intermediate section with embedded point defects, and (III) an outer section containing a set of radial wave-guides at prescribed azimuthal locations. The inner section is intended to host the source of the ultrasonic excitation. Wave propagation at frequencies inside the band-gap can be achieved by creating a network of line defects (i.e. waveguides) associated with either localized or spatially confined modes [24,25]. The waveguides are dyna-mically coupled to the inner section I by a network of point defects that are engi-neered to exhibit resonant modes at desired coupling frequencies. The dynamic response of the lens embedded in a bulk material is described by the Navier's equations. The process is illustrated on a semicircular lens made of cylindrical nickel inclusions distributed in an epoxy background [3]. The Navier's equations for the bulk material are:

$$\rho(x,y,z)\ddot{u}_i(x,y,z,t) = \partial_j\big[C_{ijkl}(x,y,z)\partial_k u_l(x,y,z)\big] \tag{1}$$

where $\rho(x,y,z)$ and $C_{ijkl}(x,y,z)$ are the space-dependent density and elastic tensor, respec-tively. For the sake of simplicity, in the numerical simulations, we consider only the shear vertical (SV) mode (i.e., particle displacement parallel to the axis of the inclusions) that is uncoupled from other bulk modes. From Equation (1), the corresponding governing equation for the SV mode becomes:

$$\rho(x,y)\frac{\partial^2 u_3}{\partial t^2} = \frac{\partial \tau_{13}}{\partial x} + \frac{\partial \tau_{23}}{\partial y} \tag{2}$$

$$\tau_{13} = C_{44}(x,y)\frac{\partial u_3}{\partial x} \tag{3}$$

$$\tau_{23} = C_{44}(x,y)\frac{\partial u_3}{\partial y} \tag{4}$$

Equations (2)–(4) can be solved by using the FDTD [26] approach that yields the following set of algebraic equations:

$$u_3(i,j,t+1) = 2u_3(i,j,t) - u_3(i,j,t-1) + \frac{\Delta t^2}{\rho(i,j)\Delta x}$$

$$\left[\tau_{13}\left(i+\frac{1}{2},j,t\right) - \tau_{13}\left(i-\frac{1}{2},j,t\right) + \tau_{23}\left(i,j+\frac{1}{2},t\right) - \tau_{23}\left(i,j-\frac{1}{2},t\right) \right]$$

$$\tau_{13}\left(i+\frac{1}{2},j,t\right) = \frac{C_{44}\left(i+\frac{1}{2},j,t\right)}{\Delta x}[u_3(i+1,j,t) - u_3(i,j,t)]$$

$$\tau_{13}(i,j+1/2,t) = \frac{C_{44}(i,j+1/2,t)}{\Delta x}[u_3(i,j+1,t) - u_3(i,j,t)]$$

where u_3, τ_{13}, τ_{23} are the out-of-plane displacement and two components of the shear stress, while i, j, and t are the space and time step indices, respectively.

Further details on the lens configuration can be found in Ref. [3]. The different waveguides can be activated by changing the excitation frequency, therefore achieving directional ultrasonic excitation via a single transducer. Figure 1(b) shows an example of the displacement field produced by the lens using a single tone frequency excitation at $f_z = 24.95$ kHz, which corresponds to the activation frequency of the channel oriented at 120°. Perfectly matched layers [27] are also used all around the computational domain so to avoid backscattering from the boundaries.

The performance of the different lens designs will be evaluated based on two main metrics: (1) the angle of aperture of the main beam, and (2) the side lobe amplitude ratio U_{s_l}. The angle of aperture is defined as the central angle $\angle AOB$, including the portion of the beam where the amplitude drops less than 90% with respect to the peak value in the propagation direction (Figure 1(b) red arrow). The side lobes amplitude ratio U_{s_l} is defined as $U_{s_l} = U_{pm}/U_{ps_l}$, where U_{pm} and U_{ps_l} represent the peak value of the displacement field for the main and the side lobes, respectively. A zero value of this ratio indicates that no side lobes exist.

Results from the above simulations clearly indicate that as the ultrasonic beam exits the channel, it undergoes a diffraction process that ultimately produces a large beam aperture ($\theta \approx 101°$ in the present example). Under these conditions, the waveguide behaves not unlike a point source through a slit, therefore drastically reducing the ability of the lens in generating directional excitation. This characteristic was found to be one of the main limitations of this lens design. New or alternative design approaches are needed to limit this diffraction effect and to improve the overall performance of the lens.

3.   Design based on GRIN inclusions: geometric acoustic analysis

In order to alleviate the effects of diffraction, we explore possible modifications to the design of the lens that are based on the use of GRIN inclusions. The ability of GRIN materials (or, equivalently, of ABH tapers) to bend incoming waves is determined by the existence of a phase velocity gradient produced by the local inhomogeneity. In the case of a geometric ABH taper, the characteristics of the profile are identified by two main parameters: the residual thickness h_r and the coefficient m of the exponential term. The former determines the lower bound of the local phase velocity while the latter determines how rapidly, in space, the phase velocity changes (i.e., it determines the spatial gradient of the phase velocity). In order to study the wave propagation in inhomogeneous media, it is convenient to use a geometric acoustic approach which provides a clear assessment of the wave trajectory when affected by the inhomogeneity. The trajectory of the individual ray for the uncoupled elastic wave is given by:

$$\frac{d\boldsymbol{x}}{dt} = c^2 \boldsymbol{s} \tag{6}$$

$$\frac{d\boldsymbol{s}}{dt} = -\frac{1}{c}\nabla\boldsymbol{s} \tag{7}$$

where $\boldsymbol{x}$ is the position vector along the ray trajectory, $\boldsymbol{s}$ is the slowness vector that can be expressed as $\boldsymbol{s} = \frac{\boldsymbol{n}}{c}$, where c is the local phase velocity, and $\boldsymbol{n}$ is the unit vector normal to the wavefront. Once the phase velocity distribution and the initial conditions (in terms of the source location and type) are specified, Equations (6)–(7) can be numerically integrated to obtain the ray trajectories across the inhomogeneous medium. In the case under study, the phase velocity c_0 of the background area is constant while the phase velocity c of the tapered area satisfies the relation $\frac{c}{c_0} = \sqrt{\frac{h(r)}{h_0}} = \sqrt{\frac{h_r + \varepsilon r^m}{h_0}}$.

We observe that similar considerations are applicable if we assume the GRIN inclusion made of an exponentially variable elastic modulus as $\frac{c}{c_0} = \sqrt{\frac{G(r)}{G_0}} = \sqrt{\frac{G_r + \varepsilon r^m}{G_0}}$. This latter approach produces in bulk materials phase velocity profiles that are equivalent, in principle, to those produced by an exponential-like taper in thin-walled structures.

Geometric acoustics was used to guide the design of the GRIN inhomogeneity and perform a preliminary assessment of the performance versus some of the design parameters. We observe that the use of a large taper coefficient m would result in a very steep variation of the phase velocity therefore increasing the reflection coefficient and preventing a progressive bending of the wavefronts. In fact, a high taper coefficient produces an inclusion mostly divided into two main regions: an outer region with low impedance mismatch with respect to the background, and an inner region with large impedance mismatch. For this reason, we fixed the taper coefficient to $m = 2.2$, which is a typical value for ABH [13–15] satisfying the smoothness criterion [15,18] and producing a smooth variation of the phase velocity, and explored the effect of a variable residual thickness h_r.

The exponential-like taper was first simulated in bulk materials by tailoring the elastic properties, in particular the local shear modulus. Figure 2(b)–(e) shows the displacement fields, obtained by FDTD simulations, where a single circular taper with fixed radius $r = 4.32$ cm and taper exponent $m = 2.2$ is placed at the exit of the waveguide. We explored the performance of tapers having different values of the lower bound of the shear modulus G_r. In Figure 2(b)–(e), the shear modulus G_r increases from left to right (i.e. from (b) to (e)) according to the following predefined values $G_r = \frac{1}{9}, \frac{3}{9}, \frac{5}{9}, \frac{7}{9} G_0$. Consequently, also the phase velocity gradient increases (from right to left). Figure 2(a) provides the reference case where no taper is placed at the exit of the channel. The insets in the upper part of each figure show the corresponding ray acoustic solutions that are in good agreement with the full wave solutions. As expected from Equations (5)–(6), larger phase velocity gradients provide more bending ability showing how an omnidirectional source

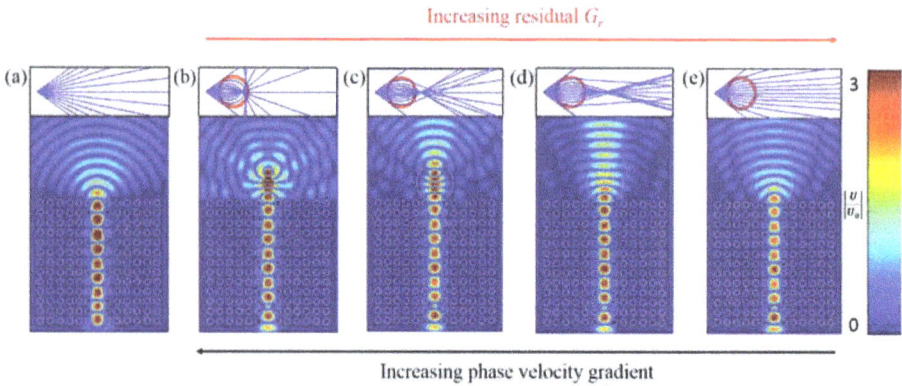

Figure 2. Magnitude of the displacement field produced by FDTD simulations showing the response of an acoustic waveguide filtered through a graded index inclusion (red circle). (a) typical diffraction pattern occurring at the exit of a waveguide without graded inclusion. (b)–(e) Magnitude of the displacement field obtained by placing a single GRIN taper at the exit of the waveguide. The tapers have all the same radius and taper coefficient while the residual shear modulus G_r at the center increases from (b) to (e). The inset provides the corresponding geometric acoustic solution which helps understanding the effect of the graded index inclusion on the ray trajectory.

can be progressively scattered (b), focused (c), or collimated (d) by different taper designs.

4. Functionally graded arrays of GRIN inclusions

The results shown in the previous section illustrated the main operating principle of the graded inclusions. In fact, a GRIN inclusion operates much like an optical lens which can scatter, focus, or collimate waves depending on its geometric design. Following a similar analogy, we can expect that an array of GRIN inclusions can be used, much like optical multi-lens devices, to produce a progressive alignment of the ultrasonic beam further improving the level of control on the excitation.

We consider a distribution of three ABH tapers forming a sequence of functionally graded inclusions $r_i = (0.2880i - 0.1440)*a$ for $i = 1, 2, 3$ (Figure 3(a)). The geometric parameters are: $r_1 = 0.432a$, $r_2 = 0.72a$, $r_3 = 1.008a$, $\Delta R_1 = 1.44a$, and $\Delta R_2 = 2.016a$, where $a = 0.03$ cm is the lattice constant which defines the unit cell of the lens geometry [3]. The residual thickness ratios for all the holes are set to $\frac{h_r}{h_0} = \frac{5}{9}$, the exponential taper is $m = 2.2$ and therefore the coefficient can be determined as $\varepsilon = \frac{4h_0}{9r_0^{2.2}}$. Equations (5)–(6) were numerically integrated to identify the trajectories of the rays emanated by a point source placed in front of the first ABH. From the ray-tracing results (Figure 3(b)), it can be seen that the graded ABH sequence provides larger flexibility in creating focused or collimated beams from initially omnidirectional sources. Note that the functionally graded array also provides control on the beam width allowing either expansion or compression (depending on the gradient of the radius profile r_i) of the incident beam. The beam width at the exit of the graded array can be, at most, the size of the ABH diameter. Figure 3(b) shows that, in the selected design, the beam width increases from $w_1 = 2.72$ cm to $w_2 = 3.96$ cm.

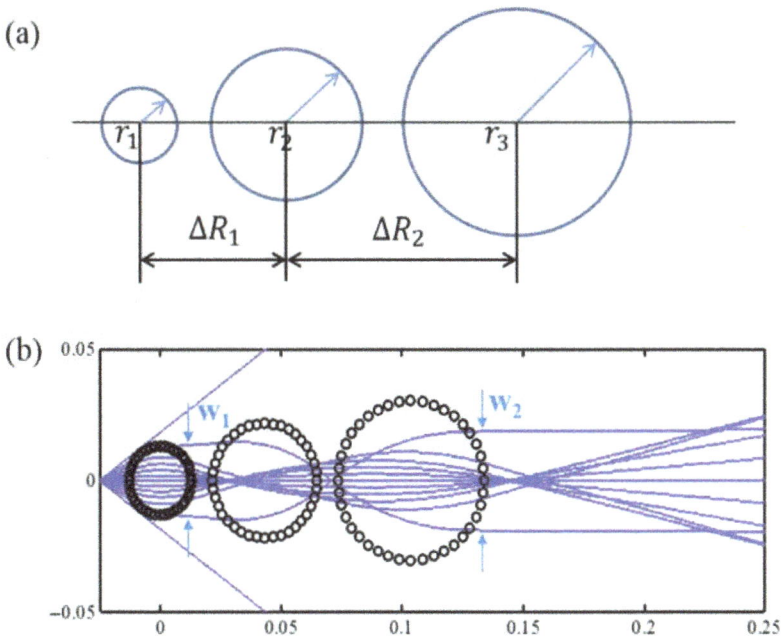

Figure 3.   (a) Notional schematic of the functionally graded ABH array. Three ABHs of different radius are used to provide progressive alignment and beam enlargement. (b) Results from the geometric acoustic analysis showing the trajectories of the rays emanating from an omnidirectional radiating (point) source placed in front of the ABH array. Results clearly show the confinement effect of the tapers as well as the magnification of the beam width.

Note that the above results for either single or multiple tapers were not obtained following an optimization process, and therefore these results are not indicative of the maximum performance of the GRIN inclusions.

5.   Application to the acoustic lens

As initially mentioned, the main objective of this study is to integrate the concept of graded ABH tapers into the design of the acoustic lens in order to improve the focusing and collimation performance. Here below, this concept is first investigated for bulk materials and then extended to thin-walled structures.

5.1.   Functionally graded tapers in bulk materials

We consider the semicircular lens as inSection 2 and place a graded array of three circular tapers (see Figure 4(a), gray circles) at the exit of the 120° waveguide. The array has the same parameters as that discussed in the previous section. The first hole of the array was centered on the outer perimeter of the lens. The dynamic response of the ABH-enhanced lens was evaluated by using a single-tone harmonic input at frequency f_z = 24.95 kHz (i.e., the activation frequency of the channel). Results are shown in Figure 4(b) and (c). A direct comparison of the dynamic response of the enhanced lens (Figure 4(b)) versus the conventional design is presented in Figure 1(b).

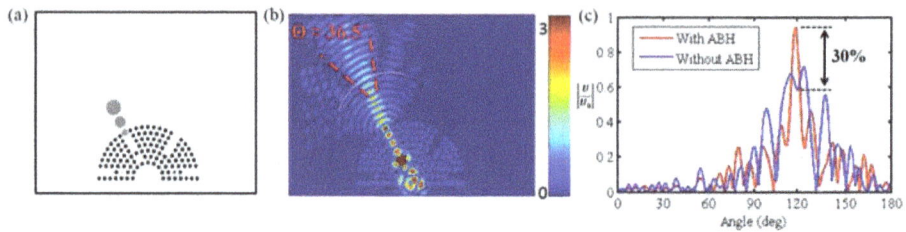

Figure 4. (a) Schematic of the ABH-enhanced lens design showing the location of the functionally graded array of tapers (gray circles). (b) Response of the ABH-enhanced lens when driven by a single-tone harmonic excitation at f_z = 24.95 kHz. (c) Directivity plot comparing the performance of the conventional (blue line) and the ABH-enhanced lens (red line). The direct comparison of the directivity plots shows the drastic reduction of the side lobes (about 40%) and of the angle of aperture as well as the increased beam intensity (about 30%) in the selected direction of excitation.

Results show large improvement in terms of beam formation and an evident reduction in the beam aperture, which is now $\theta \approx 36.5°$ (about 65% improvement with respect to the original drop-channel design). The performance of the two designs is also compared in Figure 4(c) in terms of directivity plots. The plots are extracted from the displacement field produced at a distance L = 34.5 cm (see dashed white line) from the center of the lens which coincides with the location of the acoustic source. The plot shows the magnitude of the displacement field normalized by the amplitude at the source location U_0. The comparison of the directivity functions clearly indicates the improvement in terms of beam formation as well as a drastic reduction of the side lobes. The improved control on the beam aperture results in a consistent increase (about 30%) of the vibrational energy delivered in the selected direction of excitation. The reduction of the side lobes can be quantified using the ratio $U_{s_l} = U_{pm}/U_{ps_l}$, where U_{pm} and U_{ps_l} represent the peak value of the displacement field for the main and the side lobes, respectively. For the bulk material, the proposed ABH-enhanced design delivers a ratio $U_{s_l} = 0.47$ that represents a reduction of about 40% with respect to the case without lens (Figure 1(b)) that achieved $U_{s_l} = 0.78$.

It is expected that by a proper optimization of the design parameters, further reduction of the side lobes as well as a fully collimated beam could be achieved.

5.2. Geometrically graded tapers in thin-walled structures

It was previously pointed out that fabricating structural materials with smoothly varying elastic properties (such as density and elastic moduli) is a challenging task. In addition, due to integrity and durability issues it is often preferred to avoid the use of multiple material interfaces. Thin-walled structures offer an interesting alternative approach for the integration of graded inclusions. In these structural components, GRIN inhomogeneities can be obtained exploiting the concept of ABH geometric taper. According to the ABH design, the inhomogeneous phase velocity distribution can be achieved by tailoring the local thickness according to an exponential-like profile, as discussed in Section 3.

The concept is illustrated using a rectangular thin silicon plate where a single drop-channel is realized following the same design procedure previously used for the bulk acoustic lens. In this case, the inclusions of the lens were obtained via through-holes (following the approach in Ref. [3]) arranged in a square lattice geometry.

In particular, the plate was built out of a square array of 11×11 circular through-holes having radius ratio $r/a = 0.46$, thickness $h_0/a = 0.4$, and lattice constant $a = 10$ cm. The corresponding infinite and perfectly periodic plate based on this configuration yields a narrow full bandgap [28] between $\Delta f = 28$–31 kHz, as shown in Figure 6(a). A drop-channel was created removing an entire array of inclusions in the center so as to create defect modes inside the bandgap that could be exploited to support the directional excitation. The channel was coupled to the area hosting the acoustic source by using one point cavity defect implemented via a missing hole. The defect exhibited a fundamental resonance frequency at $f = 29$ kHz.

A graded ABH taper was then located in front of the channel to achieve beam control. The ABH taper had radius $r = 25$ cm, residual thickness $h_r = 1.77$ cm and a taper coefficient $m = 2.2$. The schematics of the two thin plates with and without the taper are shown in Figure 5(a) and (b).

In order to test the ABH taper design, we solved the plate model by finite elements using the commercial software Comsol Multiphysics (Comsol 4.3, COMSOL Inc., Burlington, MA, USA). Both plates were excited by a harmonic uniformly distributed boundary load in the out-of-plane direction tuned to the cavity mode frequency and applied at the right boundary of the plate. Perfectly matched layers were used all around the boundary of the plate to reduce backscattering and to improve the beam visualization.

Figure 5 shows the performance of the ABH-graded design by direct comparison with the conventional design. The response of the two lenses is shown in terms of normalized out-of-plane displacement fields. Similarly to what observed for the bulk material design, the ABH-graded taper is extremely effective in controlling the

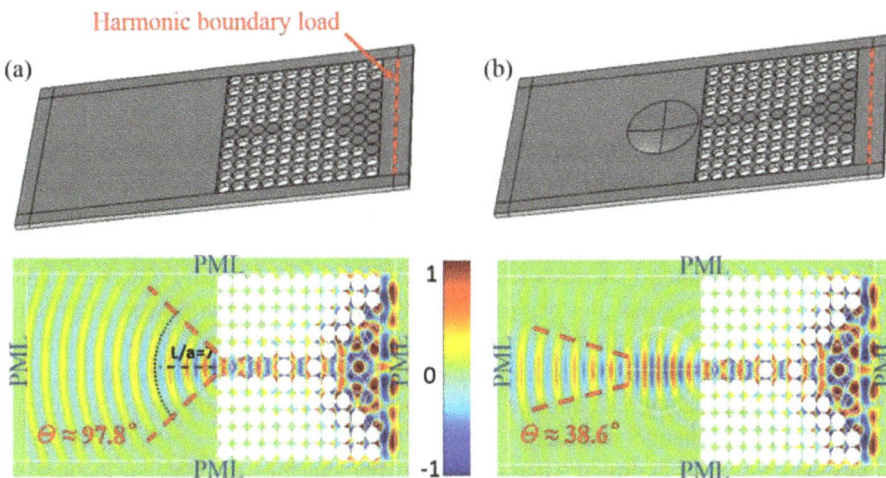

Figure 5. Numerical investigation on the performance of the (a) conventional and (b) ABH-enhanced lens. The top figures show the geometric design of the host structure with the embedded lens and the ABH taper. Results are shown in terms of out-of-plane displacement amplitude maps when the lens is excited with a harmonic force load at the frequency of the cavity mode $f = 29$ kHz. The direct comparison shows the focusing ability of the ABH taper that drastically reduces the angle of aperture and increases the energy transferred in the channel direction.

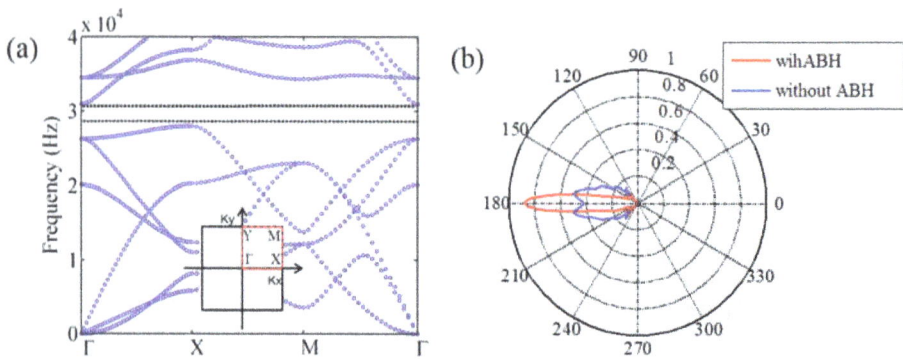

Figure 6.   (a) Shows the band structures of the periodic air/silicon system used to build the lens in the thin plate structure. Results indicate the existence of a full bandgap in the range $\Delta f = 28\text{–}31$ kHz (black dotted lines). (b) Directivity plots at $L/a = 7$ from the exit of the drop-channel. The comparison between the response of the lens without (blue line) and with (red line) the ABH shows a drastic reduction of the side lobes and a large increase in the beam intensity in the channel direction.

formation of the beam and reducing the diffraction at the exit of the channel. The aperture of the beam is reduced from $\theta \approx 97.8°$ to $\theta \approx 38.6°$ (about 60%). The directivity plots (Figure 6(b)), extracted from the response at distance $L/a = 7$ from the exit section of the channel, indicate a drastic reduction of the side lobes. The ratio U_{S_l} drops from 0.54 for the case without lens to approximately 0.12 (side lobes have almost completely disappeared) in the ABH-enhanced case. We also observe an outstanding increase in the amount of energy transferred in the channel direction, as shown by the nondimensional field U/U_0 that increases from 0.43 to 0.86 (about 100%).

6.   Conclusions

We presented an approach to improve the performance of structure-embedded acoustic lenses by exploiting a network of engineered local inhomogeneities. The conventional drop-channel lens design was able to achieve selective beam-forming and beam-steering via a single ultrasonic transducer but suffered from diffraction effects that severely reduced its performance in terms of directivity and energy intensity. GRIN inclusions implemented either via material or geometric inhomogeneity were proven to provide a viable solution and to drastically improve the performance of the lens. We numerically demonstrated that either single or multiple inclusions allow an effective control of the beam aperture, drastically reducing the side lobes and consistently improving the intensity of the beam in the channel direction.

   For the bulk material design where the lens is implemented via material inhomogeneity, the beam aperture was reduced of about 65% while the side lobes ratio U_{S_l} dropped of about 40% from 0.78 (without lens) to 0.47 (with lens).

   In the case of the thin-walled structure, the beam control was achieved via ABH tapers. This design provided a reduction of the beam aperture of about 60% and of the side lobes ratio from 0.54 to about 0.12, that is, an almost complete elimination of the side lobes.

The proposed design is of particular interest for applications on thin-walled structures where the graded inhomogeneity can be achieved via geometric ABH tapers. These tapers allow a high degree of control on the beam properties while involving minimal fabrication complexity. The ABH-enhanced drop-channel lens can have critical implications in ultrasonic-based SHM systems. In particular, the outstanding reduction in the number of transducers (only one is needed versus the several dozen required by the PA technology) would reduce considerably the system complexity and the probability of malfunctions and false alarms.

The largely increased directionality of the lens will drastically improve the ability to selectively scan a structure for damage delivering more effectively ultrasonic energy in a predefined direction and reducing backscattering from unwanted sources.

References

[1] A. Lamberti and F. Semperlotti, *Detecting closing delaminations in laminated composite plates using nonlinear structural intensity and time reversal mirrors*, Smart Mat. Struct. 22 (2013), pp. 125006. doi:10.1088/0964-1726/22/12/125006.

[2] L. Yu and V. Giurgiutiu, *In situ 2-D piezoelectric wafer active sensors arrays for guided wave damage detection*, Ultrasonics. 48 (2008), pp. 117.

[3] H. Zhu and F. Semperlotti, *A passively tunable acoustic metamaterial lens for selective ultrasonic excitation*, J. Appl. Phys. 116 (2014), pp. 094901. doi:10.1063/1.4894279.

[4] F. Semperlotti and H. Zhu, *Achieving selective interrogation and sub-wavelength resolution in thin plates with embedded metamaterial acoustic lenses*, J. Appl. Phys. 116 (2014), pp. 054906. doi:10.1063/1.4892017.

[5] Y. Pennec, B. Djafari-Rouhani, J. Vasseur, H. Larabi, A. Khelif, A. Choujaa, S. Benchabane, and V. Laude, *Acoustic channel drop tunneling in a phononic crystal*, Appl. Phys. Lett. 87 (2005), pp. 261912. doi:10.1063/1.2158019.

[6] S. Fan, P. Villeneuve, J. Joannopoulos, M. Khan, C. Manolatou, and H. Haus, *Theoretical analysis of channel drop tunneling processes*, Phys. Rev. B. 59 (1999), pp. 15882–15892. doi:10.1103/PhysRevB.59.15882.

[7] D.W. Prather, *Photonic Crystals – Theory, Applications, and Fabrication*, Wiley, Hoboken, NJ, 2009.

[8] V. Romero-García, A. Cebrecos, R. Picó, V.J. Sánchez-Morcillo, L.M. Garcia-Raffi, and J.V. Sánchez-Pérez, *Wave focusing using symmetry matching in axisymmetric acoustic gradient index lenses*, Appl. Phys. Lett. 103 (2013), pp. 264106. doi:10.1063/1.4860535.

[9] J. Zhao, R. Marchal, B. Bonello, and O. Boyko, *Efficient focalization of antisymmetric Lamb waves in gradient-index phononic crystal plates*, Appl. Phys. Lett. 101 (2012), pp. 261905. doi:10.1063/1.4773369.

[10] Y. Wu, Y. Lai, and Z.-Q. Zhang, *Effective medium theory for elastic metamaterials in two dimensions*, Phys. Rev. B. 76 (2007), pp. 205313. doi:10.1103/PhysRevB.76.205313.

[11] T.T. Wu, M.J. Chiou, Y.C. Lin, and T. Ono, *Design and fabrication of a gradient-index phononic quartz plate lens*, Proceedings SPIE 8994, Photonic and Phononic Properties of Engineered Nanostructures, IV SPIE, San Diego, CA, 2014, pp. 89940G.

[12] J. Zhao, Y. Pan, and Z. Zhong, *Analytical solution for SH wave propagating through a graded plate of metamaterial*, Front. Mech. Eng. 6 (2011), pp. 301–307.

[13] J.G. Yu, F.E. Ratolojanahary, and J.E. Lefebvre, *Guided waves in functionally graded viscoelastic plates*, Composite Struct. 93 (2011), pp. 2671–2677.

[14] V. Krylov and F. Tilman, *Acoustic 'black holes' for flexural waves as effective vibration dampers*, J. Sound Vib. 274 (2004), pp. 605–619. doi:10.1016/j.jsv.2003.05.010.

[15] C. Pekeris, *Theory of propagation of sound in a half-space of variable sound velocity under conditions of formation of a shadow zone*, J. Am. Acous. Soc. 18 (1946), pp. 295. doi:10.1121/1.1916366.

[16] M. Mironov, *Propagation of a flexural wave in a plate whose thickness decreases smoothly to zero in a finite interval*, Sov. Phys. Acoust. 34 (1988), pp. 318.

[17] D. O'Boy, V. Krylov, and V. Kralovic, *Damping of flexural vibrations in rectangular plates using the acoustic black hole effect*, J. Sound Vib. 329 (2010), pp. 4672–4688. doi:10.1016/j. jsv.2010.05.019.

[18] L. Zhao, S.C. Conlon, and F. Semperlotti, *Broadband energy harvesting using acoustic black hole structural tailoring*, Smart Mat. Struct. 23 (2014), 065021 (9pp.). doi:10.1088/ 0964-1726/23/6/065021

[19] P.A. Feurtado, S.C. Conlon, and F. Semperlotti, *A normalized wavenumber variation parameter for acoustic black hole design*, J. Am. Acous. Soc. EL. 136 (2014), pp. 148–152.

[20] A. Climente, D. Torrent, and J. Sánchez-Dehesa, *Gradient index lenses for flexural waves based on thickness variations*, Appl. Phys. Lett. 105 (2014), pp. 064101. doi:10.1063/ 1.4893153.

[21] H. Zhu and F. Semperlotti, *Phononic thin plates with embedded acoustic black holes*, arXiv:1410.1833. (2014).

[22] Z. Liu, X. Zhang, Y. Mao, Y. Zhu, Z. Yang, C. Chan, and P. Sheng, *Locally resonant sonic materials*, Science. 289 (2000), pp. 1734–1736. doi:10.1126/science.289.5485.1734.

[23] H. Sanchis-Alepuz, Y.A. Kosevich, and J. Sanchez-Dehesa, *Acoustic analogue of electronic Bloch oscillations and resonant Zener tunneling in ultrasonic superlattices*, Phys. Rev. Lett. 98 (2007), pp. 134301. doi:10.1103/PhysRevLett.98.134301.

[24] M.D.E.F. Torres, D. Garcia-Pablos, and N. Garcia, Sonic band gaps in finite elastic media: Surface states and localization phenomena in linear and point defects, Phys. Rev. Lett. 82 (1999), pp. 3054. doi:10.1103/PhysRevLett.82.3054.

[25] M. Kafesaki, M. Sigalas, and N. Garcia, *Frequency modulation in the transmittivity of wave guides in elastic-wave band-gap materials*, Phys. Rev. Lett. 85 (2000), pp. 4044.

[26] A. Taflove, *The Finite-Difference Time-Domain Method*, Artech House, Boston, MA, 1998.

[27] F. Collino and C. Tsogka, *Application of the perfectly matched absorbing layer model to the linear elastodynamic problem in anisotropicheterogeneous media*, Geophysics. 66 (2001), pp. 294–307. doi:10.1190/1.1444908.

[28] S. Mohammadi, A.A. Eftekhar, A. Khelif, H. Moubchir, R. Westafer, W.D. Hunt, and A. Adibi, *Complete phononic bandgaps and bandgap maps in two-dimensional silicon phononic crystal plates*, Electronics Lett. 43 (2007), pp. 898–899.

10

Elastic metamaterials and dynamic homogenization

Ankit Srivastava*

Department of Mechanical, Materials, and Aerospace Engineering, Illinois Institute of Technology, Chicago, IL 60616, USA

In this paper, we review the recent advances which have taken place in the understanding and applications of acoustic/elastic metamaterials. Metamaterials are artificially created composite materials which exhibit unusual properties that are not found in nature. We begin with presenting arguments from discrete systems which support the case for the existence of unusual material properties such as tensorial and/or negative density. The arguments are then extended to elastic continuums through coherent averaging principles. The resulting coupled and nonlocal homogenized relations, called the Willis relations, are presented as the natural description of inhomogeneous elastodynamics. They are specialized to Bloch waves propagating in periodic composites and we show that the Willis properties display the unusual behavior which is often required in metamaterial applications such as the Veselago lens. We finally present the recent advances in the area of transformation elastodynamics, charting its inspirations from transformation optics, clarifying its particular challenges, and identifying its connection with the constitutive relations of the Willis and the Cosserat types.

Keywords: metamaterials; dynamic homogenization; periodic composites; cloaking

1. Introduction

Metamaterials are artificially designed composite materials which can exhibit properties that cannot be found in nature. These properties can be electronic, magnetic, acoustic, or elastic and have, of late, come to include static [1] properties. In the context of acoustic metamaterials, these properties refer to the bulk modulus and density, and for elastic metamaterials, they refer to the moduli (bulk, shear, and anisotropic) and density of a designed composite material. As such, they are used for the fine-tuned, predominantly frequency-dependent control of the trajectory and dissipation characteristics of acoustic and stress waves. These materials have found natural applications in the research areas of cloaking, imaging, and noise and vibration control. The primary driver in acoustic metamaterials research has been research in the area of photonic metamaterials. As such, many of the conclusions drawn from the photonics research directly apply to acoustic waves and acoustic metamaterials due to the essential similarity of the governing equations in the two cases. Realizing analogous results for elastic metamaterials is complicated by the fact that the governing equation for elasticity admits both longitudinal and shear wave solutions which are capable of exchanging energy between each other.

*Email: ankit.srivastava@iit.edu

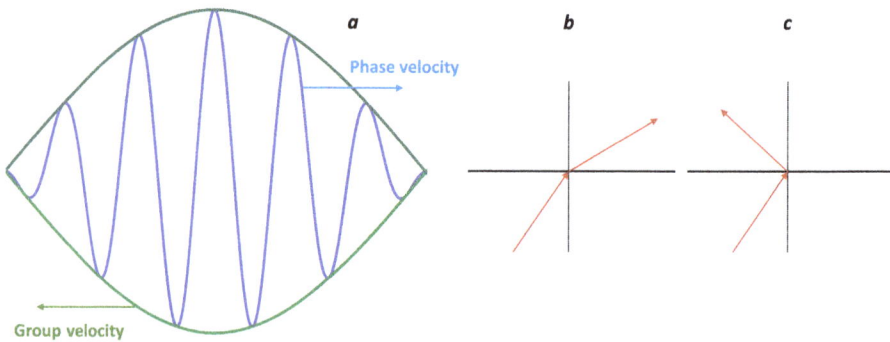

Figure 1. LHMs and negative refraction, a. LHMs are characterized by anti-parallel phase and group-velocity directions, b. refraction at an interface between two normal materials, c. refraction at an interface between a normal material and an LHM.

However, even in the case of elastic metamaterials some general ideas have been borrowed from photonic metamaterials research.

There are two broad directions from which the research area of metamaterials can be approached. The first direction seeks to find the uses and applications of those materials which exhibit unnatural material properties such as negative density and moduli (negative ϵ and μ in the case of electromagnetism). In this approach, the question of the existence of such a material is secondary and the applications themselves are of primary importance. Within this approach, researchers have conjectured that materials with simultaneously negative material properties will exhibit such exotic phenomena as negative refraction, reversed Doppler effect, and reversed Cherenkov radiation [2]. Such materials have been termed left-handed materials (LHMs) and are characterized by the quality that a wave of appropriate frequency traveling through such a material will display anti-parallel phase and group-velocity directions (Figure 1). There are other scenarios, however, which allow for the existence of anti-parallel waves such as guided waves [3] and negative group-velocity bands in photonic and phononic crystals (see also [4–6]). There exists the possibility of achieving negative refraction in such media [7,8]. In fact, it has been shown recently that negative energy refraction can be accompanied by positive phase-velocity refraction, and conversely that positive energy refraction can be accompanied by negative phase- velocity refraction [9]. LHMs, on the other hand, can be uniquely attributed a negative index of refraction through arguments of causality [10]. This unique characteristic makes them suitable for applications in creating flat lenses which can beat the diffraction limit in imaging applications [11]. If, in addition to LHMs, one could find materials with any desired material property then it becomes theoretically possible to design perfect cloaks which render an object enclosed within it completely invisible. The actual design of the cloak is approached through clever coordinate transformations which preserve the form of the governing equation making it possible to identify the desired change of wave trajectories with transformed material properties. This approach to designing cloaks has been applied to electromagnetic waves [12–16], acoustic waves [17], and elastic waves [18,19].

The second broad direction in metamaterials research seeks to find those material microstructures which will display the unusual properties required for the application areas discussed earlier. No naturally occurring homogeneous material displays LHM properties of negative density and stiffness [20]. Furthermore, naturally occurring materials offer only a very limited spectrum of density and moduli which is not enough to realize

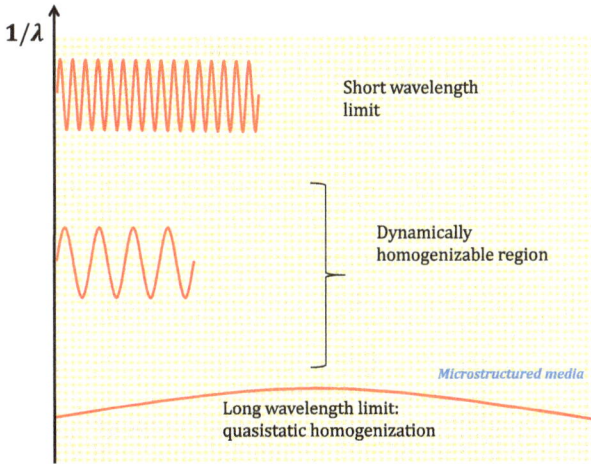

Figure 2.   Homogenizable region.

cloaking and other trajectory control applications. In order to achieve the properties required in the applications discussed earlier, researchers have sought to design heterogeneous materials at appropriate length scales which exhibit desirable effective properties [21].

Figure 2 shows three broad regions in which wave phenomena can be studied. In the low-frequency region, the predominant wavelength, λ, is much larger than the micro-structural length scale of the material in which it is traveling. In this region, wave characteristics are nondispersive and propagation is controlled by the static averages of material properties. The traditional area of static homogenization is appropriate for determining density, $\rho(x)$, and stiffness tensor, $C(x)$, which control wave behavior in this regime [22]. At the other end of the scale, the wavelength of the wave is on the same scale as or shorter than the length scale of the microstructure. Wave behavior in this regime is dominated by scattering at material interfaces and the heterogeneous material cannot be effectively defined by average homogenized properties. This is the case when one considers the optical branches in phononics or photonics [23]. Between these two scales lies a regime where the heterogeneous material may still be defined by homo-genized effective properties but those properties must take into consideration the disper-sive nature of wave propagation. This effectively means that in this regime the homogenized material properties which control wave propagation will need to be fre-quency-dependent and, therefore, static homogenization techniques are not sufficient anymore. The primary problem, therefore, is one of relating the microstructure of a composite to the frequency-dependent effective properties which will adequately represent wave phenomena in it and which are useful to the applications discussed above.

2.   Emergence of negative and tensorial material properties

Metamaterial applications naturally require frequency dependence and additional tensorial complexity (e.g. tensorial density) of material properties. The concept of dispersive (frequency-dependent) material properties naturally arises when laws of motion are enforced at scales below which additional heterogeneity, capable of dynamics, exist. Furthermore, this process of homogenization can give rise to tensorial forms of those material properties which are traditionally taken as scalar, such as density. Consider, for

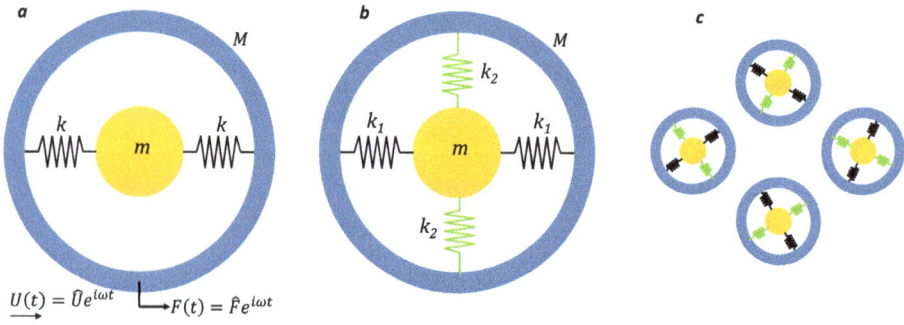

Figure 3. Frequency-dependent effective mass, a. one dimension, b and c. tensorial frequency-dependent effective mass.

instance, the dynamics of the one-dimensional composite system shown in Figure 3a [24,25]. If we relate the harmonic macroscale force to the harmonic macroscale acceleration, we find:

$$\hat{F} = -\omega^2 M^{\text{eff}}\,\hat{U}; \ M^{\text{eff}}(\omega) = M + \frac{m\omega_0^2}{\omega_0^2 - \omega^2}; \ \omega_0 = \sqrt{2k/m} \tag{1}$$

showing that by insisting on the applicability of Newton's second law at the macroscale, we inevitably end up with a frequency-dependent effective mass. The appearance of mass dispersion directly results from the act of subsuming the microstructural dynamic effects into a homogenized macroscale description. M^{eff} increases to infinity as ω approaches ω_0 (resonance condition) and becomes negative beyond that. By extending the idea to two and three dimensions and by incorporating springs of different moduli, it is clear that the effective mass can not only be made frequency-dependent but also anisotropic in nature. In two dimensions (Figure 3b), the relevant equations become:

$$\hat{F}_i = -\omega^2 M_{ij}^{\text{eff}}\,\hat{U}_j; \ M_{ij}^{\text{eff}}(\omega) = \left[M + \frac{m\omega_{[j]}^2}{\omega_{[j]}^2 - \omega^2} \right]\delta_{ij}; \ \omega_{[j]} = \sqrt{2k_j/m}; \ i,j = 1,2 \tag{2}$$

where δ_{ij} is the delta function. In the above, the off-diagonal terms of M^{eff} are zero. They can be made nonzero, thereby coupling the displacements, U_j, by incorporating the springs along axes which are not oriented along the axes of the orthogonal coordinate system [24] (Figure 3c).

The assumption of harmonicity may seem restrictive in the above examples. In general, the time-dependent macroscale displacement vector, $\mathbf{U}(t)$, would be related to the time-dependent macroscale force vector, $\mathbf{F}(t)$, via a kernel, $\mathbf{H}(t)$, through the convolution operator, $\mathbf{F} = \mathbf{H} * \mathbf{U}$. However, it was shown by Milton et al. [24] that the models of Figure 3 are sufficiently rich to approximate the kernel under fairly unrestrictive conditions. By periodically repeating the microstructure shrinking the unit cell size to zero while appropriately scaling the relevant quantities, one can obtain the justification for the existence of a frequency-dependent mass density tensor $\rho(\omega)$. Analogously to the frequency-dependent mass of Equation (1), the frequency-dependent density is expected to be negative in some frequency range, the exact range depending upon the resonant behavior of the microstructure. If the same composite can also be made to exhibit negative modulus in the frequency range where its density is negative, then it would serve as an acoustic/elastic LHM with applicability to the metamaterial applications discussed above.

It must be mentioned that in the examples of Figure 3, mass can only become negative in a homogenized effective sense and that a resonance is essential for achieving the negative behavior. However, effective mass can be tensorial and anisotropic in nature even away from the resonance (Equation 2). This requirement of resonance being important for achieving negative effective properties is seen in photonic metamaterials. A photonic LHM requires simultaneously negative dielectric permittivity (ϵ) and magnetic permeability (μ). Plasmon resonances in metals give rise to a frequency-dependent dielectric permittivity which has a similar functional form as Equation (1). Below the plasmon frequency ϵ is negative, however, at very low frequencies the effect of the plasmon is destroyed by dissipation. Pendry et al. [26] showed that the plasmon frequencies can be reduced by introducing artificial resonances through a periodic assembly of thin metallic wires. It was further shown [27] that a periodic array of nonmagnetic, resonant conducting units displays an effective μ which assumes negative values above the resonance. By combining the above, Smith et al. [21] proposed a periodic array of interspaced conducting nonmagnetic split ring resonators and continuous wires which exhibited a frequency region where both ϵ^{eff} and μ^{eff} were simultaneously negative.

The general ideas from photonic metamaterials research have inspired researchers to propose composite designs which use local mechanical resonances to exhibit negative effective density and moduli. In some cases, researchers have even proposed negative designs which do not use resonances. The proposed designs began with Liu et al. suggesting a design based on silicone-coated lead spheres embedded in epoxy matrix [28,29]. It was subsequently suggested that acoustic/elastic LHMs could be constructed by mixing two structures that independently exhibit negative density and modulus [30]. Negative modulus could be achieved through a periodic array of Helmholtz resonators [31,32], and the negative density could be achieved through an array of thin membranes [33]. The combination of these two structures was shown to exhibit LHM properties [34,35] Since then, researchers have proposed more complicated acoustic/elastic LHM architectures. These include multiply resonant microstructures that lead to density and some components of the modulus tensor simultaneously becoming negative [36], periodic array of curled perforations [37,38], tunable piezoelectric resonator arrays [39], anisotropic LHMs by arranging layers of perforated plates [40], etc. It is clear that the materials which would satisfy the requirements for LHM properties will need to be dispersive since there does not exist any material with quasistatic LHM properties. This consequence places rather strong constraints on the broadband applicability of metamaterials to applications such as perfect cloaking and superlensing. Additionally, since LHM properties result from internal resonances, they are inevitably accompanied with large dissipation, thereby, limiting their use in practical applications. Some proposals have been made to compensate for the losses through active gain [41]. However, there exist compelling theoretical arguments from causality, which limit the potential benefit of such methods [42].

3. Dynamic homogenization

The theoretical support to the field of metamaterials is provided through dynamic homogenization techniques that relate the microstructure of a composite to its frequency-dependent dynamic effective properties. These effective properties must adequately represent the wave behavior in the composite within desired frequency ranges. The majority of research interest in the area of metamaterials is restricted to periodic composites, which display highly dispersive wave behavior [43–48]. These periodic composites admit Bloch waves as solutions and many different numerical algorithms have been developed for calculating the dispersive properties of these waves. These include the popular plane wave expansion method [49], the finite difference time domain method [50], the multiple scattering method [51], variational methods [52,53], secondary expansions [54], etc.

3.1. *Averaging techniques*

There exist different ways by which effective constitutive parameters for wave propagation in metamaterials may be defined. The most common route to determining these parameters is by the use of retrieval methods [55–58], where the assumption is that local effective properties may be used to define periodic composites such as metamaterials. These properties are determined by analyzing the complex reflection and transmission coefficients of a finite slab of the composite. The retrieval method, which began with electromagnetic metamaterials, was subsequently extended to acoustic metamterials [59]. However, the retrieval method, while simple in principle, produces effective parameters which fail to satisfy basic passivity and causality properties [60] and exhibits other anti-resonant artifacts [61–63]. An excellent review [64] points out that the majority of metamaterial homogenization studies published in the last decade failed to respect basic causality and passivity properties and in some cases also violated the second law of thermodynamics. Another broad technique for determining effective properties is through the Coherent Potential Approximation method [65] and its enhancements [66,67]. The metamaterial under study is embedded in a matrix, which has the properties of the effective media. These properties are then determined by minimizing wave scattering in the surrounding matrix and as such only apply in the long wavelength limit. Recent efforts which employ a similar idea of matching the surface responses of the structural unit of a metamaterial with a homogenized medium show applicability beyond the long wavelength limit [68]. The focus here is on the dynamic homogenization techniques which are geared toward extracting the effective dynamic properties of periodic acoustic/elastic composites which are applicable in the long wavelength limit and beyond. These properties are expected to represent Bloch wave propagation in such periodic composites and, as such, can be termed Bloch wave homogenization techniques. It must be noted that they are different from asymptotic homogenization [69–71] techniques which have traditionally been used for the calculation of the Bloch wave band structure. Asymptotic homogenization was also limited in application to describing only the behavior of the fundamental Bloch mode at low frequencies [72,73]. Recently, however, it has been extended to higher frequencies and higher Bloch modes [74–77] (see also [78]). Additionally, there have been other efforts to bridge the scales for the study of dispersive systems based upon variational formulations [79], micromechanical techniques [80], Fourier transform of the elastodynamic equations [81], and strain projection methods [82].

3.2. *Ensemble averaging*

The pioneering work in the area of homogenization of inhomogeneous electromagnetics/elastodynamics was done by Willis [83–86]. Here, we present the basic ideas that have led to the current state of the art in the area of dynamic homogenization. We consider a volume Ω in which the equations of motion and kinematic relations are given point-wise by:

$$\sigma_{ij,j} + f_i = \dot{p}_i; \quad \varepsilon_{ij} = \frac{1}{2}(u_{i,j} + u_{j,i}),$$

(3)

where $\boldsymbol{\sigma}$, $\mathbf{f}$, $\boldsymbol{\varepsilon}$, $\mathbf{p}$, and $\mathbf{u}$ are the space and time-dependent stress tensor, body force vector, strain tensor, momentum vector, and displacement vector, respectively. The pointwise constitutive relations are:

$$\sigma_{ij} = C_{ijkl}\,\varepsilon_{kl}; \quad p_i = \rho \dot{u}_i,$$

(4)

with the usual symmetries for $\mathbf{C}$. Equations (3 and 4) are supplemented with appropriate boundary conditions on $\partial\Omega$ and initial conditions at $t = 0$. We seek to average the equations of motion and find the constitutive parameters that relate the averaged field variables $\langle\boldsymbol{\sigma}\rangle$, $\langle\boldsymbol{\varepsilon}\rangle$, $\langle\mathbf{p}\rangle$, and $\langle\dot{\mathbf{u}}\rangle$. The homogenization procedure can be derived for random composites and subsequently specialized to the periodic case [87]. Random composites represent families whose physical properties vary not only with position $\mathbf{x}$ but also with a parameter α. α is a member of a sample space, $\mathcal{A}$, over which a probability measure $\mathcal{P}$ is defined. In essence, we solve many problems over Ω where Ω is defined, for each problem, by materials with properties $\mathbf{C}(\mathbf{x},\alpha), \rho(\mathbf{x},\alpha)$ (Figure 4a). Each problem has the same body force, initial conditions, and boundary conditions. This leads to the pointwise solution $\mathbf{u}(\mathbf{x},\alpha)$ (and other derived field variables) for each problem. Now ensemble averages are defined over the sample space:

$$\langle\phi\rangle(\mathbf{x}) = \int_{\mathcal{A}} \phi(\mathbf{x},\alpha)\mathcal{P}(d\alpha)$$

(5)

Ensemble averaging the equation of motion (Equation (3)) we get:

$$\langle\sigma\rangle_{ij,j} + f_i = \langle\dot{p}\rangle_i; \quad \langle\varepsilon\rangle_{ij} = \frac{1}{2}(\langle u\rangle_{i,j} + \langle u\rangle_{j,i})$$

(6)

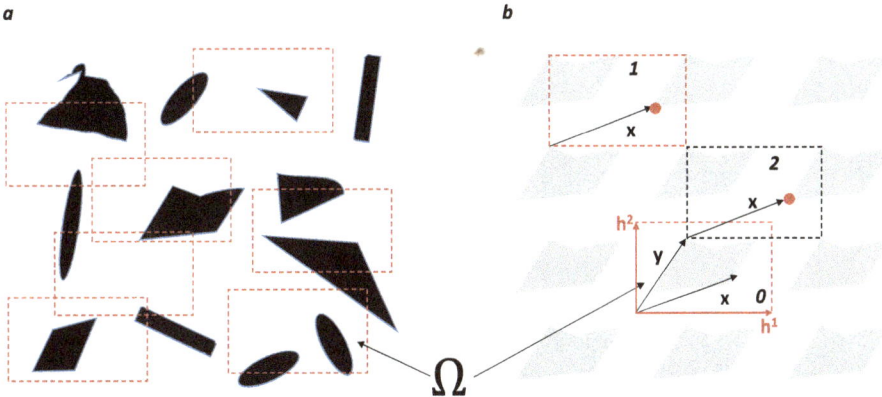

Figure 4.   Ensemble averaging: a. random composite, b. periodic composite.

which could be solved if the ensemble averages of stress and momentum could be related to the ensemble averages of strain and velocity through appropriate homogenized relations. It should be noted that such relations cannot be directly derived from averaging Equation (4).

3.3. Homogenized properties

At this point, a comparison medium is introduced with homogeneous properties $(\mathbf{C}^0, \rho^0)$ which transforms Equation (4) to:

$$\sigma_{ij} = C^0_{ijkl}\,\varepsilon_{kl} + \Sigma_{ij}; \quad p_i = \rho^0 \dot{u}_i + P_i, \tag{7}$$

where $\mathbf{\Sigma}, \mathbf{P}$ are stress and momentum polarizations (see [88–91] for polarizations). The above could be averaged and the required homogenized relation extracted if the ensemble averages of $\mathbf{\Sigma}, \mathbf{P}$ could be determined. Using Equation (7) in the equation of motion we have:

$$\left(C^0_{ijkl}\,u_{k,l}\right)_{,j} + f_i + \Sigma_{ij,j} - \dot{P}_i = \rho^0 \ddot{u}_i \tag{8}$$

Taking into consideration the initial and boundary conditions, the solution to Equation (8) can be written as

$$u_i(\mathbf{x},t) = u_i^0(\mathbf{x},t) + \int_t \int_\Omega G^0_{ij}(\mathbf{x},t;\mathbf{x}',t')\left[\Sigma_{jk,k}(\mathbf{x}',t') - \dot{P}_j(\mathbf{x}',t')\right]\,d\mathbf{x}'dt' \tag{9}$$

where $\mathbf{u}^0$ is the solution to

$$\left(C^0_{ijkl}\,u_{k,l}\right)_{,j} + f_i = \rho^0 \ddot{u}_i \tag{10}$$

with the same boundary conditions as in (Equation (8)) and $\mathbf{G}^0$ is the Green's function of the comparison medium satisfying:

$$C^0_{jikl}\,G^0_{kp,jl} + \delta_{ip}\delta(\mathbf{x}-\mathbf{x}')\delta(t-t') = \rho^0\delta_{ik}\,\ddot{G}^0_{kp} \tag{11}$$

with appropriate homogeneous boundary conditions on $\partial\Omega$. Integration by parts of (Equation (9)) formally gives:

$$\mathbf{u} = \mathbf{u}^0 - \mathbf{S}^0 \otimes \mathbf{\Sigma} - \mathbf{M}^0 \otimes \mathbf{P} \tag{12}$$

where $\otimes$ represents convolution in space and time as shown in Equation (9). After appropriate space and time differentials we have:

$$\mathbf{\varepsilon} = \mathbf{\varepsilon}^0 - \mathbf{S}^0_x \otimes \mathbf{\Sigma} - \mathbf{M}^0_x \otimes \mathbf{P}$$

$$\dot{\mathbf{u}} = \dot{\mathbf{u}}^0 - \mathbf{S}^0_t \otimes \mathbf{\Sigma} - \mathbf{M}^0_t \otimes \mathbf{P} \tag{13}$$

where $\mathbf{S}^0_x, \mathbf{M}^0_x, \mathbf{S}^0_t, \mathbf{M}^0_t$ are integral operators [87]. The above can be used to express $\mathbf{\Sigma}, \mathbf{P}$ in terms of $\langle\varepsilon\rangle, \langle\dot{\mathbf{u}}\rangle$ after eliminating $\mathbf{C}^0, \dot{\mathbf{u}}^0$ (see [24] for details). This relation is formally given:

$$\boldsymbol{\Sigma} = \mathbf{T}_{11} \otimes \langle \varepsilon \rangle + \mathbf{T}_{12} \otimes \langle \dot{\mathbf{u}} \rangle$$

$$\mathbf{P} = \mathbf{T}_{21} \otimes \langle \varepsilon \rangle + \mathbf{T}_{22} \otimes \langle \dot{\mathbf{u}} \rangle \tag{14}$$

It is clear from the above that the ensemble averages of $\boldsymbol{\Sigma}, \mathbf{P}$ depend on both $\langle \varepsilon \rangle$ and $\langle \dot{\mathbf{u}} \rangle$. In conjunction with the ensemble average of Equation (7), it means that $\langle \boldsymbol{\sigma} \rangle$ and $\langle \mathbf{p} \rangle$ will depend upon both $\langle \varepsilon \rangle$ and $\langle \dot{\mathbf{u}} \rangle$. The coupled homogenized constitutive relations which naturally emerge from ensemble averaging are formally given as

$$\langle \boldsymbol{\sigma} \rangle = \mathbf{C}^{\text{eff}} \otimes \langle \varepsilon \rangle + \mathbf{S}^{\text{eff}} \otimes \langle \dot{\mathbf{u}} \rangle$$

$$\langle \mathbf{p} \rangle = \bar{\mathbf{S}}^{\text{eff}} \otimes \langle \varepsilon \rangle + \boldsymbol{\rho}^{\text{eff}} \otimes \langle \dot{\mathbf{u}} \rangle \tag{15}$$

A material which exhibits the above coupled constitutive relation, which is a generalization of the classical elastic constitutive relation, will be termed a Willis material. It is interesting to note from Norris et al. [92] that a periodic composite formed using Willis materials, under the dynamic homogenization process shown in this section, results in an effective dynamic constitutive relation, which is again of the Willis kind. This shows that the Willis constitutive relation given above is closed under homogenization whereas the classical constitutive relation is not.

3.4. Specialization to Bloch/Floquet waves in periodic composites

It should be noted that the averaged fields in Equation (15) depend upon $\mathbf{x}, t$ and that the constitutive properties appearing in (Equation (15)) are integral operators in both space and time domains. These operators are considerably simplified when applied to the case of Bloch waves in periodic composites. Ensemble averages can be specialized to the periodic case. In Figure 4b, the unit cell denoted by index 0, Ω, is characterized by base vectors $\mathbf{h}^i$. The reciprocal base vectors of the unit cell are given by

$$\mathbf{q}^1 = 2\pi \frac{\mathbf{h}^2 \times \mathbf{h}^3}{\mathbf{h}^1 \cdot (\mathbf{h}^2 \times \mathbf{h}^3)}; \quad \mathbf{q}^2 = 2\pi \frac{\mathbf{h}^3 \times \mathbf{h}^1}{\mathbf{h}^2 \cdot (\mathbf{h}^3 \times \mathbf{h}^1)}; \quad \mathbf{q}^3 = 2\pi \frac{\mathbf{h}^1 \times \mathbf{h}^2}{\mathbf{h}^3 \cdot (\mathbf{h}^1 \times \mathbf{h}^2)} \tag{16}$$

such that $\mathbf{q}^i \cdot \mathbf{h}^j = 2\pi \delta_{ij}$. The rest of the composite can be generated by repeating Ω such that the material properties have the following periodicity:

$$C_{jkmn}(\mathbf{x} + n_i \mathbf{h}^i) = C_{jkmn}(\mathbf{x}); \quad \rho(\mathbf{x} + n_i \mathbf{h}^i) = \rho(\mathbf{x}), \tag{17}$$

where n_i are integers. The infinite periodic composite, thus generated, accepts Bloch waves as solutions with a wave vector specified by $\mathbf{k} = Q_i \mathbf{q}^i$. Any field variable $\boldsymbol{\Phi}(\mathbf{x}, t)$ (stress, strain, velocity, or momentum) can be expressed as $\hat{\boldsymbol{\Phi}}(\mathbf{x}) \exp[i(\mathbf{k} \cdot \mathbf{x} - \omega t)]$ where $\hat{\boldsymbol{\Phi}}$ is Ω periodic. Now consider another realization of the composite generated by repeating $\Omega^{(1)}$ (Figure 4b). $\Omega^{(1)}$ is chosen randomly but due to the periodicity of the composite it is equivalent to some $\Omega^{(2)}$ which is translated from Ω by a vector $\mathbf{y}$ such that $\mathbf{y} \in \Omega$. This realization also accepts Bloch wave solutions of the form $\hat{\boldsymbol{\Phi}}^y(\mathbf{x}) \exp[i(\mathbf{k} \cdot \mathbf{x} - \omega t)]$ where the superscript y denotes the translated unit cell and $\hat{\boldsymbol{\Phi}}^y(\mathbf{x})$ is again Ω periodic. As can be seen from Figure 4b any Ω periodic quantity $\hat{\Phi}^{y}(\mathbf{x})$ is equivalent to $\hat{\Phi}(\mathbf{x} + \mathbf{y})$. The medium translated by $\mathbf{y}$ can be regarded as one of a statistical ensemble if $\mathbf{y}$ is assumed to be uniformly distributed over Ω. Ensemble averaging of the field variables is, therefore, equivalent to:

$$\langle \boldsymbol{\Phi} \rangle (\mathbf{x}) = \int_{\Omega} \hat{\boldsymbol{\Phi}}^{y}(\mathbf{x}) \, e^{i(\mathbf{k} \cdot \mathbf{x} - \omega t)} d\mathbf{y} = \left[\int_{\Omega} \hat{\boldsymbol{\Phi}}(\mathbf{x} + \mathbf{y}) d\mathbf{y} \right] e^{i(\mathbf{k} \cdot \mathbf{x} - \omega t)} = \langle \hat{\boldsymbol{\Phi}} \rangle \, e^{i(\mathbf{k} \cdot \mathbf{x} - \omega t)} \qquad (18)$$

which is the usual unit cell averaging *carried over the Ω periodic parts of the field variables and not the full field variables*. It can be shown that when the averaging is performed over the Ω periodic parts of the field variables then the resulting effective properties automatically satisfy the dispersion relation of the periodic composite [93]. For periodic composites, the homogenized effective constitutive relation (Equation (15)) can be written in the tensorial form [94]:

$$\langle \hat{\sigma} \rangle_{i} = C_{ijkl}^{\mathrm{eff}} \langle \hat{\varepsilon} \rangle_{kl} + S_{ijk}^{\mathrm{eff}} \langle \hat{u} \rangle_{k}$$

$$\langle \hat{p} \rangle_{i} = \bar{S}_{ijk}^{\mathrm{eff}} \langle \hat{\varepsilon} \rangle_{jk} + \rho_{ij}^{\mathrm{eff}} \langle \hat{u} \rangle_{j} \qquad (19)$$

where $\mathbf{C}^{\mathrm{eff}}, \mathbf{S}^{\mathrm{eff}}, \bar{\mathbf{S}}^{\mathrm{eff}}, \boldsymbol{\rho}^{\mathrm{eff}}$ are functions of $\mathbf{k}, \omega$. Furthermore, the constitutive tensors display the following additional symmetries:

$$C_{ijkl}^{\mathrm{eff}} = C_{jikl}^{\mathrm{eff}} = C_{ijlk}^{\mathrm{eff}} = (C_{klij}^{\mathrm{eff}})^{*}$$

$$\bar{S}_{ijk}^{\mathrm{eff}} = (S_{jki}^{\mathrm{eff}})^{*}$$

$$\rho_{ij}^{\mathrm{eff}} = (\rho_{ji}^{\mathrm{eff}})^{*} \qquad (20)$$

where * represents a complex conjugate. The above relations hold generally for Bloch waves in three-dimensional linear periodic composites and are in congruence with effective dynamic properties for electromagnetic Bloch/Floquet waves [95]. In one dimension these relations simplify further [96]:

$$\langle \hat{\sigma} \rangle = C^{\mathrm{eff}} \langle \hat{\varepsilon} \rangle + S^{\mathrm{eff}} \langle \hat{u} \rangle$$

$$\langle \hat{p} \rangle = \bar{S}^{\mathrm{eff}} \langle \hat{\varepsilon} \rangle + \rho^{\mathrm{eff}} \langle \hat{u} \rangle \qquad (21)$$

where it can be shown that C^{eff} and ρ^{eff} are real and $\bar{S}^{\mathrm{eff}} = (S^{\mathrm{eff}})^{*}$. Moreover, $\bar{S}^{\mathrm{eff}}, S^{\mathrm{eff}} \to 0$ as $\omega \to 0$ whereas C^{eff} and ρ^{eff} approach their quasistatic homogenized limits as $\omega \to 0$. Explicit calculations of effective properties in three- and one-dimensional periodic composites are provided in Refs [92,94,96].

3.5. Nonuniqueness of the homogenized relations and LHM properties

Equation (21) are homogenized effective dynamic constitutive relations *of one kind* for Bloch wave propagation in one-dimensional periodic composites. It will be shown later that this form of the constitutive relations may be important for acoustic/elastic cloaking applications. However, these relations are not unique and can be transformed into a form which is more directly applicable to LHMs. The nonuniqueness of the effective relations results directly from the fact that they involve integral operators in the space and time domains of fields which are derived essentially from the same displacement field (Equation (15)). It is possible to transform $\mathbf{C}^{\mathrm{eff}}$ to $\mathbf{C}^{\mathrm{eff}} + \hat{\mathbf{C}}^{\mathrm{eff}}$ and $\mathbf{S}^{\mathrm{eff}}$ to $\mathbf{S}^{\mathrm{eff}} + \hat{\mathbf{S}}^{\mathrm{eff}}$ with appropriate conditions on $\hat{\mathbf{C}}^{\mathrm{eff}}, \hat{\mathbf{S}}^{\mathrm{eff}}$ such that the first equation of Equation (15) is preserved. Similarly, the second equation of Equation (15) could be preserved under appropriate transformations to $\bar{\mathbf{S}}^{\mathrm{eff}}, \rho^{\mathrm{eff}}$. These conditions are described in Ref. [97], where it is also pointed out that the nonuniqueness disappears if one assumes the existence of an inelastic strain in the body. For the electromagnetic case a similar result was proven by Fietz and Shvets [98], where they showed that the analogous electromagnetic effective dynamic constitutive relation can be determined uniquely only in the presence of a magnetic monopole current. In the absence of inelastic strain, there exists considerable freedom in the definitions of the effective parameters (see [99] for instance).

As a consequence of the nonuniqueness of the effective dynamic constitutive relations, Equation (21), in one dimension, can be transformed so as to subsume the effects of $S^{\mathrm{eff}}, \bar{S}^{\mathrm{eff}}$ into a modified set of effective modulus and density parameters [93–100]:

$$\langle \hat{\sigma} \rangle = \bar{C}^{\mathrm{eff}} \langle \hat{\varepsilon} \rangle; \quad \langle \hat{p} \rangle = \bar{\rho}^{\mathrm{eff}} \langle \hat{u} \rangle \tag{22}$$

The modified parameters $\bar{C}^{\mathrm{eff}}, \bar{\rho}^{\mathrm{eff}}$ are functions of frequency and automatically satisfy the dispersion relation of the composite:

$$\sqrt{\frac{\bar{C}^{\mathrm{eff}}}{\bar{\rho}^{\mathrm{eff}}}} = \frac{\omega}{k} \tag{23}$$

where k is the one-dimensional wavenumber.

Figure 5 shows effective properties calculated for a three-phase composite where the central phase (M_3) is stiff and heavy and can resonate due to the light and compliant M_2 phase. This unit cell is the one-dimensional equivalent of the locally resonant structure

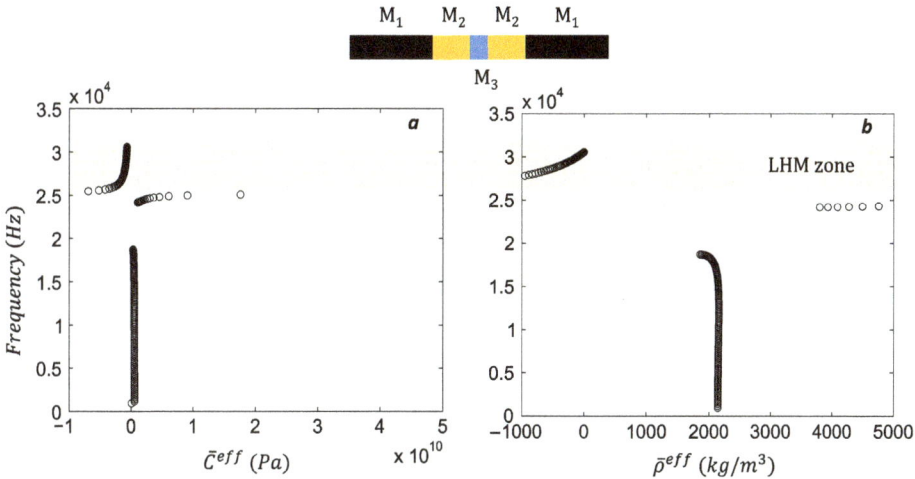

Figure 5.   Effective properties for an internally resonant one-dimensional unit cell, a. $\bar{C}^{\mathrm{eff}}$, b. $\bar{\rho}^{\mathrm{eff}}$.

considered by Liu et al. [28]. The geometrical and material properties of the unit cell are provided in Ref. [101]. These properties are calculated from Equation (22) over the first two branches of the phononic band structure of the composite. It is clear that for this unit cell both $\bar{C}^{\text{eff}}$ and $\bar{\rho}^{\text{eff}}$ are simultaneously negative in the frequency region denoted by the yellow rectangle in Figure 5. This frequency region corresponds to the Veselago left-hand metamaterial zone (see also [102]).

3.6. *Applicability of the homogenized relations*

It should be noted that Equation (15) is exact for a given elastodynamic problem on Ω if the boundary conditions of the problem are appropriately represented in the associated homogeneous problem (Equation (10)). However, this is often very complicated and the effort is to represent finite composites with homogenized properties which correspond to an infinite body. This is a valid approach in the homogenization limit [103] and is equivalent to ascribing the Bloch homogenized properties presented in section (III D) to the finite and or semi-infinite cases which appear in focusing, negative refraction, and cloaking applications. Refs [99,104,105] have studied this approximation and concluded that the approximation grows worse as frequency is increased. However, the approximation of homogenization for three-phase locally resonant unit cells is, in general, better than it is for nonresonant unit cell. Willis [105] has further studied the applicability of a modified form of the homogenized relations based upon weighted averages of the fields.

4. **Coordinate transformations and metamaterials**

In general, research in electromagnetic/acoustic/elastic wave cloaking has developed independently of the research in metamaterials/dynamic homogenization but they have converged to some common ideas and are intricately related now. Research in electromagnetic cloaking began with Pendry's observation that Maxwell's equations preserve their form under coordinate transformations, albeit with modified ϵ, μ [106–108]. Specifically, Maxwell's equations at a fixed frequency, ω, are:

$$\nabla \times \mathbf{E} + i\omega\mu\mathbf{H} = 0; \quad \nabla \times \mathbf{H} - i\omega\,\epsilon\mathbf{E} = 0$$

under the coordinate transformation $\mathbf{x}' = \mathbf{x}'(\mathbf{x})$, they retain their form:

$$\nabla' \times \mathbf{E}' + i\omega\,\mu'\mathbf{H}' = 0; \quad \nabla' \times \mathbf{H}' - i\omega\,\epsilon'\mathbf{E}' = 0$$

with the following transformed properties:

$$\mu'(\mathbf{x}') = \mathbf{A}\mu(\mathbf{x})\mathbf{A}^{\mathrm{T}}/det\mathbf{A}; \quad \epsilon'(\mathbf{x}') = \mathbf{A}\epsilon(\mathbf{x})\mathbf{A}^{\mathrm{T}}/det\mathbf{A}$$

where $A_{ij} = \partial x'_i/\partial x_j$ (and transformed fields). This observation was used to design a cloak through coordinate transformations which mapped the trajectories of electromagnetic waves in the presence of a cavity surrounded by a cloak to a homogeneous material [15,16]. This ensured that the cavity would become invisible to light waves generated by any source and incident at it from any direction. There has been considerable research activity in the field since then (see [109] for a review).

Cummer and Schurig [110] showed that there exists complete isomorphism between electromagnetic waves and acoustic waves in two dimensions. In cylindrical coordinates with z invariance and accounting for a tensorial but diagonal mass density, the acoustic wave equations and constitutive relation are given by:

$$i\omega\rho_\phi v_\phi = -\frac{1}{r}\frac{\partial p}{\partial\phi}; \quad i\omega\rho_r v_r = -\frac{\partial p}{\partial r}$$

$$i\omega p = -\frac{\lambda}{r}\left[\frac{\partial(rv_r)}{\partial r} + \frac{\partial v_\phi}{\partial\phi}\right] \tag{24}$$

where λ is the bulk modulus of the fluid, p is pressure, and $\mathbf{v}$ is vector fluid velocity. These compare with z-invariant Maxwell's equations in cylindrical coordinates for transverse-electric polarization:

$$i\omega\mu_r(-H_r) = -\frac{1}{r}\frac{\partial(-E_z)}{\partial\phi}; \quad i\omega\mu_\phi H_\phi = -\frac{\partial(-E_z)}{\partial r}$$

$$i\omega(-E_z) = -\frac{1}{zr}\left[\frac{\partial(rH_\phi)}{\partial r} + \frac{\partial(-H_r)}{\partial\phi}\right] \tag{25}$$

$\left[p, v_r, v_\phi, \rho_r, \rho_\phi, \lambda^{-1}\right] \rightleftharpoons \left[-E_z, H_\phi, -H_r, \mu_\phi, \mu_r, \varepsilon_z\right]$ clarifies the duality between Equations (24) and (25). Once the duality is established, designs for electromagnetic cloaks can be easily transformed to designs for acoustic cloaks. In the three-dimensional case Chen and Chan [111] noted that the acoustic wave equations retain their form but with transformed material properties. Specifically, the acoustic wave equation at a fixed frequency, ω, is:

$$\nabla\left[\frac{1}{\rho(\mathbf{x})}\nabla p(\mathbf{x})\right] = -\frac{\omega^2}{\kappa(\mathbf{x})}p(\mathbf{x})$$

where p is the fluid pressure and κ is the bulk modulus of the fluid. Under the coordinate transformation $\mathbf{x}' = \mathbf{x}'(\mathbf{x})$, the acoustic wave equation retains its form:

$$\nabla'\left[\frac{1}{\rho'(\mathbf{x})}\nabla' p'(\mathbf{x}')\right] = -\frac{\omega^2}{\kappa'(\mathbf{x}')}p'(\mathbf{x}')$$

with the following transformed properties:

$$\frac{1}{\rho'(\mathbf{x}')} = \mathbf{A}\frac{1}{\rho(\mathbf{x})}\mathbf{A}^{\mathrm{T}}/det\mathbf{A}; \quad \kappa'(\mathbf{x}') = \kappa(x)\,det\mathbf{A}$$

Once the invariance of the acoustic equation was noted, ideas from electromagnetic cloak design were brought to bear upon the design of acoustic cloaks. The anisotropic densities needed to realize such cloaks could be achieved by using layered fluids [112,113] (see also [114,115]), however, Norris [17] realized that such *inertial* cloaks suffer from a considerable mass penalty. He presented an alternative route to designing cloaks for acoustic waves based upon pentamode materials [17,116,117]. The existence of different routes to acoustic wave cloaking has been noted in literature [118]. For both electromagnetic and acoustic cloaking cases, the problem boils down to finding the microstructures that would exhibit those effective properties which are required by an appropriate coordinate transformation. An excellent review exists on the topic of transformational acoustics and its application to the design of acoustic cloaks [119].

For elastodynamics, however, it was found [18] that the equations change form under coordinate transformation and the cloaking ideas from electromagnetism and acoustics cannot be directly applied to elastic waves in solids. Specifically, the elastodynamic wave equations at a fixed frequency are given by

$$\nabla \cdot \boldsymbol{\sigma} = -\omega^2 \rho\, \mathbf{u}; \quad \boldsymbol{\sigma} = \mathbf{C}\nabla\mathbf{u}$$

which, under the coordinate transformation $\mathbf{x}' = \mathbf{x}'(\mathbf{x})$, transform to:

$$\nabla' \cdot \boldsymbol{\sigma}' = \mathbf{D}'\nabla'\mathbf{u}' - \omega^2 \boldsymbol{\rho}'\mathbf{u}'; \quad \boldsymbol{\sigma}' = \mathbf{C}'\nabla'\mathbf{u}' + \mathbf{S}'\mathbf{u}' \tag{26}$$

The modified constitutive parameters are given by [18]

$$C'_{pqrs} = \frac{1}{det\mathbf{A}} A_{pi} A_{qj} C_{ijkl} A_{rk} A_{sl}; \quad S'_{pqr} = \frac{1}{det\mathbf{A}} A_{pi} A_{qj} C_{ijkl} B_{rkl}$$

$$D'_{pqr} = \frac{1}{det\mathbf{A}} B_{pij} C_{ijkl} A_{qk} A_{rl}; \quad \rho'_{pq} = \frac{1}{det\mathbf{A}} \left[\rho A_{pi} A_{qi} + B_{pij} C_{ijkl} B_{qkl} \right] \tag{27}$$

where $A_{ij} = \partial x'_i/\partial x_j$ and $B_{ijk} = \partial^2 x'_i/\partial x_j \partial x_k$. This lack of invariance has made it hard to transfer the success of electromagnetic and acoustic cloaking to elastic wave cloaking. Moreover, the additional tensors $\mathbf{D}', \mathbf{S}'$ which appear in Equation (26) are highly unusual as far as conventional materials are concerned. However, these parameters are not unusual for a Willis material. In fact, Equation (15), under a fixed frequency is given by

$$\langle \boldsymbol{\sigma} \rangle = \mathbf{C}^{\text{eff}} \otimes \langle \varepsilon \rangle - i\omega\, \mathbf{S}^{\text{eff}} \otimes \langle \mathbf{u} \rangle$$

$$\langle \mathbf{p} \rangle = \bar{\mathbf{S}}^{\text{eff}} \otimes \langle \varepsilon \rangle - i\omega \rho^{\text{eff}} \otimes \langle \mathbf{u} \rangle \tag{28}$$

where $\otimes$ refers to convolution in space. With the following kinematic and dynamic relations:

$$\langle \varepsilon \rangle = \left[\nabla \langle \mathbf{u} \rangle + (\nabla \langle \mathbf{u} \rangle)^{\mathrm{T}} \right]; \quad \nabla \cdot \langle \boldsymbol{\sigma} \rangle = -i\omega \langle \mathbf{p} \rangle \tag{29}$$

Equation (28) transforms into the form of Equation (26). Moreover, it can be shown that Equation (26), under the coordinate transformation $\mathbf{x}'' = \mathbf{x}''(\mathbf{x}')$, retains its form. Therefore, it appears that if elastic materials can be defined by the Willis form then the lessons learned from transformational optics and acoustics can be applied to elastic wave cloaking. It must be mentioned here that Willis material is not the only way of achieving elastic wave cloaking. Norris [19] showed that Cosserat type materials [120] would also serve the purpose. Some other techniques which have been studied for elastic wave cloaking include using anisotropy [121] and nonlinear prestressing [122–125].

5. Conclusions

In this paper, we have approached the problem of acoustic/elastic metamaterials from two different directions, theory and applications. On the theoretical side we seek to define those effective dynamic homogenized relations which would describe the propagation characteristics of the mean wave in a composite. It is clear that the Willis constitutive relations (Equation (15)) are the natural consequence of dynamic averaging and can be safely taken to apply to the homogenized dynamic behaviour of any elastic composite. Additionally, the Willis relations are found to be closed under homogenization whereas the conventional elastodynamic constitutive relations are not. On the application side we seek to control wave trajectories in certain specific ways and search for materials which would allow us to exert such control. Inevitably, we find ourselves searching for materials with unusual properties which can, often, only be achieved in a homogenized sense. This is the common ground for theory and applications in acoustic/elastic metamaterials research. Some techniques (coordinate transformations) which have served well for the control of electromagnetic waves, when applied to acoustic/elastic waves, reinforce the idea that the Willis relations are a good description of inhomogeneous elastodynamics. This results from the fact that the Willis relations are closed under coordinate transformations. However, Willis relations are not unique and this nonuniqueness is mirrored in the practical manifestation that several different routes exist to achieve the aims of wave control in applications such as cloaking.

Acknowledgments

The author acknowledges the support of the UCSD subaward [UCSD/ONR W91CRB-10-1-0006] to the Illinois Institute of Technology (DARPA AFOSR [grant number RDECOM W91CRB-10–1–0006] to the University of California, San Diego).

Disclosure statement

No potential conflict of interest was reported by the author.

References

[1] D. Jang, L.R. Meza, F. Greer, and J.R. Greer, Fabrication and deformation of three-dimensional hollow ceramic nanostructures, Nat. Mater. 12 (2013), pp. 893–898. doi:10.1038/nmat3738

[2] V. Veselago, The electrodynamics of substances with simultaneously negative values of ε and μ, Soviet Phys. Uspekhi. 10 (1968), pp. 509–514. doi:10.1070/PU1968 v010n04ABEH003699

[3] J.D. Achenbach, *Wave Propagation in Elastic Solids: North-Holland Series in Applied Mathematics and Mechanics,* North-Holland, Amsterdam, 1987.

[4] A. Oliner and T. Tamir, Backward waves on isotropic plasma slabs, J. Appl. Phys. 33 (1962), p. 231. doi:10.1063/1.1728499

[5] I.V. Lindell, S. Tretyakov, K. Nikoskinen, and S. Ilvonen, BW media? Media with negative parameters, capable of supporting backward waves, Microw. Opt. Technol. Lett. 31 (2001), p. 129. doi:10.1002/mop.1378

[6] R. Meisels, F. Kuchar, and K. Hingerl, et al., Opt. Exp. 13 (2005), p. 8596. 10.1364/OPEX.13.008596

[7] M. Notomi, Theory of light propagation in strongly modulated photonic crystals: Refractionlike behavior in the vicinity of the photonic band gap, Phys. Rev. B. 62 (2000), pp. 10696–10705. doi:10.1103/PhysRevB.62.10696

[8] A. Sukhovich, L. Jing, and J. Page, Negative refraction and focusing of ultrasound in two-dimensional phononic crystals, Phys. Rev. B. 77 (2008), p. 14301. doi:10.1103/PhysRevB.77.014301

[9] S. Nemat-Nasser, Arxiv Preprint Arxiv:1404.5880, (2014).

[10] D.R. Smith and N. Kroll, Negative refractive index in left-handed materials, Phys. Rev. Lett. 85 (2000), pp. 2933–2936. doi:10.1103/PhysRevLett.85.2933

[11] J. Pendry, Negative refraction makes a perfect lens, Phys. Rev. Lett. 85 (2000), p. 3966. doi:10.1103/PhysRevLett.85.3966

[12] A. Greenleaf, M. Lassas, and G. Uhlmann, Anisotropic conductivities that cannot be detected by EIT, Physiol. Meas. 24 (2003), p. 413. doi:10.1088/0967-3334/24/2/353

[13] A. Greenleaf, M. Lassas, and G. Uhlmann, On nonuniqueness for Calderón's inverse problem, Math. Res. Lett. 10 (2003), p. 685. doi:10.4310/MRL.2003.v10.n5.a11

[14] U. Leonhardt, Notes on conformal invisibility devices, New. J. Phys. 8 (2006), p. 118. doi:10.1088/1367-2630/8/7/118

[15] U. Leonhardt, Optical conformal mapping, Science. 312 (2006), p. 1777. doi:10.1126/science.1126493

[16] J. Pendry, D. Schurig, and D. Smith, Controlling electromagnetic fields, Science. 312 (2006), p. 1780. doi:10.1126/science.1125907

[17] A. Norris, Acoustic cloaking theory, Proc. Royal Soc. A: Mathl, Phys. Eng. Sci. 464 (2008), p. 2411. doi:10.1098/rspa.2008.0076

[18] G. Milton, M. Briane, and J. Willis, On cloaking for elasticity and physical equations with a transformation invariant form, New J. Phys. 8 (2006), p. 248. doi:10.1088/1367-2630/8/10/248

[19] A. Norris and A. Shuvalov, Elastic cloaking theory, Wave Motion. 48 (2011), p. 525. doi:10.1016/j.wavemoti.2011.03.002

[20] J. Li and C. Chan, Double-negative acoustic metamaterial, Phys. Rev. E. 70 (2004), p. 55602. doi:10.1103/PhysRevE.70.055602

[21] D. Smith, W. Padilla, D. Vier, S. Nemat-Nasser, and S. Schultz, Composite medium with simultaneously negative permeability and permittivity, Phys. Rev. Lett. 84 (2000), p. 4184. doi:10.1103/PhysRevLett.84.4184

[22] S. Nemat-Nasser and M. Hori, *Micromechanics: Overall Properties of Heterogeneous Materials*, Elsevier, Amsterdam, 1999.

[23] C. Kittel and P. McEuen, *Introduction to Solid State Physics* Vol. 8, Wiley, New York, 1976.

[24] G. Milton and J. Willis, Proc. Royal Soc. A: Mathl, Phys. Eng. Sci. 463 (2007), p. 855.

[25] H. Huang, C. Sun, and G. Huang, On the negative effective mass density in acoustic metamaterials, Int. J. Eng. Sci. 47 (2009), p. 610. doi:10.1016/j.ijengsci.2008.12.007

[26] J. Pendry, A. Holden, W. Stewart, and I. Youngs, Extremely low frequency plasmons in metallic mesostructures, Phys. Rev. Lett. 76 (1996), p. 4773. doi:10.1103/PhysRevLett.76.4773

[27] J. Pendry, A. Holden, D. Robbins, and W. Stewart, Magnetism from conductors and enhanced nonlinear phenomena, IEEE Trans. Microw. Theory Tech. 47 (1999), p. 2075. doi:10.1109/22.798002

[28] Z. Liu, X. Zhang, Y. Mao, Y. Zhu, Z. Yang, C. Chan, and P. Sheng, Locally resonant sonic materials, Science. 289 (2000), p. 1734. doi:10.1126/science.289.5485.1734

[29] P. Sheng, X. Zhang, Z. Liu, and C. Chan, Locally resonant sonic materials, Phys. B: Conden. Matter. 338 (2003), p. 201. doi:10.1016/S0921-4526(03)00487-3

[30] Y. Ding, Z. Liu, C. Qiu, and J. Shi, Metamaterial with simultaneously negative bulk modulus and mass density, Phys. Rev. Lett. 99 (2007), p. 93904. doi:10.1103/PhysRevLett.99.093904

[31] N. Fang, D. Xi, J. Xu, M. Ambati, W. Srituravanich, C. Sun, and X. Zhang, Ultrasonic metamaterials with negative modulus, Nat. Mater. 5 (2006), p. 452. doi:10.1038/nmat1644

[32] Y. Cheng, J. Xu, and X. Liu, One-dimensional structured ultrasonic metamaterials with simultaneously negative dynamic density and modulus, Phys. Rev. B. 77 (2008), p. 45134. doi:10.1103/PhysRevB.77.045134

[33] S.H. Lee, C.M. Park, Y.M. Seo, Z.G. Wang, and C.K. Kim, Acoustic metamaterial with negative density, Phys. Lett. A. 373 (2009), p. 4464. doi:10.1016/j.physleta.2009.10.013

[34] F. Bongard, H. Lissek, and J.R. Mosig, Acoustic transmission line metamaterial with negative/zero/positive refractive index, Phys. Rev. B. 82 (2010), p. 94306. doi:10.1103/PhysRevB.82.094306

[35] S.H. Lee, C.M. Park, Y.M. Seo, Z.G. Wang, and C.K. Kim, Composite acoustic medium with simultaneously negative density and modulus, Phys. Rev. Lett. 104 (2010), p. 54301. doi:10.1103/PhysRevLett.104.054301

[36] Y. Lai, Y. Wu, P. Sheng, and Z.-Q. Zhang, Hybrid elastic solids, Nat. Mater. 10 (2011), p. 620. doi:10.1038/nmat3043

[37] Z. Liang and J. Li, Extreme acoustic metamaterial by coiling up space, Phys. Rev. Lett. 108 (2012), p. 114301. doi:10.1103/PhysRevLett.108.114301

[38] V. Romero-García, A. Krynkin, L. Garcia-Raffi, O. Umnova, and J. Sánchez-Pérez, Multi-resonant scatterers in sonic crystals: Locally multi-resonant acoustic metamaterial, J. Sound Vib. 332 (2013), p. 184. doi:10.1016/j.jsv.2012.08.003

[39] F. Casadei, T. Delpero, A. Bergamini, P. Ermanni, and M. Ruzzene, Piezoelectric resonator arrays for tunable acoustic waveguides and metamaterials, J. Appl. Phys. 112 (2012), p. 64902. doi:10.1063/1.4752468

[40] J. Christensen and F.J.G. De Abajo, Anisotropic metamaterials for full control of acoustic waves, Phys. Rev. Lett. 108 (2012), p. 124301. doi:10.1103/PhysRevLett.108.124301

[41] A.K. Popov and V.M. Shalaev, Compensating losses in negative-index metamaterials by optical parametric amplification, Opt. Lett. 31 (2006), p. 2169. doi:10.1364/OL.31.002169

[42] M. Stockman, Criterion for negative refraction with low optical losses from a fundamental principle of causality, Phys. Rev. Lett. 98 (2007), p. 177404. doi:10.1103/PhysRevLett.98.177404

[43] F. Bloch, Über die quantenmechanik der elektronen in kristallgittern, Z. Phys. 52 (1929), pp. 555–600. doi:10.1007/BF01339455

[44] S. John, et al., Strong localization of photons in certain disordered dielectric superlattices, Phys. Rev. Lett. 58 (1987), pp. 2486–2489. doi:10.1103/PhysRevLett.58.2486

[45] E. Yablonovitch, Inhibited spontaneous emission in solid-state physics and electronics, Phys. Rev. Lett. 58 (1987), pp. 2059–2062. doi:10.1103/PhysRevLett.58.2059

[46] M. Kushwaha, P. Halevi, L. Dobrzynski, and B. Djafari-Rouhani, Acoustic band structure of periodic elastic composites, Phys. Rev. Lett. 71 (1993), p. 2022. doi:10.1103/PhysRevLett.71.2022

[47] M. Sigalas and E. Economou, Band structure of elastic waves in two dimensional systems, Solid State Commun. 86 (1993), p. 141. doi:10.1016/0038-1098(93)90888-T

[48] R. Martinezsala, J. Sancho, J. Sanchez, V. Gómez, J. Llinares, and F. Meseguer, Sound attenuation by sculpture, Nature. 378 (1995), p. 241. doi:10.1038/378241a0

[49] K. Ho, C. Chan, and C. Soukoulis, Existence of a photonic gap in periodic dielectric structures, Phys. Rev. Lett. 65 (1990), p. 3152. doi:10.1103/PhysRevLett.65.3152

[50] C. Chan, Q. Yu, and K. Ho, Order-N spectral method for electromagnetic waves, Phys. Rev. B. 51 (1995), p. 16635. doi:10.1103/PhysRevB.51.16635

[51] M. Kafesaki and E. Economou, Multiple-scattering theory for three-dimensional periodic acoustic composites, Phys. Rev. B. 60 (1999), p. 11993. doi:10.1103/PhysRevB.60.11993

[52] S. Nemat-Nasser, Harmonic waves in layered composites, J. Appl. Mech. 39 (1972), p. 850. doi:10.1115/1.3422814

[53] A. Srivastava and S. Nemat-Nasser, Mixed-variational formulation for phononic band-structure calculation of arbitrary unit cells, Mech. Mater. 74 (2014), p. 67. doi:10.1016/j.mechmat.2014.03.002

[54] M. Hussein, Proc. Royal Soc. A: Mathl, Phys. Eng. Sci. 465 (2009), p. 2825.

[55] S. Obrien and J. Pendry, J. Phys.: Condens. Matter. 14 (2002), p. 6383.

[56] T. Koschny, P. Markoś, D. Smith, and C. Soukoulis, Resonant and antiresonant frequency dependence of the effective parameters of metamaterials, Phys. Rev. E. 68 (2003), p. 65602. doi:10.1103/PhysRevE.68.065602

[57] X. Chen, T.M. Grzegorczyk, B.-I. Wu, J. Pacheco Jr, and J.A. Kong, Robust method to retrieve the constitutive effective parameters of metamaterials, Phys. Rev. E. 70 (2004), p. 16608. doi:10.1103/PhysRevE.70.016608

[58] D. Smith, D. Vier, T. Koschny, and C. Soukoulis, Electromagnetic parameter retrieval from inhomogeneous metamaterials, Phys. Rev. E. 71 (2005), p. 36617. doi:10.1103/PhysRevE.71.036617

[59] V. Fokin, M. Ambati, C. Sun, and X. Zhang, Method for retrieving effective properties of locally resonant acoustic metamaterials, Phys. Rev. B. 76 (2007), p. 144302. doi:10.1103/PhysRevB.76.144302

[60] P.C. Chaumet, A. Sentenac, and A. Rahmani, Coupled dipole method for scatterers with large permittivity, Phys. Rev. E. 70 (2004), p. 36606. doi:10.1103/PhysRevE.70.036606

[61] D.A. Powell and Y.S. Kivshar, Substrate-induced bianisotropy in metamaterials, Appl. Phys. Lett. 97 (2010), p. 91106. doi:10.1063/1.3486480

[62] A. Alù, Restoring the physical meaning of metamaterial constitutive parameters, Phys. Rev. B. 83 (2011), p. 81102. doi:10.1103/PhysRevB.83.081102

[63] A. Alù, First-principles homogenization theory for periodic metamaterials, Phys. Rev. B. 84 (2011), p. 75153. doi:10.1103/PhysRevB.84.075153

[64] C. Simovski, Material parameters of metamaterials (a Review), Opt. Spectrosc. 107 (2009), p. 726. doi:10.1134/S0030400X09110101

[65] V.M. Shalaev, Electromagnetic properties of small-particle composites, Phys. Rep. 272 (1996), p. 61. doi:10.1016/0370-1573(95)00076-3

[66] Y. Wu, J. Li, Z.-Q. Zhang, and C. Chan, Effective medium theory for magnetodielectric composites: Beyond the long-wavelength limit, Phys. Rev. B. 74 (2006), p. 85111. doi:10.1103/PhysRevB.74.085111

[67] X. Hu, K.-M. Ho, C. Chan, and J. Zi, Homogenization of acoustic metamaterials of Helmholtz resonators in fluid, Phys. Rev. B. 77 (2008), p. 172301. doi:10.1103/PhysRevB.77.172301

[68] M. Yang, G. Ma, Y. Wu, Z. Yang, and P. Sheng, Homogenization scheme for acoustic metamaterials, Phys. Rev. B. 89 (2014), p. 64309. doi:10.1103/PhysRevB.89.064309

[69] A. Bensoussan, J. Lions, and G. Papanicolaou, Asymptotic Analysis for Periodic Structures Vol. 5, North Holland, New York, 1978.

[70] E. Sánchez-Palencia, Non-Homogeneous Media and Vibration Theory Vol. 127, Springer, New York, 1980. Available at http://www.springer.com/us/book/9783540100003

[71] N. Bakhvalov and G. Panasenko, Homogenisation: Averaging Processes in Periodic Media: Mathematical Problems in the Mechanics of Composite Materials, Kluwer Academic Publishers, Dordrecht, 1989. Available at http://www.springer.com/us/book/9780792300496

[72] W.J. Parnell and I.D. Abrahams, Dynamic homogenization in periodic fibre reinforced media. Quasi-static limit for SH waves, Wave Motion. 43 (2006), pp. 474–498. doi:10.1016/j.wavemoti.2006.03.003

[73] I. Andrianov, V. Bolshakov, V. Danishevs' Kyy, and D. Weichert, Proc. Royal Soc. A: Mathl, Phys. Eng. Sci. 464 (2008), p. 1181.

[74] R. Craster, S. Guenneau, and S. Adams, Mechanism for slow waves near cutoff frequencies in periodic waveguides, Phys. Rev. B. 79 (2009), p. 45129. doi:10.1103/PhysRevB.79.045129

[75] R. Craster, J. Kaplunov, and A. Pichugin, Proc. Royal Soc. A: Mathl, Phys. Eng. Sci. 466 (2010), p. 2341.

[76] T. Antonakakis, R. Craster, and S. Guenneau, Proc. Royal Soc. A: Mathl, Phys. Eng. Sci. 469 (2013), p. 20120533.

[77] T. Antonakakis, R. Craster, and S. Guenneau, Homogenisation for elastic photonic crystals and dynamic anisotropy, J. Mech. Phys. Solids 71 (2014), p. 84. doi:10.1016/j.jmps.2014.06.006

[78] G. Nagai, J. Fish, and K. Watanabe, Stabilized nonlocal model for dispersive wave propagation in heterogeneous media, Comput. Mech. 33 (2004), p. 144. doi:10.1007/s00466-003-0513-5

[79] T. McDevitt, G. Hulbert, and N. Kikuchi, An assumed strain method for the dispersive global–local modeling of periodic structures, Comput. Methods Appl. Mech. Eng. 190 (2001), p. 6425. doi:10.1016/S0045-7825(00)00184-5

[80] Z.-P. Wang and C. Sun, Modeling micro-inertia in heterogeneous materials under dynamic loading, Wave Motion. 36 (2002), p. 473. doi:10.1016/S0165-2125(02)00037-9

[81] S. Gonella and M. Ruzzene, Homogenization of vibrating periodic lattice structures, Appl. Math. Model. 32 (2008), p. 459. doi:10.1016/j.apm.2006.12.014

[82] M.I. Hussein and G.M. Hulbert, Mode-enriched dispersion models of periodic materials within a multiscale mixed finite element framework, Finite Elem. Anal. Des. 42 (2006), p. 602. doi:10.1016/j.finel.2005.11.002

[83] J. Willis, Variational and related methods for the overall properties of composites, Adv. Appl. Mech. 21 (1981), p. 1. doi:10.1016/S0065-2156(08)70330-2

[84] J. Willis, Variational principles for dynamic problems for inhomogeneous elastic media, Wave Motion. 3 ((1981), p. 1. doi:10.1016/0165-2125(81)90008-1

[85] J. Willis, The overall elastic response of composite materials, J. Appl. Mech. 50 (1983), p. 1202. doi:10.1115/1.3167202

[86] J. Willis, Variational principles and operator equations for electromagnetic waves in inhomogeneous media, Wave Motion. 6 (1984), p. 127. doi:10.1016/0165-2125(84)90009-X

[87] J. Willis, *Continuum Micromechanics*, Springer-Verlag, New York, 1997, pp. 265–290.

[88] Z. Hashin, in *Proceedings of the IUTAM Symposium Non-Homogeneity in Elasticity and Plasticity, Warsaw, Poland*, 1959.

[89] Z. Hashin and S. Shtrikman, On some variational principles in anisotropic and nonhomogeneous elasticity, J. Mech. Phys. Solids. 10 (1962), p. 335. doi:10.1016/0022-5096(62)90004-2

[90] Z. Hashin and S. Shtrikman, A variational approach to the theory of the elastic behaviour of polycrystals, J. Mech. Phys. Solids. 10 (1962), p. 343. doi:10.1016/0022-5096(62)90005-4

[91] Z. Hashin, *Theory of Mechanical Behavior of Heterogeneous Media*, Tech. Rep. University of Pennsylvania, Philadelphi Towne School of Civil and Mechanical Engineering, Philidelphia, 1963. Available at http://www.worldcat.org/title/theory-of-mechanical-behavior-of-heterogeneous-media/oclc/55660866

[92] A. Norris, A. Shuvalov, and A. Kutsenko, Proc. Royal Soc. A: Mathl, Phys. Eng. Sci. 468 (2012), p. 1629.

[93] S. Nemat-Nasser, J. Willis, A. Srivastava, and A. Amirkhizi, Homogenization of periodic elastic composites and locally resonant sonic materials, Phys. Rev. B. 83 (2011), p. 104103. doi:10.1103/PhysRevB.83.104103

[94] A. Srivastava and S. Nemat-Nasser, Proc. Royal Soc. A: Mathl, Phys. Eng. Sci. 468 (2012), p. 269.

[95] A. Amirkhizi and S. Nemat-Nasser, Microstructurally-based homogenization of electromagnetic properties of periodic media, Comptes Rendus Mecanique. 336 (2008), p. 24. doi:10.1016/j.crme.2007.10.012

[96] S. Nemat-Nasser and A. Srivastava, Overall dynamic constitutive relations of layered elastic composites, J. Mech. Phys. Solids. 59 (2011), pp. 1953–1965. doi:10.1016/j.jmps.2011.07.008

[97] J. Willis, Effective constitutive relations for waves in composites and metamaterials, Proc. Royal Soc. A: Mathl, Phys. Eng. Sci. 467 (2011), pp. 1865–1879. doi:10.1098/rspa.2010.0620

[98] C. Fietz and G. Shvets, *in Society of Photo-Optical Instrumentation Engineers (SPIE) Conference Series*, San Diego, CA, 2–6 August, Vol. 7392, 2009, 9.

[99] A. Shuvalov, A. Kutsenko, A. Norris, and O. Poncelet, Effective Willis constitutive equations for periodically stratified anisotropic elastic media, Proc. Royal Soc. A: Mathl, Phys. Eng. Sci. 467 (2011), pp. 1749–1769. doi:10.1098/rspa.2010.0389

[100] J. Willis, Exact effective relations for dynamics of a laminated body, Mech. Mater. 41 (2009), pp. 385–393. doi:10.1016/j.mechmat.2009.01.010

[101] S. Nemat-Nasser and A. Srivastava, Negative effective dynamic mass-density and stiffness: Micro-architecture and phononic transport in periodic composites, AIP Adv. 1 (2011), p. 41502. doi:10.1063/1.3675939

[102] X. Liu, G. Hu, G. Huang, and C. Sun, An elastic metamaterial with simultaneously negative mass density and bulk modulus, Appl. Phys. Lett. 98 (2011), p. 251907. doi:10.1063/1.3597651

[103] R.V. Kohn and S.P. Shipman, Magnetism and Homogenization of Microresonators, Multiscale Model. Simul. 7 (2008), p. 62. doi:10.1137/070699226

[104] A. Srivastava and S. Nemat-Nasser, Wave Motion. 51 (2014), 1045–1054.

[105] J. Willis, Arxiv Preprint Arxiv:1311.3875. (2013).

[106] E.J. Post, *Formal Structure of Electromagnetics: General Covariance and Electromagnetics*, Courier Dover Publications, Mineola, NY, 1997.

[107] A. Ward and J. Pendry, Refraction and geometry in Maxwell's equations, J. Modern Opt. 43 (1996), pp. 773–793. doi:10.1080/09500349608232782

[108] A. Ward and J. Pendry, A program for calculating photonic band structures and Green's functions using a non-orthogonal FDTD method, Comput. Phys. Commun. 112 (1998), p. 23. doi:10.1016/S0010-4655(98)00049-6

[109] H. Chen, C. Chan, and P. Sheng, Transformation optics and metamaterials, Nat. Mater. 9 (2010), p. 387. doi:10.1038/nmat2743

[110] S. Cummer and D. Schurig, One path to acoustic cloaking, New. J. Phys. 9 (2007), p. 45. doi:10.1088/1367-2630/9/3/045

[111] H. Chen and C. Chan, Acoustic cloaking in three dimensions using acoustic metamaterials, Appl. Phys. Lett. 91 (2007), p. 183518. doi:10.1063/1.2803315

[112] D. Torrent and J. S_anchez-Dehesa, Acoustic cloaking in two dimensions: A feasible approach, New. J. Phys. 10 (2008), p. 63015. doi:10.1088/1367-2630/10/6/063015

[113] A.N. Norris, Acoustic metafluids, J. Acoust. Soc. Am. 125 (2009), p. 839. doi:10.1121/1.3050288

[114] J. Mei, Z. Liu, W. Wen, and P. Sheng, Effective dynamic mass density of composites, Phys. Rev. B. 76 (2007), p. 134205. doi:10.1103/PhysRevB.76.134205

[115] D. Torrent and J. S_anchez-Dehesa, Anisotropic mass density by two-dimensional acoustic metamaterials, New J. Phys. 10 (2008), p. 23004. doi:10.1088/1367-2630/10/2/023004

[116] G.W. Milton and A.V. Cherkaev, Which elasticity tensors are realizable?, J. Eng. Mater. Technol. 117 (1995), p. 483. doi:10.1115/1.2804743

[117] N.H. Gokhale, J.L. Cipolla, and A.N. Norris, Special transformations for pentamode acoustic cloaking, J. Acoust. Soc. Am. 132 (2012), p. 2932. doi:10.1121/1.4744938

[118] J. Hu, X. Liu, and G. Hu, Constraint condition on transformation relation for generalized acoustics, Wave Motion. 50 (2013), p. 170. doi:10.1016/j.wavemoti.2012.08.004

[119] H. Chen and C.T. Chan, Acoustic cloaking and transformation acoustics, J. Phys. D. Appl. Phys. 43 (2010), p. 113001. doi:10.1088/0022-3727/43/11/113001

[120] E. Cosserat and F. Cosserat, *Theorie des corps deformables*, Hermann, Paris, 1909.

[121] A.V. Amirkhizi, A. Tehranian, and S. Nemat-Nasser, Stress-wave energy management through material anisotropy, Wave Motion. 47 (2010), p. 519. doi:10.1016/j.wavemoti.2010.03.005

[122] W.J. Parnell, Proc. Royal Soc. A: Mathl, Phys. Eng. Sci. 468 (2012), p. 563.

[123] W.J. Parnell and T. Shearer, Antiplane elastic wave cloaking using metamaterials, homogenization and hyperelasticity, Wave Motion. 50 (2013), p. 1140. doi:10.1016/j.wavemoti.2013.06.006

[124] A.N. Norris and W.J. Parnell, Hyperelastic cloaking theory: transformation elasticity with pre-stressed solids, Proc. Royal Soc. A: Mathl, Phys. Eng. Sci. 468 (2012), pp. 2881–2903.

[125] W.J. Parnell, A.N. Norris, and T. Shearer, Employing pre-stress to generate finite cloaks for antiplane elastic waves, Appl. Phys. Lett. 100 (2012), p. 171907. doi:10.1063/1.4704566

11

A nanoporous thin-film miniature interdigitated capacitive impedance sensor for measuring humidity

T. Islam[a]*, Upendra Mittal[a,b], A.T. Nimal[b] and M.U. Sharma[b]

[a]Department of Electrical Engineering, Faculty of Engineering & Technology, Jamia Millia Islamia (A Central University), Maulana Mohammed Ali Jauhar Marg, New Delhi 110025, India; [b]Solid State Physics Laboratory, Lucknow Road, Timarpur, Delhi 110054, India

This paper presents a development of a low-cost miniature humidity sensor with an interdigitated aluminium electrode connected in parallel on quartz substrate. Interdigitated capacitive device has been fabricated using the photolithography method. The aluminium electrode was covered with sensitive film of a nanoporous thin film of γ-Al$_2$O$_3$ made from novel sol–gel technique. Nanostructured thin film offers very high surface to volume ratio with distribution of micro pores for moisture detection. Pore morphologies of the film have been studied by field emission electron microscope and X-ray diffraction methods. Impedance measurement of the miniature capacitive humidity sensor toward relative humidity was investigated at room temperature by Agilent 4294A impedance analyzer (Agilent, Santa Clara, CA, USA). The device exhibits short response and recovery times and good repeatability.

Keywords: γ-Al$_2$O$_3$ film; interdigitated capacitive device; impedance; humidity sensor; sensing properties

1. Introduction

Humidity detection has always been an important task in the fields of automotive industry, food processing, meteorology, semiconductor technology, civil engineering, medicine and health care, industrial and agricultural production, environment protection, etc. Conventional techniques of humidity measurements are mechanical, chilled mirror hygrometers, wet and dry bulb psychrometers, infrared (IR) optical absorption hygrometers, capacitive and impedance type hygrometer. Various fiber optics-based techniques, that is direct spectroscopic, evanescent wave, in-fiber grating and interferometric methods, have also been used for humidity sensing. As technology improved the demand for low cost, reliable, quick response, and compact electronics circuit-based devices increased. Continuous efforts have been devoted to the research and development of humidity sensors to improve the sensing characteristics and miniaturize the size [1–4]. Recently, surface acoustic wave (SAW) humidity sensors became an attraction for researchers as they exhibit the advantages of high sensitivity, very small size, integrated electronic circuitry, easy-to-realize wireless communication, and so on [5–10]. SAW sensor with hygroscopic material coating makes excellent miniaturized humidity sensors having high sensitivity and fast response (few seconds), for real-time measurement of humidity over

---

*Corresponding author. Email: tislam@jmi.ac.in

wide dynamic range. Mainly three working mechanisms contribute to the sensor output: mass loading, acoustoelectric, and viscoelastic effects [11]. Each mechanism can be effectively controlled for making an accurate low-cost humidity sensor. SAW sensor also has advantage as it can circumvent the problems of detecting very high impedance at dry atmosphere encountered for the impedance-type humidity sensors. It is known that the SAW humidity sensors usually have a high operating frequency in order to be smaller and more sensitive [8,11,12]. However, it creates a problem in the preparation of uniform humidity sensitive films with suitable thickness having very small sensing area. Otherwise, it may add to the cost of sensor preparation and affect the consistency in mass production. In last few years, a different kind of SAW impedance sensor had been developed and used in humidity measurements or other applications [13–15]. As we know that chemically, thermally, and mechanically γ-Al$_2$O$_3$ is a highly stable material, this nanoporous metal oxide will be suitable for humidity measurement. It is possible to make the nanostructure by chemical anodization, template and low-temperature sol–gel method. Pore morphology can be tuned to fabricate humidity sensor for different working range from few traces of moisture to percentage relative humidity [16,17]. In the recent past, several works have been reported by the authors and some other groups to utilize the material for both trace and RH level humidity measurement [18,19]. Some of the commercial dew point meters for industrial applications employ thin-film porous aluminium oxide-based capacitive technique [18].

In the present work, we have prepared an interdigitated capacitive impedance humidity sensor. An interdigitated capacitive device with sensitive film of γ-Al$_2$O$_3$ with 300 MHz delay line has been fabricated with standard lithography process. Each interdigitated Al electrode (IDT) for the capacitive device is deposited on the substrate. The sensing film is deposited on the electrode by dip coating of sol–gel solution of γ-Al$_2$O$_3$. Since the sensing film on IDT forms an interdigital capacitive humidity sensor, experiments have been conducted to observe the humidity response of the impedance-type sensor and it shows good response. Effects of the sensitive film on the sensitivity, repeatability, and response time of the sensor are also examined. Proposed interdigitated capacitive sensor is in miniaturized form and it can be batch fabricated with identical characteristics using existing integrated circuit fabrication technology. Also material requirement for sensing film is very small. Thus, it is possible to fabricate the device at low cost.

2. Sol solution and microstructures of the porous γ-Al$_2$O$_3$

Coating sol solution of γ-Al$_2$O$_3$ has been prepared by hydrolyzing the mixture solution of aluminium sec. butoxide (C$_{12}$H$_{27}$AlO$_3$) and water and subsequently peptizing the colloidal solution by adding concentrated hydrochloric (HCl) acid. The solution is then refluxed and the binder polyvinyl alcohol is added. Details of preparation of sol are reported elsewhere [20,21]. To determine the pore morphology of the γ-Al$_2$O$_3$ a thin film has been prepared on ST-X quartz substrate by dip coating method and film has been sintered at 400°C for 4 h. The morphology of the γ-Al$_2$O$_3$ nanostructure film was characterized by X-ray diffraction (XRD) and field emission electron microscope (FESEM). The main object of the XRD plot is to establish the formation of thin film of γ-Al$_2$O$_3$ on quartz substrate. Figure 1(a) shows XRD results for γ-Al$_2$O$_3$ film deposited on quartz substrate while Figure 1(b) shows XRD result for γ-Al$_2$O$_3$ film deposited on alumina substrate (alpha). Figure 1(c) shows XRD result of quartz substrate only without film.When we make a thin film of γ-Al$_2$O$_3$ on Quartz, the XRD peak is obtained at 2θ of 50°. However, the XRD peak of pure quartz material

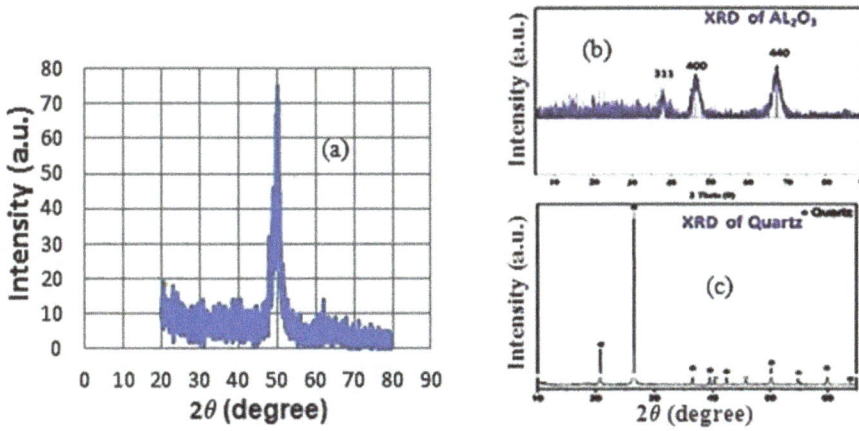

Figure 1.   (a) XRD result of γ-Al₂O₃ film deposited on quartz substrate (b) XRD result of γ-Al₂O₃ deposited on alumina substrate (alpha) (c) XRD result of quartz substrate without any film.

(without γ-Al$_2$O$_3$ film) is obtained at 2θ of 28°. This study has confirmed the formation of γ-Al$_2$O$_3$ film on Quartz substrate.

Figure 2 shows the FESEM image of the porous structure of the sample at low magnification (scale = 100 nm), which suggests that the pores are almost spherical in shape with smooth surface. Pore size has been measured with the help of FESEM at very high resolution (432.390 K X). We have observed the distribution of pores size in the range of 5–10 nm as shown on the photograph. The measurement of pore size has been done using the software available with Carl Zeiss FESEM (model-Supra 55; Carl Zeiss SMT Inc., Peabody, MA, USA) instrument. The average pore size is found to be ~10 nm.

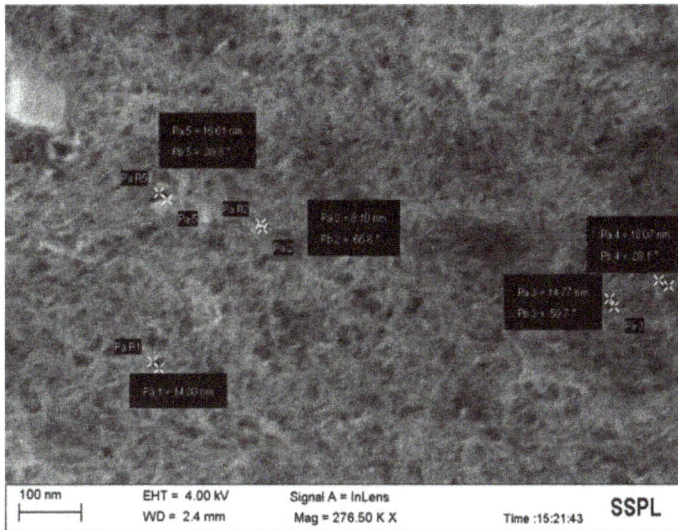

Figure 2.   Field emission scanning electron micrograph of the nanoporous structure of γ-Al$_2$O$_3$ film deposited on quartz substrate.

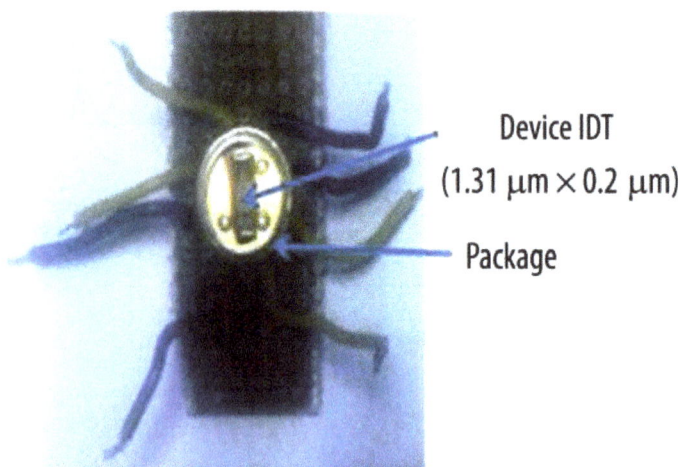

Device IDT
(1.31 μm × 0.2 μm)

Package

Figure 3.   Photograph of the thin-film interdigital capacitive humidity sensor.

2.1. Fabrication of the sensor

Performance of the capacitive impedance sensor depends on the (i) shape and dimension of the electrode, (ii) crack free γ-Al_2O_3 film on the device, (iii) thickness, and (iv) the uniformity of the film. An interdigitated electrode having 300 MHz SAW delay line device is fabricated using the photolithography process on ST-cut quartz substrate with zero temperature coefficient of delay. For the fabrication of the capacitive device, a circular quartz substrate of 3″diameter and 500 μm thickness has been used. After proper cleaning of substrate, aluminium (99.999% pure) metallization has been done using the thermal evaporation technique. Using the wet etching process aluminium electrode IDE (inter digital electrode) has been obtained on the quartz substrate. Then the substrate has been diced and packaged. The size of IDE is 1.31 μm × 0.2 μm (electrode finger width × aluminium thickness) and the capacitive chip size is 5 mm × 8 mm × 0.5 mm. The gap between two adjacent fingers is 1.31 μm.

The device is packaged with a metal shell to ensure its stability. To deposit the sensing film, the device is properly cleaned in the argon plasma for removing the undesirable impurity. The γ-Al_2O_3 film is then deposited on the electrode as well as mass loading area by dipping the device with IDE two times in the sol solution. The device electrodes covered with the sensitive film are sintered in oven at 400°C for 4 h [18]. Thickness of the film has been optimized for the improvement in the sensitivity. The γ-Al_2O_3 film has been characterized using the SOPRA optical ellipsometry system (Sopra Inc., Palo Alto, CA, USA) for thickness measurement. The film thickness is found to be in the range of few hundreds nanometer (~200 nm). The photograph of the interdigitated capacitive sensor is shown in Figure 3.

3. Determination of sensor characteristics

3.1. Experimental set up

To obtain a rapid, accurate, and reliable response of the sensor to the different concentrations of humidity in the nitrogen gas, a computer-based data acquisition system is employed. The sensor is put inside a rectangular shape steel test chamber of nearly 50 cc volume. The sensor chamber is fitted in series with a commercial Honeywell RH meter (Bombay Engineering, Kolkata, India). Humidity contained nitrogen gas is obtained by passing the dry nitrogen gas through water vapor. The amount of residual humidity in the dry N_2 gas is less than 4 ppm. The humidity level in the chamber is controlled by precision needle valve and it is measured by calibrated commercial RH meter. The accuracy of the commercial meter is $\sim \pm 2\%$ RH. The humidity level in the moist gas varied in the range of 20–97% RH at room temperature of 25°C. The leads of the sensor were connected to an impedance analyzer (Agilent 4294A; Agilent, Santa Clara, CA, USA). The impedance analyzer was interfaced to a PC through a data acquisition system. The sensor was excited by AC signal of amplitude 500 mV (r.m.s.) and the frequency of 1 kHz. Here frequency of operation is used to design the finger width of the one IDE. In the present work, the IDE formed below sensing film is used as capacitive sensor. For capacitive moisture sensing, the low signal frequency of 1 kHz is desirable for high sensitivity [1,18]. Experiments were conducted at room temperature to examine the sensor parameters such as: (i) capacitive response, (ii) dissipation factor (D), (iii) impedance response (Z), (iv) transient response for response and recovery times, and (v) repeatability of sensor response output.

3.2. Electrical characteristics of the sensor

Capacitance change with the variation of humidity is shown in Figure 4. The capacitance value increases with increase in humidity. It has been observed that if the signal frequency is increased, capacitance change with humidity becomes smaller. Dissipation factor of the capacitor is shown in Figure 5. Impedance change of the sensor with change in % RH is shown in Figure 6. Since the film is made of metal oxide at low humidity it shows very high impedance and as humidity increases, it decreases rapidly. Response and recovery times are other important parameters of the sensor which are determined from the transient response curve. The response time is the time taken by the capacitive sensor to change the

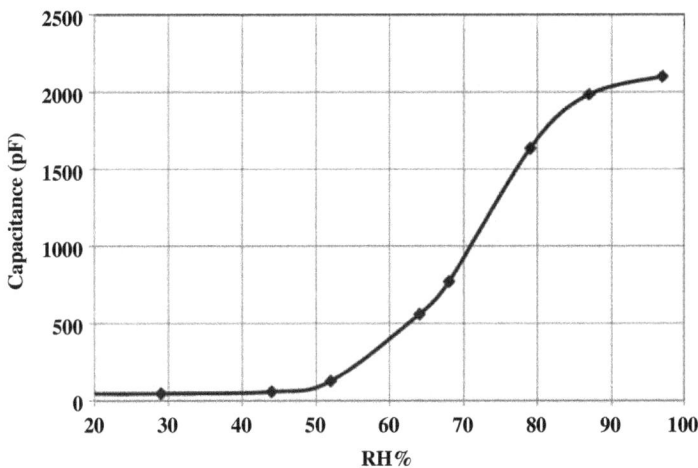

Figure 4.   Capacitance change of the thin-film capacitive sensor with variation of humidity in the range of 20–97% RH.

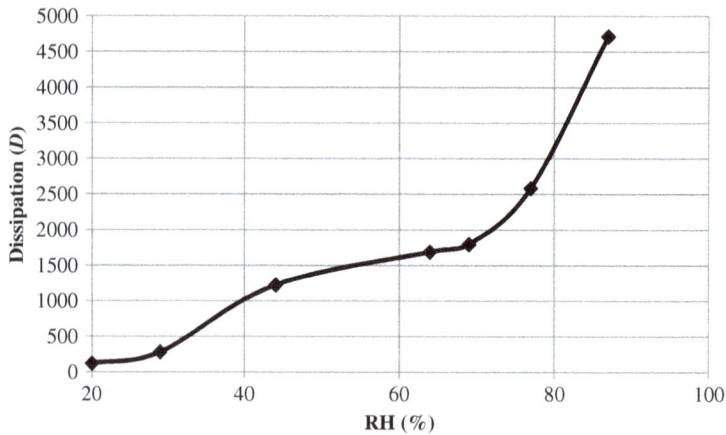

Figure 5.   Change in dissipation factor (D) with variation of humidity in the range of 20–87% RH.

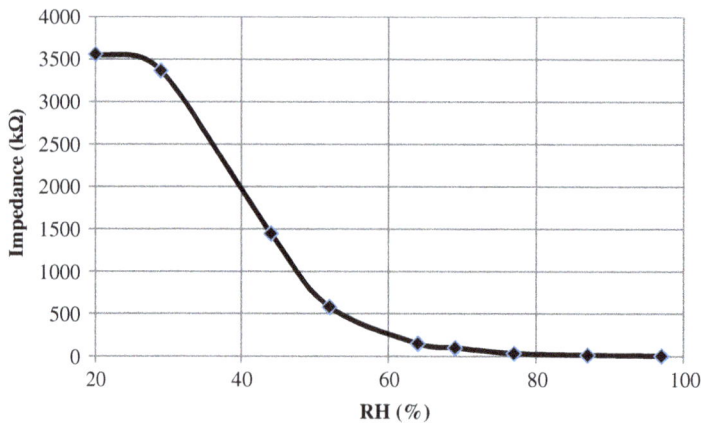

Figure 6.   Impedance (Z) response of the sensor for increase in RH from 20–97%.

output from 10% to 90% of its maximum value while the recovery time is the time taken by the sensor to return from 90% output to 10% of its initial value. For real-time application, these parameters should be as small as possible.

Figure 7 shows the transient response curve for 87–20% RH changes in humidity. The response and recovery times are approximately 15 and 75 s, respectively. The repeatability of the sensor output for the same humidity change for several cycles is shown in Figure 8. Only capacitive type humidity sensor fabricated on quartz has been tested and their repeatability is very good and its value is close to 100%.

4.   Results and discussion

Table 1 summarizes the experimental results for the miniaturized capacitive sensor. Figure 4 shows the capacitance change of the thin-film sensor with variation of humidity. Capacitance value increases monotonically in the beginning but there is a sharp change in the value above 50%. The capacitance change is almost linear in the range of 50–80% RH

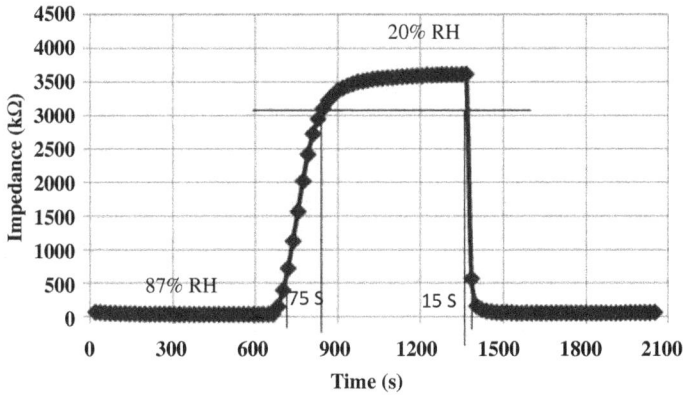

Figure 7.　Transient response of the sensor to the humidity change from 20% to 87% RH.

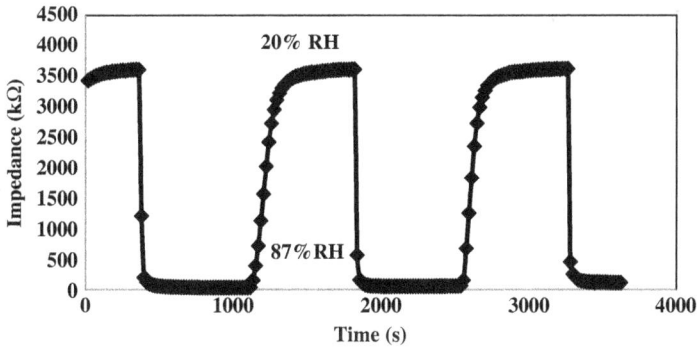

Figure 8.　Repeatability of the sensor output from 20% to 87% RH change in humidity.

Table 1.　Electrical characteristics of the sensor.

Operating moisture range	20–87%RH
Nominal capacitance	100 ± 1 pF
Sensitivity	7.2 pF/%RH
Response time (t90)	15 s (20–87%RH)
Recovery time (t90)	75 s (20–87%RH)
Nonlinearity	1.05% (max [(C/Cmax) × 100])
Temperature coefficient	0.04 pF/°C (22–85°C)
Operating frequency range	1–100 kHz (suitable frequency = 1 kHz)

and above 80% RH, it increases slowly. When the sensor is exposed to humidity, the vapor molecules initially adsorb on the surface and then condensed in the nano-order pores causing effective change in dielectric constant. But since the sensing area is very small, the capacitance change at lower humidity is small and as humidity increases enough vapor molecules condensed in the porous layer causing large change in dielectric constant. Also at higher humidity above 80%, many of the pores are already or nearly filled up causing lesser change is dielectric constant [18]. Figure 6 shows the impedance response of the sensor for increase in RH. The impedance change is due to change capacitive reactance as well as conductivity of the water adsorbed. Initially water molecules are chemically adsorbed then physically attached on the chemically adsorbed layer. The decrease in reactance is due to the increase in capacitance and the increase in conductivity is due to increase in protons' (H^+) concentration of the sensor with vapor. The adsorbed OH^- ions provide easy path for proton hopping [16]. Also the small gap between the fingers plays important role for sharp change in impedance. Because of very thin sensing layer, the response is faster and hysteresis error is found to be negligible. Hysteresis is one of the concerns of any humidity sensor working on adsorption and desorption phenomenon. Hysteresis error arises due to incomplete adsorption and desorption of water molecules, and for a good sensor it should be as small as possible. The hysteresis error depends on the pore morphology and thickness of the sensing layer [1,18]. Since the sensor is based on thin film of metal oxide, the output drift is considerably smaller [16]. Transient response of the sensor is shown in Figure 7, and the sensor shows very fast response time (15 s). The response behavior is governed by the pore morphologies, and the morphologies are governed by the average dimension of the pore, the distribution of the pores, thickness of the porous layer, and stability of the sensing film [1]. Pore morphology plays a significant role in the condensation of water in the porous matrix. Since the mean free path and size of incident moisture cluster varies with the change in humidity concentration, an optimum pore size comparable to mean free path is desirable for humidity sensing. The response also depends on the signal frequency, which is related to orientation loss of humidity. The interdigital capacitance C_{IDT} shown in Figure 9 can be determined from the following equation [22].

$$C_{IDT} = C_{UC}(N - L) \tag{1}$$

where C_{IDT} is the capacitance of the IDT, C_{UC} is the capacitance of unit cell, N is the number of electrodes, and L is the length of the electrode. C_{UC} can be written as

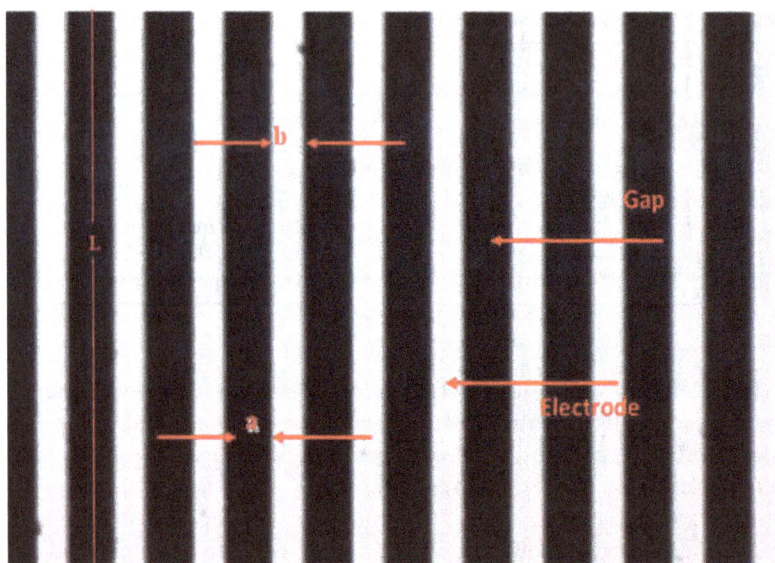

Figure 9.   The interdigital electrode structure of the capacitive impedance sensor.

$$C_{UC} = C_1 + C_2 + C_3 \tag{2}$$

The capacitance C_1 and C_2 are calculated by conformal mapping $K[x]$ of the two semi-planes with dielectric constants ε_1 and ε_2 into two parallel-plate capacitors.

$$C_1 + C_2 = \varepsilon_0 \frac{\varepsilon_1 + \varepsilon_2}{2} \frac{K\left[\left(1 - \left(\frac{a}{b}\right)^2\right)^{1/2}\right]}{K\left[\frac{a}{b}\right]} \tag{3}$$

where a is the gap between the electrodes, b is the single electrode width. The capacitance C_3 is given by

$$C_3 = \varepsilon_0 \varepsilon_3 \frac{h}{a} \tag{4}$$

where, ε_3 is the effective dielectric constant of the sensing film in the presence of water, h is the thickness of the electrode. As a (gap between the electrodes) is inversely related with C_3, decreasing the gap between the electrodes or increasing the number of fingers (actual sensing area) may increase sensitivity toward lower humidity range. Reliability of the sensor system is shown in Figure 8. As our capacitive type humidity sensor has been tested with same humidity several times and it gives the same output in every cycle as the γ-Al_2O_3 film coated capacitive device is very stable and repeatability of the sensor system is very good. It has been reported that γ-Al_2O_3 film is highly stable thermally, hence the effect of ambient temperature variation is negligible [18].

5. Conclusion

In the present work, a miniaturized interdigitated capacitive device using thin film of nanoporous alumina for humidity sensing has been fabricated. The humidity sensing film has been prepared by sol–gel technique. Since the sensing film on the IDE of the device forms an interdigital capacitive impedance sensor, initial experiments have been conducted to study the impedance response with 20–97% RH change in N_2 gas. The achieved characteristics are suitable for employing the sensor to measure humidity in the prescribed range. The response characteristics are significant, fast, and highly reproducible. However, the sensitivity can be increased in the lower humidity range by increasing the area of the interdigital electrode and sensing film of the device. The actual sensing area is only 5 mm × 8 mm.

Acknowledgments

The authors would like to thank Director, Solid State Physics Laboratory, Delhi and Mr. Lokesh Kumar from the Department of Electrical Engg., Jamia Milia Islamia University, Delhi, for providing support to carry out the research work. Authors would also like to thank Dr. Shiv Kumar, Mr. Akhilesh Pandy, and Mr. Anand Kumar from the Solid State Physics Laboratory, Delhi, for their support for ellipsometry, XRD, and FESEM measurements. The authors gratefully acknowledge the help and cooperation from their colleagues.

References

[1] Z.M. Rittersma, *Recent achievements in miniaturised humidity sensors—a review of transduction techniques*, Sens. Actuators A. 96 (2002), pp. 196–210. doi:10.1016/S0924-4247(01) 00788-9

[2] Z.Y. Wang, L.Y. Shi, F.Q. Wu, S.A. Yuan, Y. Zhao, and M.H. Zhang, *The sol–gel template synthesis of porous TiO_2 for a high performance humidity sensor*, Nanotechnology. 22 (2011), pp. 275502–275510. doi:10.1088/0957-4484/22/27/275502

[3] P. Sahoo, D.S. Oliveira, M.A. Cotta, S.D.S. Dash, A.K. Tyagi, and B. Raj, *Enhanced surface potential variation on nanoprotrusions of GaN microbelt as a probe for humidity sensing*, J. Phys. Chem. C. 115 (2011), pp. 5863–5867. doi:10.1021/jp111505m

[4] J.C. Tellis, C.A. Strulson, M.M. Myers, and K.A. Kneas, *Relative humidity sensors based on an environment-sensitive fluorophore in hydrogel films*, Anal. Chem. 83 (2011), pp. 928–932. doi:10.1021/ac102616w

[5] S. Lei, D.J. Chen, and Y.Q. Chen, *A surface acoustic wave humidity sensor with high sensitivity based on electrospun MWCNT/Nafion nanofiber films*, Nanotechnology. 22 (26) (2011), pp. 265504–265512. doi:10.1088/0957-4484/22/26/265504

[6] H.S. Hong and G.S. Chung, *Humidity sensing characteristics of Ga-doped zinc oxide film grown on a polycrystalline AlN thin film based on a surface acoustic wave*, Sens. Actuators B. 150 (2010), pp. 681–685. doi:10.1016/j.snb.2010.08.020

[7] Y. Li, P. Li, M.J. Yang, S. Lei, Y.Q. Chen, and X.S. Guo, *A surface acoustic wave humidity sensor based on electrosprayed silicon-containing polyelectrolyte*, Sens. Actuators B. 145 (2010), pp. 516–520. doi:10.1016/j.snb.2009.12.062

[8] K. Ocakoglu and S. Okur, *Humidity sensing properties of novel ruthenium polypyridyl complex*, Sens. Actuators B. 151 (2010), pp. 223–228. doi:10.1016/j.snb.2010.09.017

[9] Y. Li, M. Yang, M. Ling, and Y. Zhu, *Surface acoustic wave humidity sensors based on poly (p – diethynylbenzene) and sodium poly sulfonate*, Sens. Actuators B. 122 (2007), pp. 560–563. doi:10.1016/j.snb.2006.03.031

[10] P.R. Story, D.W. Galipeau, and R.D. Mileham, *A study of low-cost sensors for measuring low relative humidity*, Sens. Actuators B. 25 (1995), pp. 681–685. doi:10.1016/0925-4005(95) 85150-X

[11] J.D.N. Cheeke, N. Tashtoush, and N. Eddy, *Surface acoustic wave humidity sensor based on the change in the viscoelastic properties of a polymer film*, Proc. IEEE Ultrason. Symp. 1 (1996), pp. 449–452.

[12] R. Rimeika, D. Ciplys, V. Poderys, R. Rotomskis, and M.S. Shur, *Fast-response surface acoustic wave humidity sensor based on hematoporphyrin film*, Sens. Actuators B. 137 (2009), pp. 592–596. doi:10.1016/j.snb.2009.02.009

[13] A. Buvailo, Y.J. Xing, J. Hines, and E. Borguet, *Thin polymer film based rapid surface acoustic wave humidity sensors*, Sens. Actuators B. 156 (2011), pp. 444–449. doi:10.1016/j. snb.2011.04.080

[14] Q.Y. Cai, R.H. Wang, L.Y. Wu, L.H. Nie, and S.Z. Yao, *Surface acoustic wave (SAW)— impedance sensor for kinetic assay of trypsin*, Microchem. J. 55 (1997), pp. 367–374. doi:10.1006/mchj.1996.1320

[15] D.Z. Liu, K. Chen, A.F. Yin, K. Ge, L.N. Nie, and S.Z. Yao, *Theory of the surface acoustic wave/impedance sensor system and its application to the end-point determination in titrimetry*, Anal. Chim. Acta. 320 (1996), pp. 245–254. doi:10.1016/0003-2670(95)00529-3

[16] E. Traversa, *Ceramic sensors for humidity detection: the state-of-the-art and future developments*, Sens. Actuators B. 23 (1995), pp. 135–156. doi:10.1016/0925-4005(94)01268-M

[17] R.K. Nahar, *Study of the performance degradation of thin film aluminum oxide sensor at high humidity*, Sens. Actuators B. 63 (2000), pp. 49–54. doi:10.1016/S0925-4005(99)00511-0

[18] T. Islam, L. Kumar, and S.A. Khan, *A novel sol–gel thin film porous alumina based capacitive sensor for measuring trace moisture in the range of 2.5–25 ppm*, Sens. Actuators B. 173 (2012), pp. 377–384. doi:10.1016/j.snb.2012.07.014

[19] T. Islam, L. Kumar, Z. Uddin, and A. Ganguly, *Relaxation oscillator based active bridge circuit for linearly converting resistance to frequency of resistive sensor*, IEEE Sens J. 13 (2013), pp. 1507–1513. doi:10.1109/JSEN.2012.2236646

[20] B.E. Yoldas, *A transparent porous alumina gels*, Am. Ceramic Soc. Bull. 54 (3) (1975), pp. 286–288.

[21] B.E. Yoldas, *Alumina sol preparation from alkoxide*, Ceramic Bull. Sci. 54 (3) (1975), pp. 289–290.
[22] H.E. Endres and S. Drost, *Optimization of the geometry of gas sensitive interdigital capacitors*, Sens. Actuators B. 4 (1991), pp. 95–98. doi:10.1016/0925-4005(91)80182-J

12

Carbon nanotube-reinforced elastomeric nanocomposites

Bismark Mensah[a], Han Gil Kim[a], Jong-Hwan Lee[a], Sivaram Arepalli[b] and
Changwoon Nah[a]*

*[a]BK21 Haptic Polymer Composite Research Team, Department of Polymer-Nano Science and
Technology, Chonbuk National University, Jeonju 561-756, Republic of Korea; [b]Nano Science
Consultants LLC, Hampton, VA 23666-6701, USA*

This review is focused on carbon nanotube (CNT)-elastomeric polymer nanocomposites, which have attracted industrial and academic interest over the years due
to their enhanced properties. Major factors notably CNT type, surface modification, dispersion of CNT, and processing techniques that affect the physical
properties of CNT-elastomeric polymer nanocomposites are reviewed, and several
key physical properties, including tensile, electrical, and thermal properties, were
also included in this review. Some of the key challenges that undermine the
effectiveness of CNTs and their composites with elastomeric polymers, and the
potential applications of CNT-elastomeric composites are also captured.

Keywords: carbon nanotubes; elastomeric polymers; nanocomposites; surface modification; physical properties

1. Introduction

The carbon nanotubes (CNTs) can be formed by rolling of graphite sheets into a tubular
structure [1,2]. They can be classified into different forms depending on the number of
layers. They include single-walled carbon nanotubes (SWCNTs), double-walled carbon
nanotubes (DWCNTs), and multiwalled carbon nanotubes (MWCNTs) (Figure 1). As can
be seen in Figure 2, the basic characteristics of CNTs can be determined by the way of
rolling of graphite sheet leading to a specific chirality defined by the chiral (roll-up)
vector, $\vec{C}_k$, and chiral angle, θ, as can be described by the Equation [3–7]:

$$\vec{C}_k = n\vec{a}_1 + m\vec{a}_2 \tag{1}$$

where the integers n, m are the number of steps along the zigzag carbon bonds of the
hexagonal lattice and $\vec{a}_1$ and $\vec{a}_2$ are unit vectors. Two limiting cases of the geometrical
structure are possible. The zigzag ($\theta = 0°$) and armchair ($\theta = 30°$) structure is shown in
Figure 3. Generally, the property of CNTs depends on their atomic arrangement, diameter
and length, and surface morphology. The chirality of CNTs strongly affects the mechanical and electric conductivity [1,4,5,7,8].

Since the first detailed characterization of CNTs in 1991 by Iijima [9], they have
received much attention because of their unique properties. The CNTs possess high

*Corresponding author. Email: cnah@jbnu.ac.kr

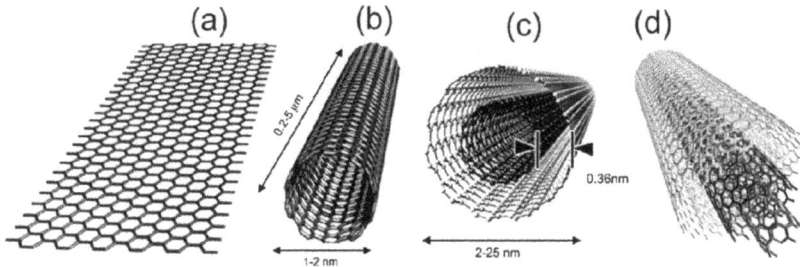

Figure 1. Structure of (a) graphite sheet, (b) single-wall (SWCNT), (c) double-wall (DWCNT), and (d) multiwall carbon nanotube (MWCNT) (available from online: http://jnm.snmjournals.org/content/48/7/1039/F1.expansion.html).

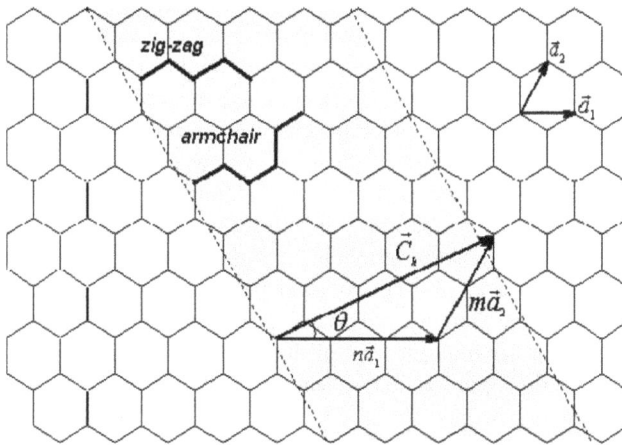

Figure 2. Schematic diagram showing how a hexagonal sheet of graphite is 'rolled' to form a carbon nanotube. Reprinted from reference [4].

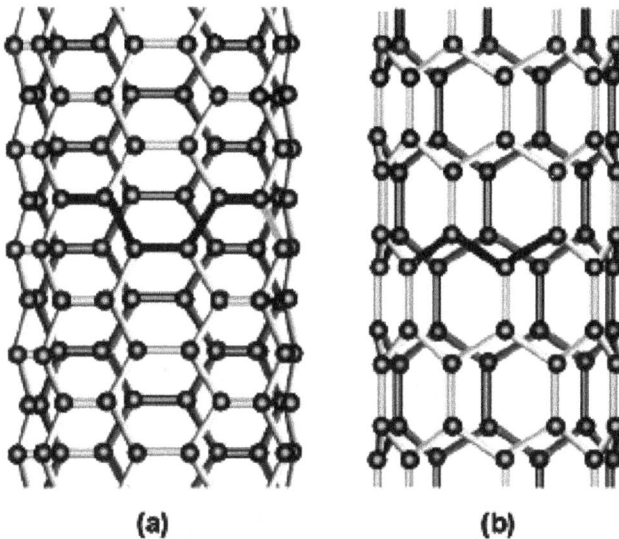

Figure 3. Illustrations of the atomic structure of (a) an armchair and (b) a ziz-zag nanotube. Reprinted from reference [4].

flexibility, low mass density, and large aspect ratio (typically: 100 ~ 1000) in addition to very high electrical and thermal conductivities. Some nanotubes are stronger than steel (about 10 ~ 100 times higher at a fraction of the weight), lighter than aluminum, and more conductive than copper [10,11]. For example, theoretical and experimental results on individual SWCNTs show extremely high tensile modulus (640 GPa to 1 TPa) and tensile strength 6 (150 ~ 180 GPa) [12]. The SWCNTs exhibit large phonon mean free path lengths that result in high thermal conductivity (theoretically > 6000 $Wm^{-1}K^{-1}$).

The superior mechanical and electrical properties of CNTs offer an exciting potential for new class of polymer composite materials to replace the traditional fillers. A number of polymers such as epoxy [13,14], polyvinylacetate (PVA) [15], polyvinylchloride (PVC) [16,17], polyethylene (PE) [18], polyamide (PA) [19], and many other polymers have been employed as matrices to prepare CNT-polymer nanocomposites. Figure 4 shows a trend of the CNT-based research activity in terms of scientific publication and patents over the years since 1992. An exponential growth of research activity indicates that the topic has become one of the hot issues recently. Figure 5 shows a similar trend of CNT-based research activity (research publication only) focusing on the elastomeric polymer matrices including rubbers. The data in Figure 5 were collected in a recent literature pool on CNT-elastomer-based compounds [20–118]. The gradual increase in the number of research publication also suggests that a great attention is now extended to the reinforcing of rubber by CNTs, even though the total number of publications is relatively lower.

Elastomeric polymers and rubbers have long been widely used for various applications including tires, seals, shock-absorbing mounts and gaskets for energy, structural, electronic, and electrical equipment [33–41]. No other polymers without elasticity can replace the unique function of the rubbery polymers. The employment of CNTs into rubbery matrices can provide further improvement of durability, strength, light weight, as well as design and processing flexibility compared with the conventional fillers like carbon blacks (CBs). Since the research works on CNT-reinforced rubber nanocomposites have been published by many researchers [41–47], there is a clear need to collect and analyze the recent technologies developed so far. The current review paper presents the streamlined technologies of various CNT-elastomer nanocomposites and their hybrid system by

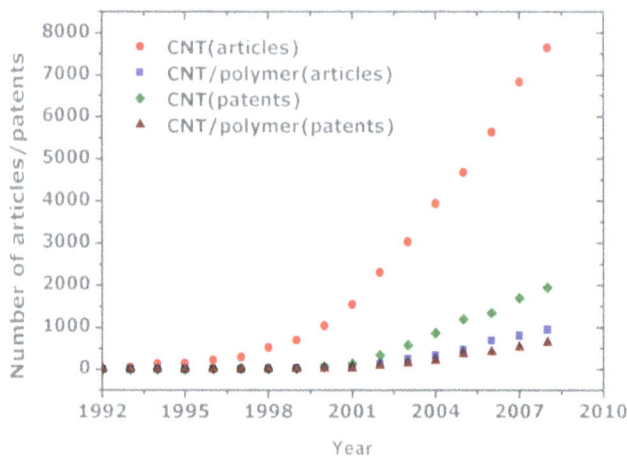

Figure 4.   A trend in the research activity of CNTs and CNT-related polymer composites over the years (available at: https://sites.google.com/site/cntcomposites/Home).

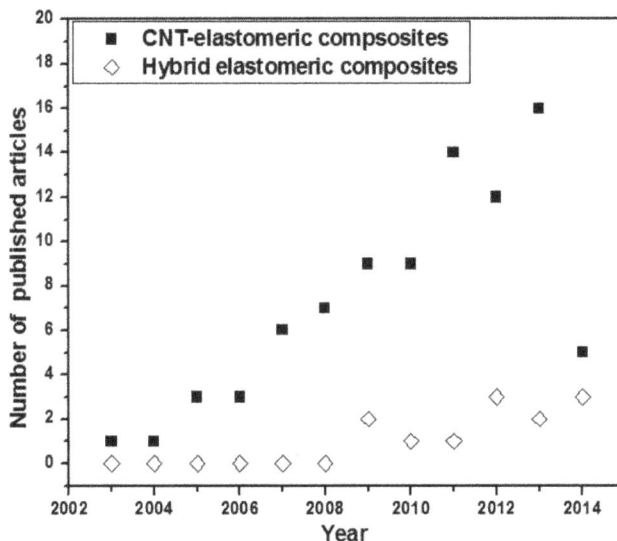

Figure 5.   A trend of the research activity of CNT-related elastomeric composites over the years.

focusing on modification of CNTs, mixing, morphology and state of dispersion of CNTs, mechanical, electrical, and thermal properties.

2. Modification of CNT

A crucial issue in fabricating CNT-polymer nanocomposites is the dispersion of individual CNT into the host polymer matrix without losing the effective interactions at the CNT–matrix interface. Individual dispersion of CNT is strongly affected by van der Waals forces of attraction between adjacent tubes and high aspect ratio, which gives rise to the formation of aggregates [3,10,12,48]. Several methods have been reported for the fabrication of CNT-elastomer composites, focusing on the successful dispersion of CNT filler in the host matrix and their effective interfacial interactions with the matrix. One of the most promising means of enhancing the dispersion of CNTs and their interfacial bonding with polymer matrix is the modification/functionalization of the CNTs [49,50,51]. The modification can be categorized into covalent and non-covalent. We focus here on a brief explanation and their relevance to processing of CNT-elastomeric nanocomposites.

Covalent functionalization is one popular method to attain modified CNTs; it deals with chemical bonding (grafting) of polymer chains to CNTs, where strong chemical bonds between nanotubes and polymers are established. There are two ways of grafting of CNTs. The 'grafting to' approach involves the synthesis of a polymer with a specific molecular weight terminated with reactive groups or radical precursors. In a subsequent reaction, the polymer chain is attached to the surface of nanotubes by addition reactions. In comparison, the 'grafting from' approach involves growing polymers from CNT surfaces via in-situ polymerization of monomers initiated by chemical species immobilized on the CNT sidewalls and CNT edges [3,48]. A notable drawback of covalent modification is the disruption of the extended π conjugation in CNTs, which can strongly affect negatively the electrical property [3]. Figure 6 presents the various approaches used to chemically modify CNTs.

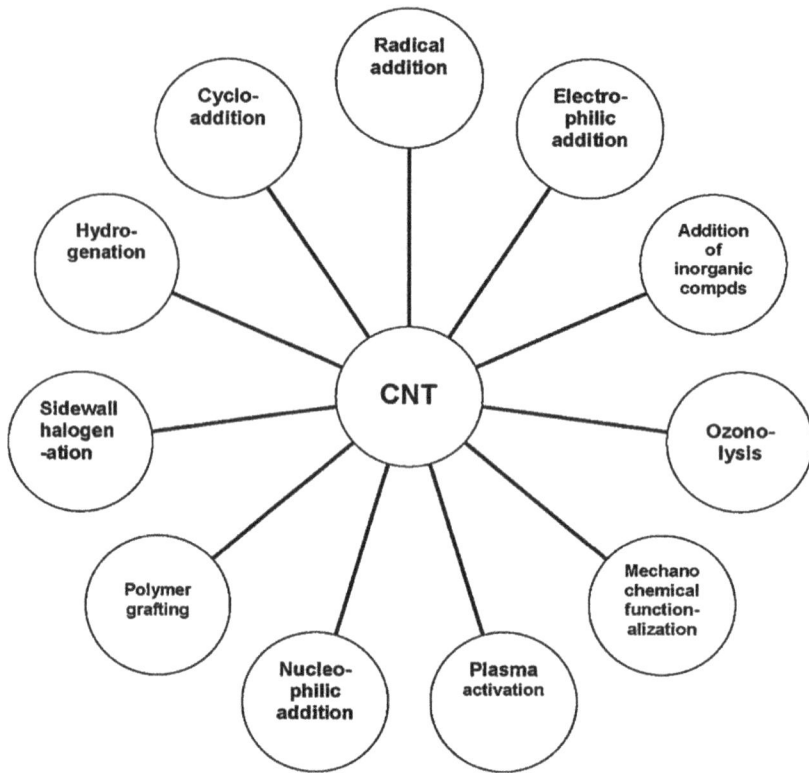

Figure 6. Various covalent functional modifications possible with carbon nanotubes. Reprinted from reference [52].

The next categorized method is the non-covalent modification, which can modify the interfacial properties of CNTs [3,53]. It is primarily concerned with the physical adsorption and/or wrapping of polymers to the surface of the CNTs. The graphitic sidewalls of CNTs provide the possibility of π–π stacking interactions with conjugated polymers, as well as organic polymers containing hetero-atoms with free electron pairs [3,54]. The use of surfactants falls in the non-covalent method and it is widely accepted. Surfactants consist of hydrophilic polar head and hydrophobic nonpolar tail. They can interact with the hydrophobic surface of CNTs to provide them a hydrophilic character, which makes the CNTs soluble in several solvents including water [55–58]. Unlike the chemical modification, the non-covalent method does not destroy the conjugated system of CNTs, leading to a less effect on the electrical performance. However, the load transferring capability might be lower due to weaker interactions at the interfaces [18,19,53].

Islam and co-workers [59] explored the dispersing power of a range of surfactants: sodium dodecylbenzenesulfonate (NaDDBS: $C_{12}H_{25}C_6H_4SO_3Na$), sodium octylbenzene sulfonate (NaOBS: $C_8H_{17}C_6H_4SO_3Na$), sodium butylbenzene sulfonate (NaBBS: $C_4H_9C_6H_4SO_3Na$), sodium benzoate ($C_6H_5CO_2Na$), sodium dodecyl sulfate (NaDS: $CH_3(CH_2)_{11}$-OSO_3Na), Triton X-100 (TX100: $C_8H_{17}C_6H_4(OCH_2CH_2)_n$-OH: $n \cong 10$), dodecyltrimethylammonium bromide (DTAB: $CH_3(CH_2)_{11}$ $N(CH_3)_3Br$), dextrin, and poly(styrene)-poly(ethylene oxide) (PS-PEO) diblock copolymer. It was shown that the stabilization of CNTs depends on the surfactant molecules that lie on

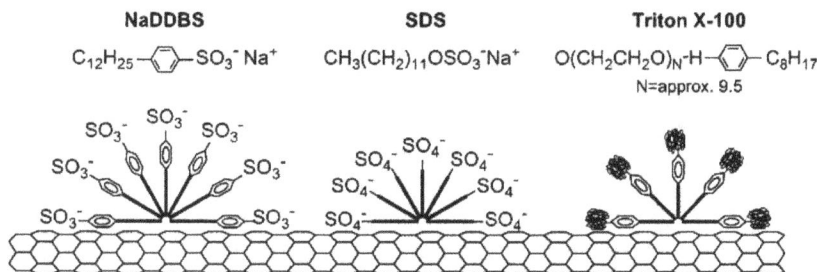

NaDDBS   **SDS**   **Triton X-100**

$C_{12}H_{25}$—◯—SO_3^- Na^+ $CH_3(CH_2)_{11}OSO_3^-Na^+$ $O(CH_2CH_2O)_N$-H—◯—C_8H_{17}

N=approx. 9.5

Figure 7. A schematic representation of adsorption behavior of surfactants onto the surface of CNTs. Reprinted from reference [59].

Table 1. Modification types of CNTs and their composites with rubber matrices.

Type of modification	Modifier	Rubber–CNT composite
Covalent	Acid treatment	NR/SBR-MWCNT [21], NR-MWCNT [60,61], NR-SWCNT [36,62],TPE-MWCNT [27], IIR-MWCNT [38], NBR/SBR-MWCNT [63], PDMS-MWCNT [47,64], CIIR-MWCNT [44], Soft epoxy [65], EPDM-MWCNT [66], SBR-CNT [67], TPU-MWCNT [68]
	Hydroxyl (-OH)	SBR/BR-MWCNT [69]
	Grafting	TPU-MWCNT [70]
	Amino/silane	NR-MWCNT [33], PDMS-MWCNT [71], BADGE/ATBN-MWCNT [72], TPU-MWCNT [73]
Non-covalent	Ionic surfactant	CR-MWCNT [34,41,74,75,76], PDMS-SWCNT [77,78], NR-MWCNT [79,80], PDMS-MWCNT [81], SBR/BR [82], NR
	Non-ionic	NR/SBR-MWCNT [83], SBR/BR-MWCNT [69]
	Catalyst	SEBS-SWCNT [39]

CNTs surface parallel to the cylindrical axis. They reported that the alkyl chain groups of a surfactant molecule adsorb flat along the length of the tube instead of the diameter. The NaDDBS and Triton X-100 (TX100) were dispersed CNTs better than SDS due to the role of benzene rings. NaDDBS was reported to be dispersed better than TX100 because of its head group and slightly longer alkyl chain. They argued that the spacing between the benzene rings on the surfactants and the CNTs surface was large enough to accommodate the SO_3^- charged groups [59]. Figure 7 is a schematic representation of how surfactants may be adsorbed onto the nanotube surface.

Table 1 summarizes the modification class of CNTs and their specific elastomeric composites appeared in the literature. The majority research works on CNT-rubber nanocomposites have been focused on the covalent modification of CNT, typically by acid treatment of CNT for processing CNT-elastomeric composites. Other covalent modification includes the amino/siliane treatment, and hydroxyl and grafting modification method. Among the non-covalent modification, the treatment by ionic surfactants was widely used by researchers. However, the choice of modification method may strongly depend on the application purpose of the resulting nanocomposites.

3. Mixing of CNT with elastomeric polymer

One of the issues in the CNT-based elastomeric polymer nanocomposites is how to achieve a homogenous good dispersion of CNTs in the polymer matrix to maximize the enhancement in various performances of the resulting composites. The major considering factors would be the CNT type and concentration, modification of CNTs, type of polymer matrix, and mixing techniques used [3,50,51]. Although, there are several mixing techniques available, most of the CNT-filled elastomeric composites mixing is widely done by using the conventional techniques such as solution mixing, melt mixing, and in-situ polymerization [15,16].

Solution mixing is one of the approaches for processing polymer nanocomposites. It involves the use of a dissolvable polymer in a solvent such as tetrahydrofuran (THF), dimethyl formamide (DMF), toluene or acetone. Then the nanofillers are also dispersed in one of these dispersible solvents by the sonication or any kind of mechanical agitation approach. The two solutions are then homogenously mixed together by mechanical mixing, magnetic agitation, or high-energy sonication. Subsequently, the CNT-polymer composites can be obtained by either vaporizing the solvent or coagulation by adding a poor solvent [25,77,78]. This method is generally considered when a better dispersion and distribution of film-type composites are needed [10]. One limitation of this method is that the polymer should be dissolved in a solvent for further processing. Badaire group [84] established that a high-power ultrasonication for a long period of time shortens the nanotube length (shorter aspect ratio), which is detrimental to the composite properties. The general guideline for the optimum sonication conditions (time, power) are not systematically established and will depend on nanotube concentration and original shape of CNTs [10]. Many CNT-rubber composites with different kinds of elastomeric matrices and CNTs have been reported via this method [25,83,85,86]. Recently, Bokobza [87] enhanced the electrical and mechanical properties of rubber based on styrene–butadiene rubber (SBR), natural rubber (NR), and ethylene–propylene–diene monomer (EPDM) with MWCNTs. In terms of mechanical properties (ultimate tensile strength), SBR obtained 1.08 MPa, NR ~ 5.88 MPa, and EPDM ~ 10.7 MPa, respectively, at 0.5 phr of filler loading. This is a clear indication that the solution mixing is sensitive to the type of matrix involved. Furthermore, the processing conditions had a strong effect on the electrical properties, which are also very sensitive to nanotube dispersion within the elastomeric matrix. A new mixing technique which can be categorized under solvent mixing approach was recently adopted by Shin et al [88] to prepare a soft, flexible, highly stretchable, and conductive composites of MWCNT and PU (poly[4,40- methylene-bis (phenyl isocyanate)-alt-1,4 butanediol/poly(butylene adipate)]). In this study, aligned MWCNT (MWCNT forest, diameter: 10 nm) were grown on iron-catalyst-coated Si wafers using a conventional chemical vapor deposition (CVD) method. Initially, the MWCNT forests were infiltrated with a PU solution in N,N-dimethyl formamide (DMF) using a simple drop-casting procedure. Later, the solvent was evaporated and a composite of MWCNT forest/PU which could be peeled off the underlying Si wafer was obtained. The prepared film was composed of a black and conductive side facing the substrate (forest side) and the other whitish and insulating side (PU side). The MWCNT/ PU material could be stretched up to about 1400% relative to the pure PU polymer (760% strain). The elongation of the material was also higher than the highly stretchable sheath-core conducting fibers (strain up to 1320%) created by wrapping CNT sheets oriented in the fiber direction on stretched rubber fiber cores prepared by Liu et al [89] and the stretchable CNT material (280% strain) for human motion detection [90]. Moreover, the

MWCNT forest/PU film obtained an overall electrical conductivity of 50–100 Sm^{-1}, Young's Modulus and ultimate strength of about 148 and 38% higher than the pure PU, respectively.

Melt blending is another method broadly accepted in fabricating the CNT-elastomeric nanocomposites [10]. CNTs are dispersed within the elastomeric polymer matrix by shearing stresses. The generated shear forces help to break CNT aggregates or prevent their reaggregations during mixing. This approach is generally noted to be simple and common for industrial-based mass production. Verge et al [91] reported on the dispersion by simple melt blending of tiny amounts of CNT in acrylonitrile–butadiene rubber (NBR). The CNT bundles were more effectively exfoliated and dispersed in the NBR with higher acrylonitrile content due to stronger interaction between the acrylonitrile and CNTs. Wu et al [92] fabricated styrene–ethylene–butylene–styrene (SEBS)/CNT composites in the presence of bis(diphenyl phosphate) (BDP) by melt mixing method and explored their flame-retardant and mechanical properties. They reported that the pretreatment of CNTs by BDP greatly improved the final dispersion in SEBS matrix. The improved CNT dispersion was reported to enhance the flame retardency and thermal stability of the composite. A double filler system such as CNT and CBs mixtures have also been reported to affect the dispersion of CNT during melt mixing process by Dang et al [28].

In-situ polymerization technique is the third traditional method for processing polymer nanocomposites. It is regarded as an alternative method, whenever both the solution and melt mixing are not suitable. With this approach, the CNTs are dispersed into a given monomer followed by polymerizing the monomers under certain reaction conditions [93]. The in-situ polymerization methods enable covalent bonding between functionalized nanotubes and the polymer matrix using various condensation reactions [48]. Higher percentage of CNTs may be easily dispersed by this method resulting in a stronger interaction with the polymer matrix. However, this method is only limited to the preparation of nanocomposites that are insoluble in solvent and thermally unstable. Majority of the scientific reports of in-situ polymerization for fabrication of thermosetting composites used epoxy as the base matrix material [94–97]. In this case, CNTs are first dispersed in a resin followed by curing the resin with a hardener under certain conditions. Nonetheless, a number of CNT-rubbery composites have been fabricated using this method and yielded good results. For example, nanocomposites based on poly(trimethylene terephthalate)-block–poly(tetra-methylene oxide) (PTT–PTMO)-segmented copolymer and COOH-functionalized SWCNTs were prepared via in-situ polymerization method by Szymczyk [98]. It was observed that the nanocomposites with low SWCNTs loading (<0.5 wt%) showed uniform dispersion of CNT in the polymer matrix. Again, the tensile strength of the nanocomposites with 0.05–0.3 wt% loading of SWCNTs obtained an improved tensile strength better than the neat PTT–PTMO copolymer without reduction in elongation at break. Roslaniec et al [99] used in-situ method to synthesize a nanocomposite based on multiblock polyester elastomers (PEE) and CNTs, and reported their enhanced mechanical and thermal properties. A key concern here is that the effectiveness of this approach may rely on several parameters such as reaction conditions, type of monomer, nature of filler and its properties, as well as the filler concentration.

The efficiency of composite processing techniques is summarized in Table 2 and Table 3, showing the influence of the techniques on the physical properties (tensile and percolation threshold) of various elastomeric polymer matrices.

In general, the melt mixing technique has been widely used to reinforce elastomers with CNTs. Recently, the combinational approach involving two or more of the individual

Table 2.   Mechanical properties of elastomeric/CNT nanocomposites.

Composite	Processing	TS increase (%)	EM increase (%)	EB increase (%)	Ref
NR + 3 phr MWCNT	Melt	~1378	~260	~(226)	[100]*
NR + 1 phr MWCNT + 29 phr silica	Melt	~38	~26	~(33)	[101]*
NR + 25 phr SWCNT	Melt	~173	~317	~(−27)	[62]ˣ
NR + 3 phr MWCNT + 40 phr silica	Melt	~2422	~380	~(264)	[100]*
NR + NBR + 2 phr MWCNT	Melt	~18	~100	~(−5)	[83]ʸ
EPDM + 3 phr MWCNT	Melt	~161	~90	~(71)	[100]*
EPDM + 3 phr MWCNT + 40 phr CB	Melt	~1587	~260	~(203)	[100]*
SBR + 5 phr MWCNT	Melt	~248	~85	~(38)	[67]ˣ
SBR + 5 phr MWCNT	Melt	~146	~157	~(10)	[102]*
SBR + 5 phr MWCNT + 20 phr EG	Melt	~296	~530	~(−3)	[102]*
SBR/BR + 5 phr MWCNT	Melt	~246	~150	~(66)	[69]ˣ
SBR + BR + 3 phr CNT + 40 phr silica	Melt	~839	~320	~(71)	[100]*
TPU + 3 wt% MWCNT-g-PTMEG	Melt	~(−4)	~4	~(−6)	[70]ˣ
SBR + 15 phr MWCNT	Solid (two-roll)	~293	~262	N/A	[63]ˣ
PDMS + 5 wt% MWCNT	Solid (two-roll)	~155	N/A	~(−21)	[64]ˣ
MVQ + 0.75 wt% MWCNT	Solvent	~38	~82	~(−46)	[103]*
MVQ + MWCNT + G (0.75 wt%)	Solvent	~109	~96	~(1)	[103]*
PDMS + 1.6 wt% SWCNT	Solvent	~14	−32	~(33)	[77]*
PDMS + 4 wt% SWCNT	Solvent	~183	~629	~(−7)	[78]ʸ
TPU + 1 wt% (MWCNT+HNT)	Solvent	~143	~35	~(68)	[68]ˣ
NR + 25 phr CNT	Solvent	~244	~594	~(−27)	[36]ˣ
SEBS + 1.5 wt% SWCNT	Solvent	~24	~154	~(−8)	[39]ʸ
NR + 1 phr MWCNT	Solvent and melt	~26	~26	~(−31)	[60]ˣ
NR + 1 wt% MWCNT	Latex	~58	~33	~(3)	[61]ˣ
SBR + 25 phr MWCNT	Latex and melt	~80	~163	~(−18)	[104]*

*untreated CNT, ˣcovalently treated CNT, ʸnon-covalently treated CNT.

processing techniques has also been adopted by many researchers [60,83,112] to achieve better dispersion of CNTs in the elastomeric polymers. Although this approach seems to give complementary advantage offered by each technique, a further study would be needed for compromising laborious procedure, high cost, and longer processing time involved.

Table 3. Percolation threshold of CNT/elastomeric composites and their hybrids compounds.

Polymer	Filler	Processing	Percolation threshold	Ref
SBR	MWCNT	Melt	$(5)^b$	[104]*
SBR/BR	MWCNT	Melt	$(<2)^a$	[69]ˣ
SBR/BR	MWCNT	Melt	$(<3)^a$	[82]ʸ
SBR/NBR	MWCNT	Melt	$(\sim0.2$ and $1)^a$	[86]*
NR/NBR	MWCNT	Melt	$(2)^b$	[83]ʸ
CIIR	MWCNT	Melt	$(6)^b$	[44]ˣ
IIR	MWCNT	Melt	$(6 \sim 8)^b$	[38]ˣ
CR	MWCNT	Melt	$(5)^b$	[76]ʸ
NBR	MWCNT	Solid (two-roll)	$(10)^b$	[63]ˣ
CR	MWCNT	Solid (two-roll)	$(3)^b$	[34]ʸ
NR	MWCNT	Solid (two-roll)	$(9 \sim 16)^a$	[105]*
SR	MWCNT	Solvent	$(<0.1)^a$	[106]*
SR	MWCNT	Solvent	$(\sim1.0)^c$	[107]*
SR	SWCNT	Solvent	$(4)^a$	[78]ʸ
PSF	MWCNT	Solvent	$(0.1 \sim 0.3)^a$	[20]*
NR	MWCNT	Solvent	$(0.5–1)^b$	[25]*
TPU	MWCNT	Solvent	$(0.5 \sim 10)^c$	[85]*
EPDM	MWCNT	Solvent	$(4)^b$	[87]*
MVQ	MWCNT	Solvent	$(1.6)^c$	[81]ʸ
MVQ	MWCNT/TRG	Solvent	$(2$ and $5)^a$	[108]*
SBR	SWCNT	Solvent	$(0.27)^a$	[109]*
SBR	MWCNT	Solvent	$(2$ and $3)^b$	[110]*
PDMS	MWCNT	Manual	$(\sim0.6)^a$	[111]*
SR	MWNCT	High-shear	$(5$ and $10)^a$	[64]ˣ

^{a}wt%, ^{b}phr, ^{c}(vol%), *untreated CNT, ˣcovalently treated CNT, ʸnon-covalently treated CNT.

In summary, although the melt mixing is widely accepted in the mass production of CNT-filled elastomeric polymer nanocomposites, there are still many arguments for the effectiveness of the melt blending of CNTs into elastomeric polymers, due to incomplete dispersion of CNTs agglomerates at milder mixing conditions and some damages of CNTs at severer shearing conditions. The solution mixing approach usually provides more uniform dispersion of CNTs, but it is however preferred on a laboratory scale. Therefore, CNT-elastomeric composites research should be further focused on finding an improved or efficient way of dispersing CNTs into polymer matrix.

4. Morphology and state of dispersion of CNT

A precise characterization skill of observing the dispersed CNTs in polymer is inevitably crucial for the better understanding of structure–property relations of polymer composites. Most of the properties of composites are strongly related to the morphology and dispersion status of CNTs [1,82]. Thus, the analytical technique which can monitor any type of properties of the polymer composites would be a tool for characterizing the morphology. Several typical techniques marked for these studies include transmission electron microscopy (TEM), scanning electron microscopy (SEM), atomic force microscopy (AFM), wide-angle x-ray diffractometry (WAXD), thermogravimetric analysis (TGA), and differential scanning calorimetry (DSC) [27,31,32,60,83,85]. The combination of the techniques is often employed to fully understand the dispersion and morphology of CNTs.

The direct observation of CNTs in the polymer matrix is one of the most powerful but difficult techniques. Several images of CNT dispersion and morphology in elastomeric polymers using TEM, SEM, and AFM analyses are shown in Figure 8 for a case study. Tsuchiya et al [109] compared the CNT dispersion in SBR between a new mixer, called 'rotation-revolution mixer', and a conventional Banbury mixer based on TEM observation. There is a clear difference in both the shape and dispersion of CNT, as can be seen in Figure 8(a) and (b). A more uniform and homogeneous dispersion was observed in their newly proposed mixer system. A similar difference in morphology and dispersion of CNTs in NR matrix is also noticeable by the presence of acid treatment as shown in Figure 8(c) and (d) [60]. The acid treatment leads to the decoration of the surface of CNTs

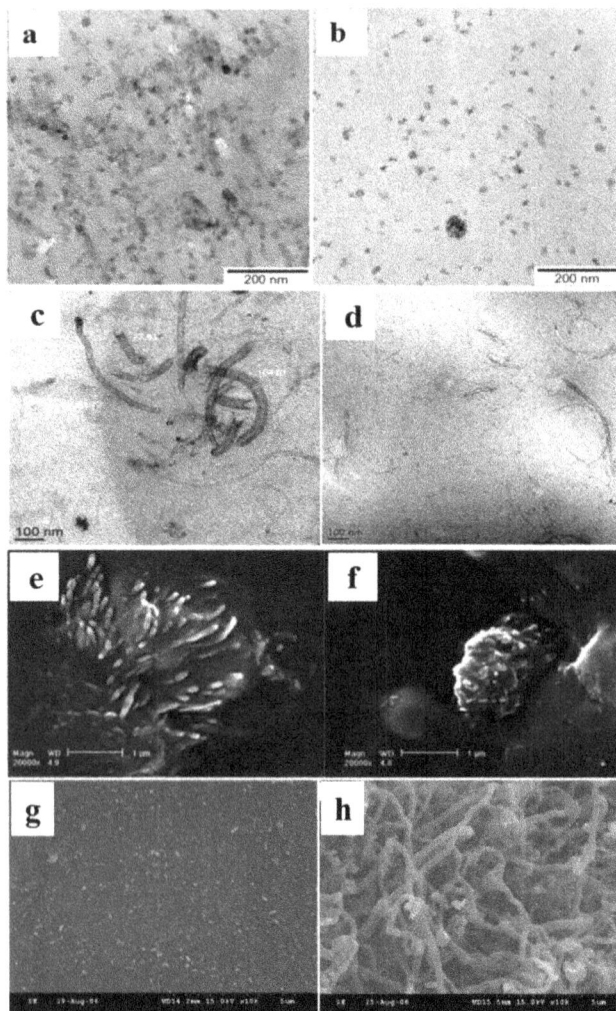

Figure 8. TEM images of SBR-MWCNT composite (0.75 phr) prepared by (a) a rotation–revolution mixer, (b) Banbury mixer [109]; TEM images of NR composite filled with 1% of (c) unmodified CNT, (d) acid-treated CNT [60]; SEM micrographs of NR composites filled with (e) acid-treated MWCNTs, (f) unmodified MWCNTs [61]; SEM images of CNT-filled rubber composite containing 7 phr CNT: (g) without stretching; (h) at 15% stretching [113].

with functional groups such as the -COOH. This in turn improves the interaction of CNTs with elastomeric polymer matrix.

The effect of acid treatment of CNTs is also confirmed using the SEM observation for the cryogenically fractured surface of NR-MWCNT composites by Peng et al [61]. In this experiment, they employed the covalently treated MWCNTs by a latex blending method. The unmodified MWCNTs still remain as agglomerates (Figure 8(f)), while the acid-modified MWCNTs show better dispersion (Figure 8(e)). It should be noted that the SEM observation generally does not give a clearer image of CNTs on the fractured surfaces of elastomeric polymers probably due to the nanosized nature of the CNTs or the mismatch in elongation between polymers and CNTs during rupture. The broken ends of CNTs tend to be embedded deeply into the bulk of polymers. In our previous study, we proposed a new approach for a clearer observation technique [113,114]. In this case, a slight compression was imposed on the polymer sheet for making the CNTs push out to the fractured surfaces during SEM observation, as shown in Figure 8(g) and (h). When the CNTs on the fractured surface of the rubber matrix were observed later, it was noticed that they have reverted into the matrix. Hence, a relatively weak interaction between the untreated CNTs and elastomeric polymers was confirmed. A similar approach was also used to observe clearly the presence of graphene oxide sheet in NBR matrix by SEM after compression of composites [115].

Another observation method is the AFM technique. But the resolution of the AFM technique has been known to be very sensitive to the hardness and smoothness of the surface of specimens. Most of the elastomeric polymers form a smoother surface. Generally, it is very difficult to obtain a successful AFM morphology (especially in topology mode) of elastomeric polymer composites, as shown in Figure 9(a) and (b). Thus, the phase mode of the AFM technique is more useful for such analysis of soft polymer-CNT composites, because the phase mode can provide the variation of surface stiffness [82]. In effect, there is great difference in stiffness between soft polymers and CNTs. Another new approach (called electronic force microscopy, EFM) was proposed in our earlier work [116]. A low current of few volts was applied onto the surface of the specimen by the AFM tip. When the tip is placed on the CNTs, higher currents are tunneling, while on the polymers, lower current is passing. The image based on the current map was quite successful to represent the dispersion of CNTs in an elastomeric polymer system, as shown in Figure 9(c) and (d).

A unique way to quantify the structures and dispersion of CNTs in an elastomer matrix is the technique based on the Raman spectroscopy. It is a widely used technique for studying structure and electronic properties of CNTs and graphitic materials [1], and the interfacial interactions between rubber matrix and MWCNTs [76,82]. Figure 10 illustrates the Raman spectra in the frequency range from 1200 to 1700 cm^{-1} for MWCNT, MWCNT in a rubber matrix, and MWCNT in a rubber matrix in the presence of ionic liquid [82]. MWCNTs show two characteristic peaks, one at 1326 cm^{-1} assigned to the D band and the other to G band at 1593 cm^{-1}. The D band is originated from disordered graphite structure and the G band is associated with tangential C–C bond stretching of the graphite-like structures. No change of the D and G band positions was observed upon incorporation of 3 phr of MWCNTs into the rubber. On other hand, a noticeable shift of G band toward the higher frequencies is been observed with ionic liquid. It is consequently established that the disentanglement of the MWCNTs and their subsequent dispersion and chemical interaction with the rubber by the ionic coupling agent might be the explanation behind the shift [82,117].

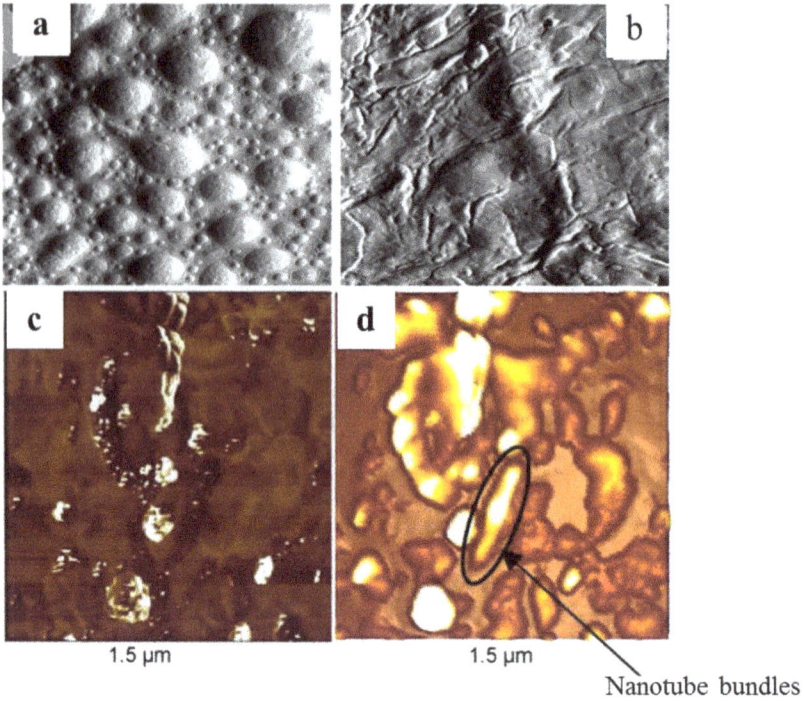

Figure 9.   AFM and EFM images of CNT dispersion in elastomeric matrix: (a) AFM images of composite with 2.8 wt% MWCNT before coagulation, (b) after coagulation of the latex beads at 60°C [24], (c) EFM phase image at 0 V, and (d) EFM phase image at 7 V of NBR-MWCNT composite [116].

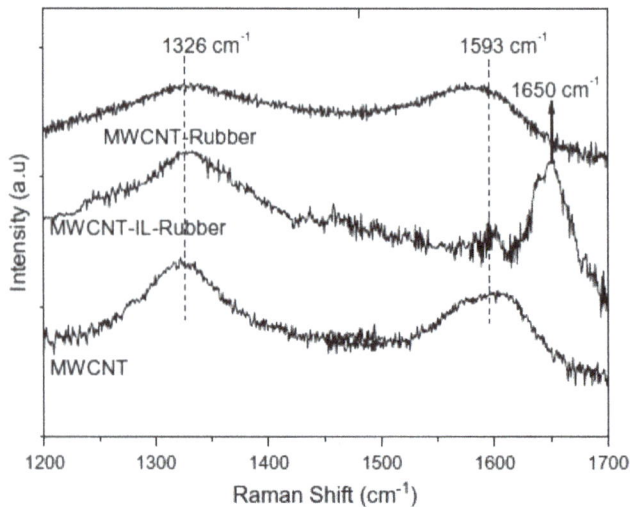

Figure 10.   Raman spectrum of MWCNT, MWCNT-rubber, and MWCNT-IL-rubber composites. IL represents an ionic liquid of 1-allyl-3methylimidazolium chloride, and the MWCNT loading was 3 phr [82].

Based on these observations, it is evident that the efficacy of CNT dispersion in an elastomer depends strongly on several factors as already mentioned. It should be emphasized here that no single technique provides sufficient analytical information related to the structural characteristics of rubber nanocomposites. Thus, it is recommended to use more than one technique to understand the structures and properties as well as the performance of the polymer nanocomposites [118].

5. Mechanical property

The physical properties of the elastomeric material strongly depend on the various parameters of the employed filler system, such as volume fraction, size and shape of particles, interactions among fillers, and filler–matrix interactions [24,119]. The interaction between the filler particles and the rubber matrix (often called as bound rubber) is generally considered to be the most important aspect [119]. The reinforcing behavior by fillers is explained by various theoretical models, including Einstein (Equation 2), Guth–Gold (Equation 3), Thomas (Equation 4), and Halpin–Tsai (Equations 5, 6) [64,119,120–122]. More extended models are available in reference [85].

$$\frac{E_f}{E_u} = 1 + 2.5\phi \tag{2}$$

$$\frac{E_f}{E_u} = 1 + 2.5\phi + 14.1\phi^2 \tag{3}$$

$$\frac{E_f}{E_u} = 1 + 2.4\phi + 10.2\phi^2 + Ae^{B\phi} \tag{4}$$

$$\frac{E_f}{E_u} = \frac{(1 + 2f\phi\eta)}{(1 - \phi\eta)} \tag{5}$$

$$\eta = \frac{(E_f/E_u - 1)}{(E_f/E_u + 2f)} \tag{6}$$

where E_f and E_u are the moduli at a given elongation of filled and unfilled polymer, ϕ is the volume fraction of fillers, f is the aspect ratio of the fillers, and A, B are constant parameters obtained experimentally.

It has been suggested that the fiber shape of nanofillers such as CNTs are the most efficient in enhancing the stiffness [123]. Previously, we reported a significant improvement in both the modulus and the strength, when the unmodified MWCNTs were added to the NR matrix (Figure 11(a)) [42].

The 100% modulus of the MWCNT-NR nanocomposites was found to be about 150% higher than that of CB-NR composites at the same loading level, even though the bound rubber content denoting filler–rubber interactions was even smaller for the MWCNT nanocomposites. Based on the SEM observation of the poor interaction between MWCNT and NR matrix, the main cause of the reinforcing mechanism of the

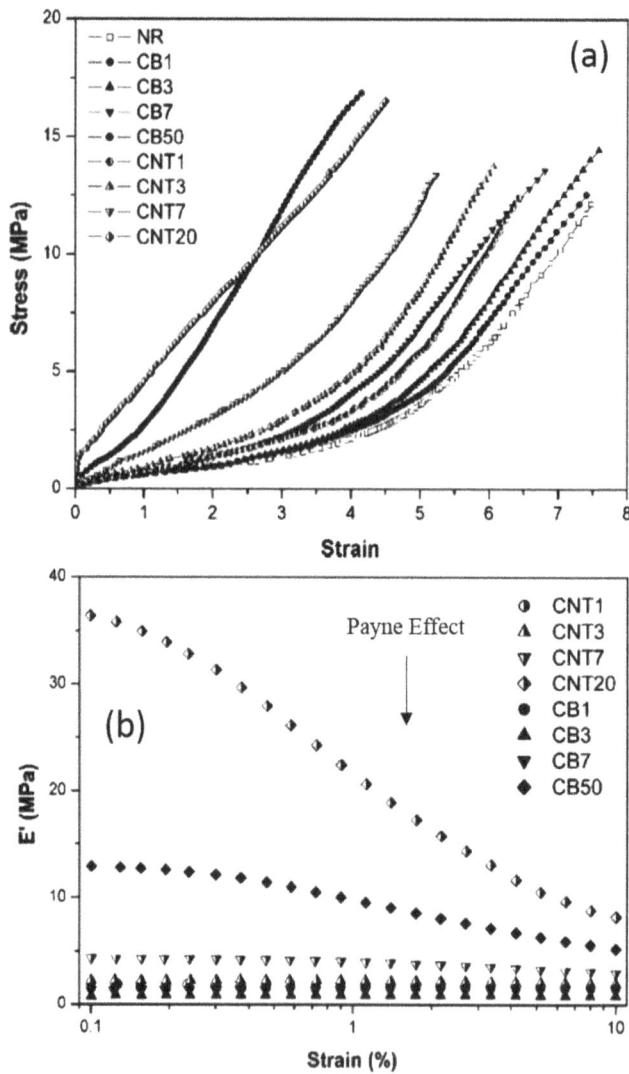

Figure 11. (a) Stress–strain curves of NR, NR/CNT, and NR/CB compounds and (b) storage modulus of NR/CNT and NR/CB compounds at different filler loadings as a function of dynamic strain amplitude [114].

MWCNT-NR nanocomposites was considered to be the effective load transfer due to the high aspect ratio of MWCNTs. The poor interactions between CNTs and rubber matrix are also contained in several literature [25,93,113,121,124]. In addition, a higher reduction in the storage modulus with increasing dynamic strain amplitude was observed for the MWCNT-NR nanocomposites than the CB-NR compounds at the same loading level (Figure 11(b)). This strongly indicates that the Payne effect is more noticeable in the nanocomposites, possibly due to the higher breakdown of the filler–filler networks among MWCNTs. A similar result was also reported later by Bokobza [25]. Another point which should be mentioned here for the mechanical reinforcement is the state of dispersion of CNTs in the rubber matrix. The surface modification of CNTs is widely accepted for the better dispersion [60,63,65,83]. For instance, a 10-times improvement in the modulus without severe deterioration of elongation at break was reported by employing a carboxylated MWCNT [24], and 58% improvement in tensile strength was reported by employing a small amount (1 wt%) of acid-modified MWCNTs in NR [61].

Recently, the hybridization of components of composites is receiving a greater attention for any synergistic effect among them. They include a dual filler system [46,101,103,108], dual matrix system [63,69,125], and ternary matrix system [126,127]. A synergistic effect of the fillers between MWCNT and graphene (G) was found in polydimethylsiloxane (PDMS) [103]. They reported a considerable improvement in tensile strength of 110% and tensile modulus of 137% by using an extremely small amount of filler loading of 0.75 wt%. Kueseng et al [83] reported another way of enhancing mechanical property based on dual rubber system. They initially prepared the master batches by mixing CNTs in NR matrix, and then the master batches were blended with NR–NBR blends of various ratios. A significant enhancement in mechanical properties was found for the master batch system than the conventional mixing approach. This was ascribed to the improved CNT dispersion, which mostly occurred at lower filler concentration between 0 and 4 phr CNT. This study clearly shows the importance of CNT dispersion in rubber matrix.

In brief, the incorporation of CNTs in elastomeric matrices characteristically increases their mechanical properties due to the enhanced load transfer capability of the fibril shape of CNTs with high aspect ratio. The proper surface modification of CNTs by either covalently or non-covalently can further improve the mechanical performance through strong chemical bonding of CNTs with the rubber molecules. A synergistic effect can also be possible by employing dual fillers and rubber system.

6. Electrical property

Since the electrical conductivity of CNTs is known to be extremely high ($\sim$18,000 Scm^{-1}) [85], a significant increase of the conductivity of elastomers can be expected when they are properly dispersed. A better dispersion of CNTs generally results in larger number of conducting paths in the rubber matrix. The critical volume fraction leading to a sharp drop in the electrical resistance is known as the percolation threshold (Figure 12).

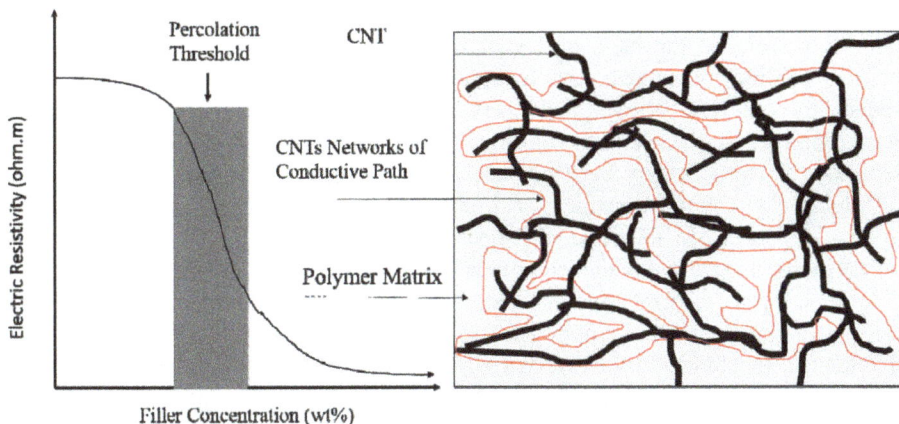

Figure 12.   An illustration of percolation threshold by network formation of CNTs in a rubber matrix.

The percolation threshold depends on the CNT type (SWCNT, MWCNT), basic features of CNTs such as amorphous carbon content, ratio of metallic to semiconductive tubes, aspect ratio, morphology, type of rubber matrix, and dispersion state and particle orientation [20,24]. Although, there are considerable disagreements in the percolation threshold values among researchers [13,15,128,129,130], it is generally agreed that the percolation threshold of CNTs can be achieved at much smaller loading (≥ 0.1 phr), when compared with conventional CB-filled elastomeric compounds [28,131].

The estimation of the electrical conductivity is generally made by measuring the electrical resistance, and it is represented in the volume resistivity (σ) given by:

$$\sigma = \frac{L}{AR} \tag{7}$$

where L and A are the length and cross-sectional area of the specimen, respectively, and R is the electrical resistance. The prediction of percolation behavior is given by the scaling law [132,133] as follows:

$$\sigma = \sigma_0(\phi - \phi_c)^t \tag{8}$$

where σ is the electrical conductivity, σ_0 is a fitting parameter, ϕ is the volume fraction of CNTs and ϕ_c is the critical percolation concentration, and t is an exponent that governs the scaling law in the vicinity of percolation. The parameters t and ϕ_c are usually determined by plotting log σ vs. log ($\phi-\phi_c$) [129]. However, the use of Equation (8) is only limited to the solid matrix composite system like polymer-CNTs composites. A dynamic colloid theory is needed for the fluid-CNTs system like CNTs colloids [134,135]. Quite a different percolation threshold was reported based on the experimental observation depending on the type of CNTs and rubber matrix. Katihabwa et al [64] reported the percolation threshold between 5 and 10 wt% of an unmodified CNT-filled PDMS compounds. Koerner et al [85] found the percolation threshold to be 0.5 ~ 10 vol% for MWCNT-filled thermoplastic polyurethane (TPU) compounds, and above the percolation threshold the conductivity reached was 1 ~ 10 Scm^{-1}. The lower percolation threshold seems to be due to the high aspect ratio of fibrous shape of CNTs.

Modification of CNTs can provide further enhancement in dispersion, and the better dispersion affects the electrical conductivity of the composite systems. Yadav et al [38] found the percolation threshold of 2 wt% by employing the non-covalent modification SWCNTs in SEBS via solvent casting method. Subramaniam et al [76] observed a good dispersion of ionic liquid-treated MWCNTs in chloroprene rubber (CR) matrix via a two-roll mill mixing. They achieved an electrical conductivity of 0.1 Scm^{-1} at the percolation threshold at 5 phr. However, a controversial result of electrical conductivity was reported by Abdullateef et al [60], when nitric acid-treated CNTs were dispersed in NR matrix with excellent dispersion. In this case, the electrical conductivity of CNTs themselves was reduced by breaking the unsaturated double bonds on the CNT's surface.

Another approach to improve the electrical conductivity is the hybridization of polymers and fillers. Yesil [127] reported the reinforcing behavior of CNTs on the recycled poly(ethylene terephthalate) (PET)–poly(ethylene naphthatalate) (PEN) blends using functional elastomers, notably ethylene-methacrylate copolymer (EMA)–glycidyl methacrylate (GMA). The miscibility between PET and PEN was greatly improved due to the compatabilizer role of the EMA–GMA, resulting in a better dispersion of CNTs in polymer matrix. This finally led to the improvement in the conductivity. It has been also reported

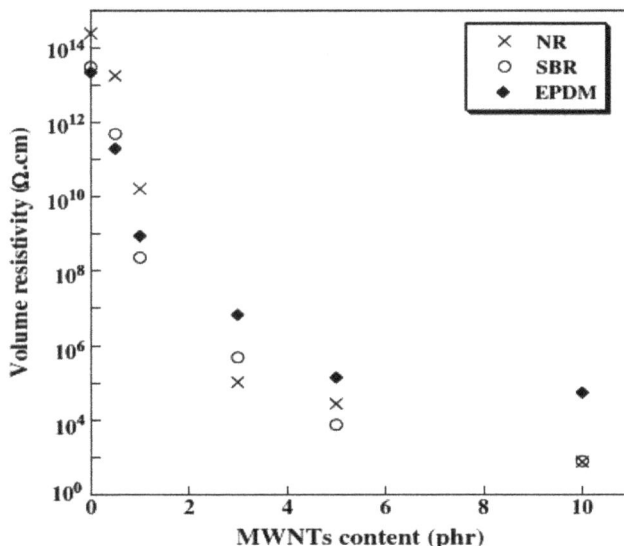

Figure 13. Dependence of CNT loading on the volume resistivity of NR, SBR, and EPDM composites [87].

that the use of dual fillers system can enhance the conductivity [108]. When the CNTs were used together with G, the dispersion and the conductivity was improved compared with that of CNT-reinforced compounds. The synergistic effect of the hybrid materials is also available for various matrix systems [63,100,103].

It should be mentioned here that the processing techniques affect the electrical conductivity as well. Bokobza [87] studied the percolation threshold of CNTs in various types of elastomers, including NR, SBR, and EPDM (Figure 13). The percolation threshold was generally found to be around 0.5 phr (0.002 in volume fraction) regardless of rubber matrix. The rotation–revolution mixer was reported to give better dispersion leading to a better conductivity than that of conventional Banbury type mixer [109]. They reported a percolation threshold less than ~1 phr for SBR-CNT composites. Table 3 summarized several factors affecting the electrical percolation threshold of representative CNT-elastomeric-based nanocomposites.

7. Thermal property

CNTs are also known to be thermally stable and have higher thermal conductivity of ~6600 $Wm^{-1}K^{-1}$ for SWCNTs [136] and 3000 $Wm^{-1}K^{-1}$ for MWCNTs [137]. When the thermally stable CNTs are dispersed in less stable rubber matrices, the thermal stability of the polymer composites can be improved depending on the loading amount of CNTs. Another reason for the thermal stability improvement of the composites is that the dispersed CNTs in rubber matrices help to dissipate the heat more quickly in the composites because of their higher thermal conductivity [138]. The heat conduction mechanism of CNTs in a rubber matrix was proposed by He et al [32] based on the molecular dynamics simulation. Meanwhile, it should be emphasized that the stability of a composite may also depend more on the chemical nature (backbone or side groups of the polymer) of the matrix material employed.

Pham et al [139] studied the thermal decomposition property of FKM reinforced with unmodified MWCNTs. The decomposition temperature gradually increased with increased MWCNTs. They pointed out two reasons for the improvement. Firstly, the presence of MWCNTs made the vulnerable sites of fluorocarbon backbone inactive by physical and chemical interactions between MWCNTs and FKM. Secondly, the MWCNTs played an antioxidant role. Perez et al [63] also investigated the effect of content and modification of MWCNTs on the thermal degradation stability of two rubber matrices, including NBR and SBR using TGA under nitrogen condition.

As summarized in Table 4 and Figure 14, the SBR-MWCNT composites showed a single drop in the weight loss curve, while the NBR-MWCNT composites showed two at the lower loading level of 0 ~ 5 phr, probably due to the characteristic feature of NBR degradation. The first loss peak of NBR-MWCNTs disappeared when the MWCNTs loading was further increased above 10 phr, where the percolation threshold was reached. They explained that the disappearance of the first loss from gaseous subproducts peak was attributed to the barrier role of percolation networks of MWCNTs.

It was reported that the modification of CNT was also effective for improving the thermal stability of NR matrix [60]. They argued that the thermal degradation improvement of ~10°C in the thermal degradation temperature by adding the acid-treated CNTs was attained, while no noticeable improvement for unmodified CNTs occurred. Shang et al [140] also studied the significance of CNT treatment on the thermal behavior of PDMS, as depicted in the TGA and their derivative curves in Figure 14.

The chitosan hydrochloric acid salt (CSCl)-treated MWCNTs improved thermal degradation stability of the SR composite with a one-step weight loss process. When non-pretreated MWCNTs were added into SR, the composite showed a three-step weight loss process and the thermal stability did not improve as well. They claimed this phenomenon indicated that the untreated MWCNTs damaged the internal structure of PDMS to some extent. More research results on the modification effect of CNTs on the thermal stability are reported [41,70,72].

Although several approaches were reported by researchers regarding the thermal stability boost of CNT-polymer-based composites, yet it seems that good dispersion is a common prerequisite of the formation of CNT networks, leading to a slowdown of thermal degradation. The proper selection of the rubber matrix would also enhance the thermal stability of the composite. For instance, FKM is known to show a superior thermal stability compared with many other rubbers due to the presence of fluorine in

Table 4.   TGA results for the SBR and NBR reinforced with multiwall carbon nanotubes under N_2 atmosphere [63].

| CNT (phr) | SBR | | NBR | | | |
| | First weight loss | | First weight loss | | Second weight loss | |
	Maximum (°C)	Residue (wt%)	Maximum (°C)	Residue (wt%)	Maximum (°C)	Residue (wt%)
0	447.4	5.0	423.9	64.9	445.7	9.2
2	450.0	6.4	424.1	64.1	446.5	12.0
5	451.3	8.8	425.8	61.7	425.8	11.8
10	454.8	12.9			443.5	17.7
15	455.1	16.3			452.7	20.6

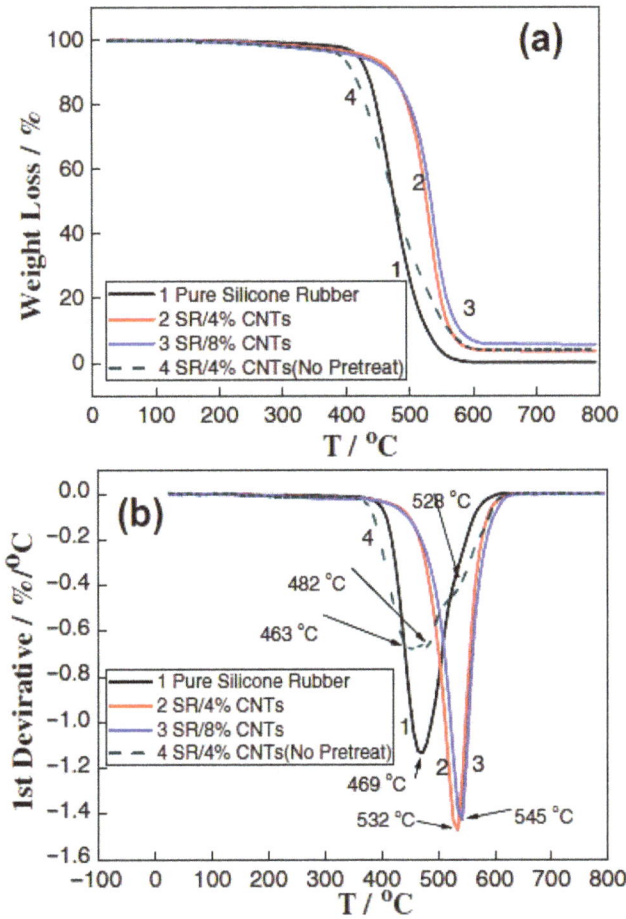

Figure 14.   TGA thermograms (a) and corresponding DTG curve (b) of pure silicone rubber (SR) and SR/MWCNTs nanocomposites [140].

their backbone. Thus, the highly polar C–F bonds and helical structure of the main chain impart high temperature stability. It was also reported earlier that NBR containing a higher amount of acrylonitrile content tends to show higher thermal stability in compounds with CNTs [91,141].

8.   Applications

The unique properties of CNTs makes them superior compared to conventional fillers such as carbon blacks and silicas which require higher filler loading to boost the physico-mechanical properties of polymers [3,9,142]. Thus, the multipurpose nature of rubber-CNT-based nanocomposites makes them potential candidates for widespread applications in various fields. The characteristic features of flexibility, stretchability, conductivity and robustness of rubber-CNT nanocomposites brands them highly applicable in pressure, and strain sensors [47,71,90], super-capacitors, and charge-storage capacitors[38,41,77], thermally stable and electrical cable materials [60,92], oil/fuel hoses and gaskets [105], and microwave, structural applications in the aerospace industry [143], etc. It is therefore obvious that rubber-CNT could enjoy unlimited applications if researchers across the globe further engage deeply into the studies of functionalization, processing techniques, structure–property–dynamics relation of elastomers, and the physics of interactions of CNT–elastomer, as well as the interactions that occur among their hybrid systems.

9. Concluding remarks

The CNT-reinforced elastomeric nanocomposites have received a gradually increased attention. Various functional properties, including mechanical, electrical, and thermal property of rubbery materials, have been altered even at a small amount of CNTs, when the proper dispersion of CNTs is achieved with the help of surface treatment. This review article has addressed the recent research issues focusing on the surface treatment of CNTs, processing, state of dispersion, and several properties of the CNT-reinforced elastomeric polymer nanocomposites.

The accepted method of CNT modification can be classified into covalent and non-covalent type. The covalent modification method, notably by acid treatment, was widely adopted for CNT-rubber nanocomposites. In the non-covalent modification, the ionic surfactants were widely selected. The proper choice of modification method seems to be crucial for the application purpose of the resulting nanocomposites.

The mixing of CNTs into rubber matrices was done by solution mixing, melt mixing, and in-situ polymerization. Even though the melt mixing was widely used in a large-scale production, the solution mixing was selected for more complete dispersion with lower damages of CNTs. There should be further challenging ways for finding more efficient mixing techniques in view of either quality of dispersion or surface damages. The characterization skills of CNTs and the state of dispersion in the rubber matrices have been successfully developed recently. They include TEM, SEM, AFM, WAXD, TGA, DSC, and others. Raman spectroscopy is newly proposed as the characterization technique for CNTs.

The incorporation of CNTs in elastomeric matrices characteristically enhances the mechanical properties due to their higher load transfer capability of the fibril shape with high aspect ratio, although there is a poor interaction between rubber and CNTs. A further improvement is expected with a proper surface modification of CNTs. The percolation threshold of the electrical conductivity of CNTs is comparatively lower than those of CBs due to their easier susceptibility for making CNT network owing to their fibril structure. Care should be given to the selection of the modification type of CNTs, since it greatly affects the electrical conductivity of CNTs themselves. There are two types of mechanisms for the improvement of the stability of CNT-reinforced rubber composites. One is from their originally higher thermal stability (when properly dispersed). The thermally stable CNTs can replace the less stable rubber matrix throughout the volume. Another is from the improved thermal conductivity which results in fast dissipation of accumulated heat inside the bulk polymer matrix.

Disclosure statement

No potential conflict of interest was reported by the authors.

Funding

This work was supported by both the BK 21 Plus and the research fund [grant number 20135010300700] from Korea Institute of Energy Technology Evaluation and Planning (KETEP).

References

[1] V. Singh, D. Joung, L. Zhai, S. Das, S.I. Khondaker, and S. Seal, *Graphene based materials: Past, present and future*, Prog. Mater. Sci. 56 (2011), pp. 1178–1271.

[2] H. Kim, A.A. Abdala, and C.W. Macosko, *Graphene/polymer nanocomposites*, Macromolecules. 43 (2010), pp. 6515–6530.

[3] M. Moniruzzaman and K.I. Winey, *Polymer nanocomposites containing carbon nanotubes*, Macromolecules. 39 (2006), pp. 5194–5205.

[4] E.T. Thostenson, Z.F. Ren, and T.W. Chou, *Advances in the science and technology of carbon nanotubes and their composites: a review*, Compos. Sci. Technol. 61 (2001), pp. 1899–1912.

[5] P.R. Bandaru, *Electrical properties and applications of carbon nanotube structures*, J. Nanosci. Nanotechnol. 7 (2007), pp. 1239–1267.

[6] J.W.G. Wildoer, L.C. Venema, A.G. Rinzler, R.E. Smalley, and C. Dekker, *Electronic structure of atomically resolved carbon nanotubes*, Nature. 391 (1998), pp. 59–62.

[7] J. Bernholc, D. Brenner, M.B. Nardelli, V. Meunier, and C. Roland, *Mechanical and electrical properties of nanotubes*, Annu. Rev. Mater. Res. 32 (2002), pp. 347–375.

[8] M.F. Yu, *Fundamental mechanical properties of carbon nanotubes: Current understanding and the related experimental studies*, J. Eng. Mater. T. Asme. 126 (2004), pp. 271–278.

[9] S. Iijima, *Helical microtubes of graphitic carbon*, Nature. 354 (1991), pp. 56–58.

[10] J.H. Du, J. Bai, and H.M. Cheng, *The present status and key problems of carbon nanotube based polymer composites*, Express Polym Lett. 1 (2007), pp. 253–273.

[11] W.A. de Heer, *Nanotubes and the pursuit of applications*, Mrs Bull. 29 (2004), pp. 281–285.

[12] T. Uchida and S. Kumar, *Single wall carbon nanotube dispersion and exfoliation in polymers*, J. Appl. Polym. Sci. 98 (2005), pp. 985–989.

[13] J. Sandler, M.S.P. Shaffer, T. Prasse, W. Bauhofer, K. Schulte, and A.H. Windle, *Development of a dispersion process for carbon nanotubes in an epoxy matrix and the resulting electrical properties*, Polymer. 40 (1999), pp. 5967–5971.

[14] F.H. Gojny, J. Nastalczyk, Z. Roslaniec, and K. Schulte, *Surface modified multi-walled carbon nanotubes in CNT/epoxy-composites*, Chem. Phys. Lett. 370 (2003), pp. 820–824.

[15] M.S.P. Shaffer and A.H. Windle, *Fabrication and characterization of carbon nanotube/poly (vinyl alcohol) composites*, Adv. Mater. 11 (1999), pp. 937–941.

[16] F. Li, H.M. Cheng, S. Bai, G. Su, and M.S. Dresselhaus, *Tensile strength of single-walled carbon nanotubes directly measured from their macroscopic ropes*, Appl. Phys. Lett.77 (2000), pp. 3161–3163.

[17] A.B. Dalton, S. Collins, E. Munoz, J.M. Razal, V.H. Ebron, J.P. Ferraris, J.N. Coleman, B.G. Kim, and R.H. Baughman, *Super-tough carbon-nanotube fibres - These extraordinary composite fibres can be woven into electronic textiles*, Nature. 423 (2003), pp. 703–703.

[18] K. Lozano, S.Y. Yang, and R.E. Jones, *Nanofiber toughened polyethylene composites*, Carbon. 42 (2004), pp. 2329–2331.

[19] J.K.W. Sandler, S. Pegel, M. Cadek, F. Gojny, M. Van Es, J. Lohmar, W.J. Blau, K. Schulte, A.H. Windle, and M.S.P. Shaffer, *A comparative study of melt spun polyamide-12 fibres reinforced with carbon nanotubes and nanofibres*, Polymer. 45 (2004), pp. 2001–2015.

[20] J.O. Aguilar, *Influence of carbon nanotube clustering on the electrical conductivity of polymer composite films*, Express Polym. Lett. 4 (2010), pp. 292–299.

[21] M. Ahmadi and A. Shojaei, *Cure kinetic and network structure of NR/SBR composites reinforced by multiwalled carbon nanotube and carbon blacks*, Thermochim. Acta. 566 (2013), pp. 238–248.

[22] U. Basuli, T.K. Chaki, and S. Chattopadhyay, *Thermomechanical and rheological behaviour of polymer nanocomposites based on ethylene–methyl acrylate (EMA) and multiwalled carbon nanotube (MWNT)*, Plast. Rubber Compos. 40 (2011), pp. 213–222.

[23] U. Basuli, T.K. Chaki, D.K. Setua, and S. Chattopadhyay, *A comprehensive assessment on degradation of multi-walled carbon nanotube-reinforced EMA nanocomposites*, J. Therm. Anal. Calorim. 108 (2011), pp. 1223–1234.

[24] S. Bhattacharyya, C. Sinturel, O. Bahloul, M.-L. Saboungi, S. Thomas, and J.-P. Salvetat, *Improving reinforcement of natural rubber by networking of activated carbon nanotubes*, Carbon. 46 (2008), pp. 1037–1045.

[25] L. Bokobza, *Multiwall carbon nanotube-filled natural rubber: Electrical and mechanical properties*, Express Polym Lett. 6 (2012), pp. 213–223.

[26] L. Bokobza, *A Raman investigation of carbon nanotubes embedded in a soft polymeric matrix*, J. Inorg. Organomet. Polym. Mater. 22 (2011), pp. 629–635.

[27] G. Broza, *Thermoplastic elastomers with multi-walled carbon nanotubes: Influence of dispersion methods on morphology*, Compos. Sci. Technol. 70 (2010), pp. 1006–1010.

[28] Z.-M. Dang, K. Shehzad, J.-W. Zha, A. Mujahid, T. Hussain, J. Nie, and C.-Y. Shi, *Complementary percolation characteristics of carbon fillers based electrically percolative thermoplastic elastomer composites*, Compos. Sci. Technol. 72 (2011), pp. 28–35.

[29] A. Fakhru'l-Razi, M.A. Atieh, N. Girun, T.G. Chuah, M. El-Sadig, and D.R.A. Biak, *Effect of multi-wall carbon nanotubes on the mechanical properties of natural rubber*, Compos. Struct. 75 (2006), pp. 496–500.

[30] J. Fritzsche, H. Lorenz, and M. Klüppel, *CNT based elastomer-hybrid-nanocomposites with promising mechanical and electrical properties*, Macromol. Mater. Eng. 294 (2009), pp. 551–560.

[31] M. Hemmati, A. Narimani, H. Shariatpanahi, A. Fereidoon, and M.G. Ahangari, *Study on morphology, rheology and mechanical properties of thermoplastic elastomer polyolefin (TPO)/carbon nanotube nanocomposites with reference to the effect of polypropylene-grafted-maleic anhydride (PP-g-MA) as a compatibilizer*, Int. J. Polym. Mater. 60 (2011), pp. 384–397.

[32] Y. He and Y. Tang, *Thermal conductivity of carbon nanotube/natural rubber composite from molecular dynamics simulations*, J. Theor. Comput. Chem. 12 (2013), pp. 1350011.

[33] A. Shanmugharaj, J. Bae, K. Lee, W. Noh, S. Lee, and S. Ryu, *Physical and chemical characteristics of multiwalled carbon nanotubes functionalized with aminosilane and its influence on the properties of natural rubber composites*, Compos. Sci. Technol. 67 (2007), pp. 1813–1822.

[34] D. Steinhauser, *Influence of ionic liquids on the dielectric relaxation behavior of CNT based elastomer nanocomposites*, Express Polym. Lett. 6 (2012), pp. 927–936.

[35] G. Sui, W.H. Zhong, X.P. Yang, and Y.H. Yu, *Curing kinetics and mechanical behavior of natural rubber reinforced with pretreated carbon nanotubes*, Mater. Sci. Eng: A. 485 (2008), pp. 524–531.

[36] G. Sui, W.H. Zhong, X.P. Yang, Y.H. Yu, and S.H. Zhao, *Preparation and properties of natural rubber composites reinforced with pretreated carbon nanotubes*, Polym. Advan. Technol. 19 (2008), pp. 1543–1549.

[37] M.A.A. Tarawneh, S.H. Ahmad, R. Rasid, S.Y. Yahya, S.A.R. Bahri, S. Ehnoum, K.Z. Ka, and L.Y. Seng, *Mechanical properties of thermoplastic natural rubber (TPNR) reinforced with different types of carbon nanotube*, Sains. Malays. 40 (2011), pp. 725–728.

[38] T.T.N. Dang, S.P. Mahapatra, V. Sridhar, J.K. Kim, K.J. Kim, and H. Kwak, *Dielectric properties of nanotube reinforced butyl elastomer composites*, J. Appl. Polym. Sci. 113 (2009), pp. 1690–1700.

[39] S.K. Yadav, S.S. Mahapatra, and J.W. Cho, *Tailored dielectric and mechanical properties of noncovalently functionalized carbon nanotube/poly(styrene-b-(ethylene- co-butylene)-b-styrene) nanocomposites*, J. Appl. Polym. Sci. 129 (2013), pp. 2305–2312.

[40] G. Scherillo, M. Lavorgna, G.G. Buonocore, Y.H. Zhan, H.S. Xia, G. Mensitieri, and L. Ambrosio, *Tailoring assembly of reduced graphene oxide nanosheets to control gas barrier properties of natural rubber nanocomposites*, Acs Appl. Mater. Inter. 6 (2014), pp. 2230–2234.

[41] K. Subramaniam, A. Das, D. Steinhauser, M. Klüppel, and G. Heinrich, *Effect of ionic liquid on dielectric, mechanical and dynamic mechanical properties of multi-walled carbon nanotubes/polychloroprene rubber composites*, Eur. Polym. J. 47 (2011), pp. 2234–2243.

[42] Y. Pan, L. Li, S.H. Chan, and J. Zhao, *Correlation between dispersion state and electrical conductivity of MWCNTs/PP composites prepared by melt blending*, Compos. Part A- Appl. S. 41 (2010), pp. 419–426.

[43] J. Huang and D. Rodrigue, *Equivalent continuum models of carbon nanotube reinforced polypropylene composites*, Mater. Design. 50 (2013), pp. 936–945.

[44] S.K. Tiwari, R.N.P. Choudhary, and S.P. Mahapatra, *Relaxation behavior of chlorobutyl e elastomer nanocomposites: Effect of temperature, multiwalled carbon nanotube and frequency*, J. Polym Res. 20 (2013), pp. 176.

[45] H. Koerner, G. Price, N.A. Pearce, M. Alexander, and R.A. Vaia, *Remotely actuated polymer nanocomposites - stress-recovery of carbon-nanotube-filled thermoplastic elastomers*, Nat. Mater. 3 (2004), pp. 115–120.

[46] A. Boonmahitthisud and S. Chuayjuljit, *NR/XSBR nanocomposites with carbon black and carbon nanotube prepared by latex compounding*, J. Metals Mater. Miner. 22 (2012), pp. 77–85.

[47] X. Wenjun and M.G. Allen, *Fabrication of patterned carbon nanotube (CNT)/elastomer bilayer material and its utilization as force sensors.* in Solid-State Sensors, Actuator and Microsystems Conf. 2009, Transducers, IEE, Denver, Colorado, USA, 2009, pp. 2242–2245.

[48] Z. Spitalsky, D. Tasis, K. Papagelis, and C. Galiotis, *Carbon nanotube–polymer composites: Chemistry, processing, mechanical and electrical properties*, Prog. Polym. Sci. 35 (2010), pp. 357–401.

[49] T.K. Das and S. Prusty, *Graphene-based polymer composites and their applications*, Polym. Plast. Technol. Eng. 52 (2013), pp. 319–331.

[50] S. Bal and S.S. Samal, *Carbon nanotube reinforced polymer composites - A state of the art*, Bull. Mater. Sci. 30 (2007), pp. 379–386.

[51] R. Andrews and M.C. Weisenberger, *Carbon nanotube polymer composites*, Curr. Opin. Solid St. M. 8 (2004), pp. 31–37.

[52] N. Roy, R. Sengupta, and A.K. Bhowmick, *Modifications of carbon for polymer composites and nanocomposites*, Prog. Polym. Sci. 37 (2012), pp. 781–819.

[53] C.-Y. Hu, Y.-J. Xu, S.-W. Duo, R.-F. Zhang, and M.-S. Li, *Non-covalent functionalization of carbon nanotubes with surfactants and polymers*, J. Chinese Chem. Soc. 56 (2009), pp. 234–239.

[54] L.A. Girifalco, M. Hodak, and R.S. Lee, *Carbon nanotubes, buckyballs, ropes, and a universal graphitic potential*, Phys. Rev. B. 62 (2000), pp. 13104–13110.

[55] S. Bandow, A.M. Rao, K.A. Williams, A. Thess, R.E. Smalley, and P.C. Eklund, *Purification of single-wall carbon nanotubes by microfiltration*, J. Phys. Chem. B. 101 (1997), pp. 8839–8842.

[56] G.S. Duesberg, M. Burghard, J. Muster, and G. Philipp, *Separation of carbon nanotubes by size exclusion chromatography*, Chem. Commun. 3 (1998), pp. 435–436.

[57] V. Krstic, G.S. Duesberg, J. Muster, M. Burghard, and S. Roth, *Langmuir-Blodgett films of matrix-diluted single-walled carbon nanotubes*, Chem. Mater. 10 (1998), pp. 2338–2340.

[58] M.I.H. Panhuis, C. Salvador-Morales, E. Franklin, G. Chambers, A. Fonseca, J.B. Nagy, W.J. Blau, and A.I. Minett, *Characterization of an interaction between functionalized carbon nanotubes and an enzyme*, J. Nanosci. Nanotechnol. 3 (2003), pp. 209–213.

[59] M.F. Islam, E. Rojas, D.M. Bergey, A.T. Johnson, and A.G. Yodh, *High weight fraction surfactant solubilization of single-wall carbon nanotubes in water*, Nano Lett. 3 (2003), pp. 269–273.

[60] A.A. Abdullateef, S.P. Thomas, M.A. Al-Harthi, S.K. De, S. Bandyopadhyay, A.A. Basfar, and M.A. Atieh, *Natural rubber nanocomposites with functionalized carbon nanotubes: Mechanical, dynamic mechanical, and morphology studies*, J. Appl. Polym. Sci. 125 (2012), pp. E76–E84.

[61] Z. Peng, C. Feng, Y. Luo, Y. Li, and L.X. Kong, *Self-assembled natural rubber/multi-walled carbon nanotube composites using latex compounding techniques*, Carbon. 48 (2010), pp. 4497–4503.

[62] G. Sui, W. Zhong, X. Yang, and S. Zhao, *Processing and material characteristics of a carbon-nanotube-reinforced natural rubber*, Macromol. Mater. Eng. 292 (2007), pp. 1020–1026.

[63] L.D. Perez, M.A. Zuluaga, T. Kyu, J.E. Mark, B.L. Lopez, and P. Tandon, *Preparation, characterization, and physical properties of multiwall carbon nanotube/elastomer composites*, Polym. Eng. Sci. 49 (2009), pp. 866–874.

[64] A. Katihabwa, W. Wencai, J. Yi, Z. Xiuying, L. Yonglai, and Z. Liqun, *Multi-walled carbon nanotubes/silicone rubber nanocomposites prepared by high shear mechanical mixing*, J. Reinf. Plast. Compos. 30 (2011), pp. 1007–1014.

[65] S.B. Jagtap, *Preparation and characterization of rubbery epoxy/multiwall carbon nanotubes composites using amino acid salt assisted dispersion technique*, Express Polym. Lett. 7 (2013), pp. 329–339.

[66] F. Barroso-Bujans, R. Verdejo, M. Perez-Cabero, S. Agouram, I. Rodriguez-Ramos, A. Guerrero-Ruiz, and M.A. Lopez-Manchado, *Effects of functionalized carbon nanotubes in peroxide crosslinking of diene elastomers*, Eur. Polym. J. 45 (2009), pp. 1017–1023.

[67] Y.S. Park, M. Huh, S.J. Kang, S.I. Yun, and K.H. Ahn, *Effect of CNT diameter on physical properties of styrene-butadiene rubber nanocomposites*, Carbon Lett. 10 (2009), pp. 320–324.

[68] L. Jiang, C. Zhang, M. Liu, Z. Yang, W.W. Tjiu, and T. Liu, *Simultaneous reinforcement and toughening of polyurethane composites with carbon nanotube/halloysite nanotube hybrids*, Compos. Sci. Technol. 91 (2014), pp. 98–103.

[69] A. Das, K.W. Stöckelhuber, R. Jurk, M. Saphiannikova, J. Fritzsche, H. Lorenz, M. Klüppel, and G. Heinrich, *Modified and unmodified multiwalled carbon nanotubes in high performance solution-styrene–butadiene and butadiene rubber blends*, Polymer. 49 (2008), pp. 5276–5283.

[70] Y.H. Zhan, R. Patel, M. Lavorgna, F. Piscitelli, A. Khan, H.S. Xia, H. Benkreira, and P. Coates, *Processing of polyurethane/carbon nanotubes composites using novel minimixer*, Plast. Rubber Compos. 39 (2010), pp. 400–410.

[71] M.-J. Jiang, Z.-M. Dang, and H.-P. Xu, *Giant dielectric constant and resistance-pressure sensitivity in carbon nanotubes/rubber nanocomposites with low percolation threshold*, Appl. Phys. Lett. 90 (2007), pp. 042914.

[72] J. Zhang, Y. Wang, X. Wang, G. Ding, Y. Pan, H. Xie, Q. Chen, and R. Cheng, *Effects of amino-functionalized carbon nanotubes on the properties of amine-terminated butadiene-acrylonitrile rubber-toughened epoxy resins*, J. Appl. Polym. Sci. 131 (2014), pp. 40472.

[73] R. Zhang, H. Deng, R. Valenca, J. Jin, Q. Fu, E. Bilotti, and T. Peijs, *Strain sensing behaviour of elastomeric composite films containing carbon nanotubes under cyclic loading*, Compos. Sci. Technol. 74 (2013), pp. 1–5.

[74] H.H. Le, X.T. Hoang, A. Das, U. Gohs, K.W. Stoeckelhuber, R. Boldt, G. Heinrich, R. Adhikari, and H.J. Radusch, *Kinetics of filler wetting and dispersion in carbon nanotube/ rubber composites*, Carbon. 50 (2012), pp. 4543–4556.

[75] F.F. Semeriyanov, A.I. Chervanyov, R. Jurk, K. Subramaniam, S. König, M. Roscher, A. Das, K.W. Stöckelhuber, and G. Heinrich, *Non-monotonic dependence of the conductivity of carbon nanotube-filled elastomers subjected to uniaxial compression/decompression*, J. Appl. Phys. 113 (2013), pp. 103706.

[76] K. Subramaniam, A. Das, and G. Heinrich, *Development of conducting polychloroprene rubber using imidazolium based ionic liquid modified multi-walled carbon nanotubes*, Compos. Sci. Technol. 71 (2011), pp. 1441–1449.

[77] K. Oh, J.Y. Lee, -S.-S. Lee, M. Park, D. Kim, and H. Kim, *Highly stretchable dielectric nanocomposites based on single-walled carbon nanotube/ionic liquid gels*, Compos. Sci. Technol. 83 (2013), pp. 40–46.

[78] T.A. Kim, H.S. Kim, S.S. Lee, and M. Park, *Single-walled carbon nanotube/silicone rubber composites for compliant electrodes*, Carbon. 50 (2012), pp. 444–449.

[79] S. Cantournet, M.C. Boyce, and A.H. Tsou, *Micromechanics and macromechanics of carbon nanotube-enhanced elastomers*, J. Mech. Phys. Solids. 55 (2007), pp. 1321–1339.

[80] D. Ponnamma, S.H. Sung, J.S. Hong, K.H. Ahn, K.T. Varughese, and S. Thomas, *Influence of non-covalent functionalization of carbon nanotubes on the rheological behavior of natural rubber latex nanocomposites*, Eur. Polym. J. 53 (2014), pp. 147–159.

[81] M.-J. Jiang, Z.-M. Dang, S.-H. Yao, and J. Bai, *Effects of surface modification of carbon nanotubes on the microstructure and electrical properties of carbon nanotubes/rubber nanocomposites*, Chem. Phys. Lett. 457 (2008), pp. 352–356.

[82] A. Das, K.W. Stöckelhuber, R. Jurk, J. Fritzsche, M. Klüppel, and G. Heinrich, *Coupling activity of ionic liquids between diene elastomers and multi-walled carbon nanotubes*, Carbon. 47 (2009), pp. 3313–3321.

[83] P. Kueseng, P. Sae-oui, and N. Rattanasom, *Mechanical and electrical properties of natural rubber and nitrile rubber blends filled with multi-wall carbon nanotube: Effect of preparation methods*, Polym. Test. 32 (2013), pp. 731–738.

[84] S. Badaire, P. Poulin, M. Maugey, and C. Zakri, *In situ measurements of nanotube dimensions in suspensions by depolarized dynamic light scattering*, Langmuir. 20 (2004), pp. 10367–10370.

[85] H. Koerner, W. Liu, M. Alexander, P. Mirau, H. Dowty, and R.A. Vaia, *Deformation–morphology correlations in electrically conductive carbon nanotube—Thermoplastic polyurethane nanocomposites*, Polymer. 46 (2005), pp. 4405–4420.

[86] C. Kummerlöwe, N. Vennemann, E. Yankova, M. Wanitschek, C. Größ, T. Heider, F. Haberkorn, and A. Siebert, *Preparation and properties of carbon nanotube composites with nitrile- and styrene-butadiene rubbers*, Polym. Eng. Sci. 53 (2013), pp. 849–856.

[87] L. Bokobza, *Enhanced electrical and mechanical properties of multiwall carbon nanotube rubber composites*, Polym. Advan. Technol. 23 (2012), pp. 1543–1549.

[88] M.K. Shin, J. Oh, M. Lima, M.E. Kozlov, S.J. Kim, and R.H. Baughman, *Elastomeric conductive composites based on carbon nanotube forests*, Adv. Mater. 22 (2010), pp. 2663–2667.

[89] Z.F. Liu, S. Fang, F.A. Moura, J.N. Ding, N. Jiang, J. Di, M. Zhang, X. Lepró, D.S. Galvão, C.S. Haines, N.Y. Yuan, S.G. Yin, D.W. Lee, R. Wang, H.Y. Wang, W. Lv, C. Dong, R.C. Zhang, M.J. Chen, Q. Yin, Y.T. Chong, R. Zhang, X. Wang, M.D. Lima, R. Ovalle-Robles, D. Qian, H. Lu, and R.H. Baughman, *Hierarchically buckled sheath-core fibers for superelastic electronics, sensors, and muscles*, Science. 349 (2015), pp. 400–404.

[90] T. Yamada, Y. Hayamizu, Y. Yamamoto, Y. Yomogida, A. Izadi-Najafabadi, D.N. Futaba, and K. Hata, *A stretchable carbon nanotube strain sensor for human-motion detection*, Nat. Nano. 6 (2011), pp. 296–301.

[91] P. Verge, S. Peeterbroeck, L. Bonnaud, and P. Dubois, *Investigation on the dispersion of carbon nanotubes in nitrile butadiene rubber: Role of polymer-to-filler grafting reaction*, Compos. Sci. Technol. 70 (2010), pp. 1453–1459.

[92] Z. Wu, H. Wang, X. Tian, X. Ding, M. Xue, H. Zhou, and K. Zheng, *Mechanical and flame-retardant properties of styrene–ethylene–butylene–styrene/carbon nanotube composites containing bisphenol A bis(diphenyl phosphate)*, Compos. Sci. Technol. 82 (2013), pp. 8–14.

[93] J.N. Coleman, U. Khan, W.J. Blau, and Y.K. Gun'ko, *Small but strong: A review of the mechanical properties of carbon nanotube-polymer composites*, Carbon. 44 (2006), pp. 1624–1652.

[94] L.S. Schadler, S.C. Giannaris, and P.M. Ajayan, *Load transfer in carbon nanotube epoxy composites*, Appl. Phys. Lett. 73 (1998), pp. 3842–3844.

[95] J. Zhu, J.D. Kim, H.Q. Peng, J.L. Margrave, V.N. Khabashesku, and E.V. Barrera, *Improving the dispersion and integration of single-walled carbon nanotubes in epoxy composites through functionalization*, Nano Lett. 3 (2003), pp. 1107–1113.

[96] X.Y. Gong, J. Liu, S. Baskaran, R.D. Voise, and J.S. Young, *Surfactant-assisted processing of carbon nanotube/polymer composites*, Chem. Mat. 12 (2000), pp. 1049–1052.

[97] Z. Wang, X. Yang, Q. Wang, H.T. Hahn, S.-G. Lee, K.-H. Lee, and Z. Guo, *Epoxy resin nanocomposites reinforced with ionized liquid stabilized carbon nanotubes*, Int. J. Smart Nano Mater. 2 (2011), pp. 176–193.

[98] A. Szymczyk, *Poly(trimethylene terephthalate-block-tetramethylene oxide) elastomer /single-walled carbon nanotubes nanocomposites: Synthesis, structure, and properties*, Appl. Polym. Sci. 126 (2012), pp. 796–807.

[99] Z. Roslaniec, G. Broza, and K. Schulte, *Nanocomposites based on multiblock polyester elastomers (PEE) and carbon nanotubes (CNT)*, Compos. Interf. 10 (2003), pp. 95–102.

[100] H. Lorenz, J. Fritzsche, A. Das, K.W. Stöckelhuber, R. Jurk, G. Heinrich, and M. Klüppel, *Advanced elastomer nano-composites based on CNT-hybrid filler systems*, Compos. Sci. Technol. 69 (2009), pp. 2135–2143.

[101] H. Ismail, A.F. Ramly, and N. Othman, *Effects of silica/multiwall carbon nanotube hybrid fillers on the properties of natural rubber nanocomposites*, J. Appl. Polym. Sci. 128 (2013), pp. 2433–2438.

[102] A. Das, G.R. Kasaliwal, R. Jurk, R. Boldt, D. Fischer, K.W. Stöckelhuber, and G. Heinrich, *Rubber composites based on graphene nanoplatelets, expanded graphite, carbon nanotubes and their combination: A comparative study*, Compos. Sci. Technol. 72 (2012), pp. 1961–1967.

[103] B. Pradhan and S.K. Srivastava, *Synergistic effect of three-dimensional multi-walled carbon nanotube-graphene nanofiller in enhancing the mechanical and thermal properties of high-performance silicone rubber*, Polym. Int. 63 (2014), pp. 1219–1228.

[104] S. Schopp, R. Thomann, K.-F. Ratzsch, S. Kerling, V. Altstädt, and R. Mülhaupt, *Functionalized graphene and carbon materials as components of styrene-butadiene rubber nanocomposites prepared by aqueous dispersion blending*, Macromol. Mater. Eng. 299 (2014), pp. 319–329.

[105] F. Deng, M. Ito, T. Noguchi, L. Wang, H. Ueki, K.-I. Niihara, Y.A. Kim, M. Endo, and Q.-S. Zheng, *Elucidation of the reinforcing mechanism in carbon nanotube/rubber nanocomposites*, ACS Nano. 5 (2011), pp. 3858–3866.

[106] C.H. Liu and S.S. Fan, *Nonlinear electrical conducting behavior of carbon nanotube networks in silicone elastomer*, Appl. Phys. Lett. 90 (2007), pp. 041905.

[107] M. Norkhairunnisa, A. Azizan, M. Mariatti, H. Ismail, and L. Sim, *Thermal stability and electrical behavior of polydimethylsiloxane nanocomposites with carbon nanotubes and carbon black fillers*, J. Compos. Mater. 46 (2011), pp. 903–910.

[108] H. Hu, L. Zhao, J. Liu, Y. Liu, J. Cheng, J. Luo, Y. Liang, Y. Tao, X. Wang, and J. Zhao, *Enhanced dispersion of carbon nanotube in silicone rubber assisted by graphene*, Polymer. 53 (2012), pp. 3378–3385.

[109] K. Tsuchiya, A. Sakai, T. Nagaoka, K. Uchida, T. Furukawa, and H. Yajima, *High electrical performance of carbon nanotubes/rubber composites with low percolation threshold prepared with a rotation–revolution mixing technique*, Compos. Sci. Technol. 71 (2011), pp. 1098–1104.

[110] L. Bokobza, *Mechanical, electrical and spectroscopic investigations of carbon nanotube-reinforced elastomers*, Vib. Spectrosc. 51 (2009), pp. 52–59.

[111] J.-B. Lee and D.-Y. Khang, *Electrical and mechanical characterization of stretchable multi-walled carbon nanotubes/polydimethylsiloxane elastomeric composite conductors*, Compos. Sci. Technol. 72 (2012), pp. 1257–1263.

[112] B. Pradhan and S.K. Srivastava, *Layered double hydroxide/multiwalled carbon nanotube hybrids as reinforcing filler in silicone rubber*, Compos. Part A: Appl. Sci. Manuf. 56 (2014), pp. 290–299.

[113] C. Nah, J.Y. Lim, R. Sengupta, B.H. Cho, and A.N. Gent, *Slipping of carbon nanotubes in a rubber matrix*, Polym. Int. 60 (2011), pp. 42–44.

[114] C. Nah, J.Y. Lim, B.H. Cho, C.K. Hong, and A.N. Gent, *Reinforcing rubber with carbon nanotubes*, J. Appl. Polym. Sci. 118 (2010), pp. 1574–1581.

[115] B. Mensah, S. Kim, S. Arepalli, and C. Nah, *A study of graphene oxide-reinforced rubber nanocomposite*, Appl. Polym. Sci. 131 (2014), pp. 40640.

[116] M.A. Kader, D. Choi, S.K. Lee, and C. Nah, *Morphology of conducting filler-reinforced nitrile rubber composites by electrostatic force microscopy*, Polym. Test. 24 (2005), pp. 363–366.

[117] Y. Sato, K. Hasegawa, Y. Nodasaka, K. Motomiya, M. Namura, N. Ito, B. Jeyadevan, and K. Tohji, *Reinforcement of rubber using radial single-walled carbon nanotube soot and its shock dampening properties*, Carbon. 46 (2008), pp. 1509–1512.

[118] H.K. Joseph, *Polymer Nanocomposites: Processing, Characterization, and Applications*, McGraw-Hill, USA, 2006, pp. 92 (272).

[119] L. Bokobza, *The reinforcement of elastomeric networks by fillers*, Macromol. Mater. Eng. 289 (2004), pp. 607–621.

[120] P. Mallick, *Fiber-Reinforced Composites*, M. Dekker, New-York, NY, 1993, pp. 130.

[121] L. Bokobza, *Multiwall carbon nanotube elastomeric composites: A review*, Polymer. 48 (2007), pp. 4907–4920.

[122] T.P. Selvin, J. Kuruvilla, and T. Sabu, *Mechanical properties of titanium dioxide-filled polystyrene microcomposites*, Mater. Lett. 58 (2004), pp. 281–289.

[123] F. Ramsteiner and R. Theysohn, *On the tensile behavior of filled composites*, Compos. Part A: Appl. Sci. Manuf. 15 (1984), pp. 121–128.

[124] C.H. Liu, H. Huang, Y. Wu, and S.S. Fan, *Thermal conductivity improvement of silicone elastomer with carbon nanotube loading*, Appl. Phys. Lett. 84 (2004), pp. 4248.

[125] S. Araby, Q. Meng, L. Zhang, H. Kang, P. Majewski, Y. Tang, and J. Ma, *Electrically and thermally conductive elastomer/graphene nanocomposites by solution mixing*, Polymer. 55 (2014), pp. 201–210.

[126] H.H. Le, M.N. Sriharish, S. Henning, J. Klehm, M. Menzel, W. Frank, S. Wiessner, A. Das, K.W. Stoeckelhuber, G. Heinrich, and H.J. Radusch, *Dispersion and distribution of carbon nanotubes in ternary rubber blends*, Compos. Sci. Technol. 90 (2014), pp. 180–186.

[127] S. Yesil, *Effect of carbon nanotube reinforcement on the properties of the recycled poly (ethylene terephthalate)/poly(ethylene naphthalate) (r-PET/PEN) blends containing functional elastomers*, Mater Design. 52 (2013), pp. 693–705.

[128] R. Andrews, D. Jacques, M. Minot, and T. Rantell, *Fabrication of carbon multiwall nanotube/polymer composites by shear mixing*, Macromol. Mater. Eng. 287 (2002), pp. 395–403.

[129] C.A. Martin, J.K.W. Sandler, M.S.P. Shaffer, M.K. Schwarz, W. Bauhofer, K. Schulte, and A. H. Windle, *Formation of percolating networks in multi-wall carbon-nanotube-epoxy composites*, Compos. Sci. Technol. 64 (2004), pp. 2309–2316.

[130] N. Grossiord, P.J.J. Kivit, J. Loos, J. Meuldijk, A.V. Kyrylyuk, P. Van Der Schoot, and C.E. Koning, *On the influence of the processing conditions on the performance of electrically conductive carbon nanotube/polymer nanocomposites*, Polymer. 49 (2008), pp. 2866–2872.

[131] K.P. Sau, D. Khastgir, and T.K. Chaki, *Electrical conductivity of carbon black and carbon fibre filled silicone rubber composites*, Angew. Makromol. Chem. 258 (1998), pp. 11–17.

[132] P.C. Ma, B.Z. Tang, and J.-K. Kim, *Effect of CNT decoration with silver nanoparticles on electrical conductivity of CNT-polymer composites*, Carbon. 46 (2008), pp. 1497–1505.

[133] W. Bauhofer and J.Z. Kovacs, *A review and analysis of electrical percolation in carbon nanotube polymer composites*, Compos. Sci. Technol. 69 (2009), pp. 1486–1498.

[134] J.J. Hernández, M.C. García-Gutiérrez, A. Nogales, D.R. Rueda, M. Kwiatkowska, A. Szymczyk, Z. Roslaniec, A. Concheso, I. Guinea, and T.A. Ezquerra, *Influence of preparation procedure on the conductivity and transparency of SWCNT-polymer nanocomposites*, Compos. Sci. Technol. 69 (2009), pp. 1867–1872.

[135] L. Chang, K. Friedrich, L. Ye, and P. Toro, *Evaluation and visualization of the percolating networks in multi-wall carbon nanotube/epoxy composites*, J. Mater. Sci. 44 (2009), pp. 4003–4012.

[136] S. Berber, Y.K. Kwon, and D. Tomanek, *Unusually high thermal conductivity of carbon nanotubes*, Phys. Rev. Lett. 84 (2000), pp. 4613–4616.

[137] P. Kim, L. Shi, A. Majumdar, and P.L. McEuen, *Thermal transport measurements of individual multiwalled nanotubes*, Phys. Rev. Lett. 87 (2001), pp. 215502.

[138] S.T. Huxtable, D.G. Cahill, S. Shenogin, L.P. Xue, R. Ozisik, P. Barone, M. Usrey, M.S. Strano, G. Siddons, M. Shim, and P. Keblinski, *Interfacial heat flow in carbon nanotube suspensions*, Nat. Mater. 2 (2003), pp. 731–734.

[139] T.T. Pham, V. Sridhar, and J.K. Kim, *Fluoroelastomer-MWNT nanocomposites-1: Dispersion, morphology, physico-mechanical, and thermal properties*, Polym. Compos. 30 (2009), pp. 121–130.

[140] S. Shang, L. Gan, and M.C.-W. Yuen, *Improvement of carbon nanotubes dispersion by chitosan salt and its application in silicone rubber*, Compos. Sci. Technol. 86 (2013), pp. 129–134.

[141] B. Likozar and Z. Major, *Morphology, mechanical, cross-linking, thermal, and tribological properties of nitrile and hydrogenated nitrile rubber/multi-walled carbon nanotubes composites prepared by melt compounding: The effect of acrylonitrile content and hydrogenation*, Appl. Surf. Sci. 257 (2010), pp. 565–573.

[142] G. Mathew, J.M. Rhee, Y.S. Lee, D.H. Park, and C. Nah, *Cure kinetics of ethylene acrylate rubber/clay nanocomposites*, J. Ind. Eng. Chem. 14 (2008), pp. 60–65.

[143] F.X. Qin, C. Brosseau, and H.X. Peng, *Microwave properties of carbon nanotube/microwire/rubber multiscale hybrid composites*, Chem. Phys. Lett. 579 (2013), pp. 40–44.

13

Energy harvesting performance of viscoelastic polyacrylic dielectric elastomers

Junshi Zhang[a,b], Yongquan Wang[a,c]*, Hualing Chen[a,c] and Bo Li[a,c]

[a]State Key Laboratory for Strength and Vibration of Mechanical Structures, Xi'an Jiaotong University, Xi'an 710049, China; [b]School of Aerospace, Xi'an Jiaotong University, Xi'an 710049, China; [c]School of Mechanical Engineering, Xi'an Jiaotong University, Xi'an 710049, China

Viscoelasticity dissipates the mechanical energy, leading to a reduction of energy conversion efficiency in both dielectric elastomer (DE) actuators and generators. By measuring the uniaxial tension-recovery experiments of very-high-bond-based DE, this article quantitatively presents the effect of viscoelasticity on energy harvesting performance of DE generators. By employing a DE strip energy harvester with constant surface charge, an analytical model is established to calculate the generated electrical energy and energy conversion efficiency. Numerical results demonstrate that viscoelasticity has a significant influence on DE energy harvesting performance.

Keywords: dielectric elastomer; viscoelasticity; energy harvesting

1. Introduction

Dielectric elastomers (DEs) exhibit large dimensional changes upon electrical stimulation: expansion in area and shrinkage in thickness [1–5]. Due to their high-strain responsive performance together with other desirable properties like low cost and high energy density, a variety of promising applications of DE have been extensively developed, such as artificial muscles [6], soft robots [7], loudspeaker [8], tunable lenses [9], and energy generators [10–12]. In 2001, Pelrine et al. [13] first presented DE generators, demonstrating that mechanical energy can be transformed into electrical energy by using DE under certain conditions. Since then, different kinds of configurations and concepts in energy harvesting were reported, with the objective to achieve high electrical energy output and energy conversion efficiency. Koh et al. [10] theoretically investigated the DE energy harvesting process and proposed a method to calculate the maximum energy that can be converted during one cycle. Kaltseis et al. [11] reported the generated electrical energy density per cycle can be 0.102 J/g and the corresponding energy conversion efficiency can be 7.5%. By using the equi-biaxial loading, significant improvements in electrical energy density (0.56 J/g) and energy conversion efficiency (27%) were achieved by Huang et al. [12]; however, the energy density is still small compared with theoretical predictions with a value of 1 J/g. The reason is due to the strong viscoelasticity of the DE materials.

Viscoelasticity (especially the very-high-bond (VHB)) greatly affects their performance [14–19]. The viscoelastic resistance dissipates the mechanical energy, leading to a

*Corresponding author. Email: yqwang@mail.xjtu.edu.cn

reduction of the energy conversion efficiency of DE generators [15,20]. Previously, experiments have shown that the stress–stretch curve of VHB strongly depends on the stretch curve [21–23]. When the DE films are used in energy harvesting, it will experience successive tension-recovery cycles with the external source of mechanical energy. Thus, viscoelasticity, intrinsically being represented by hysteresis, will undoubtedly have a significant impact on the amount of the generated energy and the energy conversion efficiency. Up to now, there have been a number of reports which mainly focus on the mechanism and optimization on DE energy harvesting [10–12,15,20,24–26], while, most of the studies [10,11,24–26] only take DE as hyper-elastic continuum materials, without taking into account the viscoelasticity and induced loss of mechanical energy.

2. Experimental details

In this article, we simply adopt the uniaxial tensile configuration to quantitatively study the effect of viscoelasticity on energy harvesting performance of DE. We define that the thickness direction is direction 3, the tension direction is direction 1, and another direction is direction 2. DE material used is VHB4910 film manufactured by 3 M, and the initial length, width and thickness of the VHB membrane are $L_1 = 75$ mm, $L_2 = 20$ mm, and $L_3 = 1$ mm, respectively. The stretch λ is defined as the ratio between the deformed size and the undeformed size. Uniaxial tension-recovery experiments are carried out at different stretch rates (from 0.029 s^{-1} to 0.71 s^{-1}, the stretch rate is defined as $\dot{\lambda} = d\lambda/dt$, where t denotes the time), and the Tytron 250 microforce testing system is employed as the experimental set-up, as shown in Figure 1.

Figure 2 plots parts of the nominal stress–stretch curve of DE at several different stretch rates. As seen in Figure 2, the stretch rate has a great effect on the stress–stretch curve of DE material, which is consistent with the previous reported works [21–23]. For the same stretch, both tension stress and recovery stress increase with the increasing

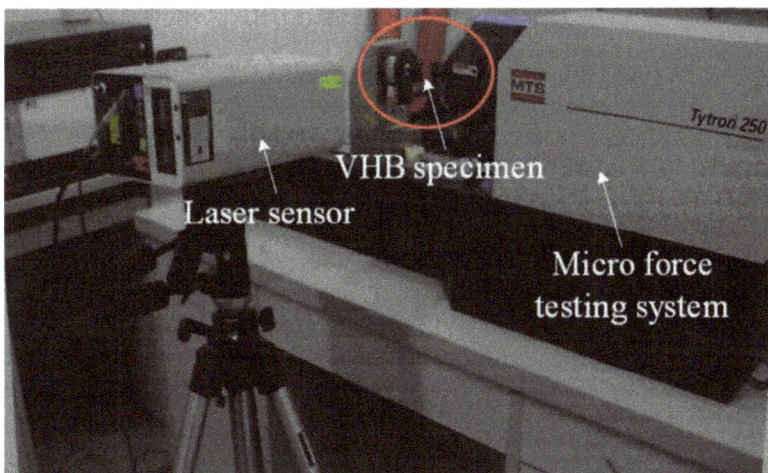

Figure 1.   Experimental set-up for uniaxial test: the Tytron 250 microforce testing system [18].

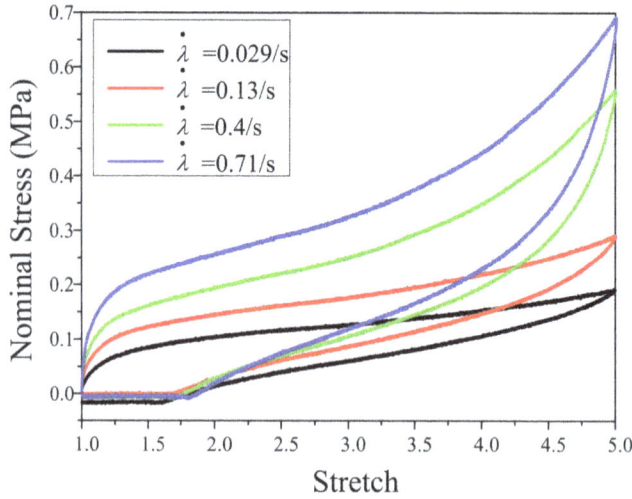

Figure 2.   The nominal stress–stretch curves of DE membrane under several stretch rates including both the tension and recovery stress.

stretch rate. Due to viscoelasticity, the recovery stress appears a noticeable hysteresis compared with the tension stress, bringing out a large internal dissipation.

3. Method

In principle, a basic cycle requires two batteries: the low-voltage one supplies charge to the DE generator and the high-voltage one stores charge. The energy harvesting cycle of DE was reported by Koh et al. [10], including four states during the process. From state 1 to state 2, subject to the external mechanical load, the DE generator extends, reaching the maximum capacitance. From state 2 to state 3, first, the low-voltage battery is applied on the DE film to output charge, then, the low-voltage battery is disconnected to maintain the charge on the DE surface as a constant. From state 3 to state 4, due to the decrease of external load, the film recovers, reducing the value of capacitance and increasing the voltage simultaneously, under the condition of a constant charge. From state 4 to state 1, when the DE generator reaches the condition of loss of tension, the DE generator is connected to the high-voltage battery, allowing the charge to transfer to the high-voltage battery.

The generated electrical energy results from the work done by the Maxwell stress. In order to calculate the energy easily, the Maxwell stress in direction 3 can be equivalent to the directions 1 and 2, based on the theory developed by Suo [4]. Thus, as shown in Figure 3 (a), $\sigma_{M1} = \sigma_{M2} = -\sigma_{M3}$ can be obtained, where σ_{M3} denotes the Maxwell stress in directions 3, σ_{M1} and σ_{M2} represent the equivalent Maxwell stresses in directions 1 and 2.

Under the condition of uniaxial tension with the application of voltage, the relationship of the stretches in directions 1 and 2 can be related as [27]

$$\lambda_2^2 = \frac{\sqrt{1 + \left(Q/\left(A_0\sqrt{\mu\varepsilon_r\varepsilon_0}\right)\right)^2}}{\lambda_1} \tag{1}$$

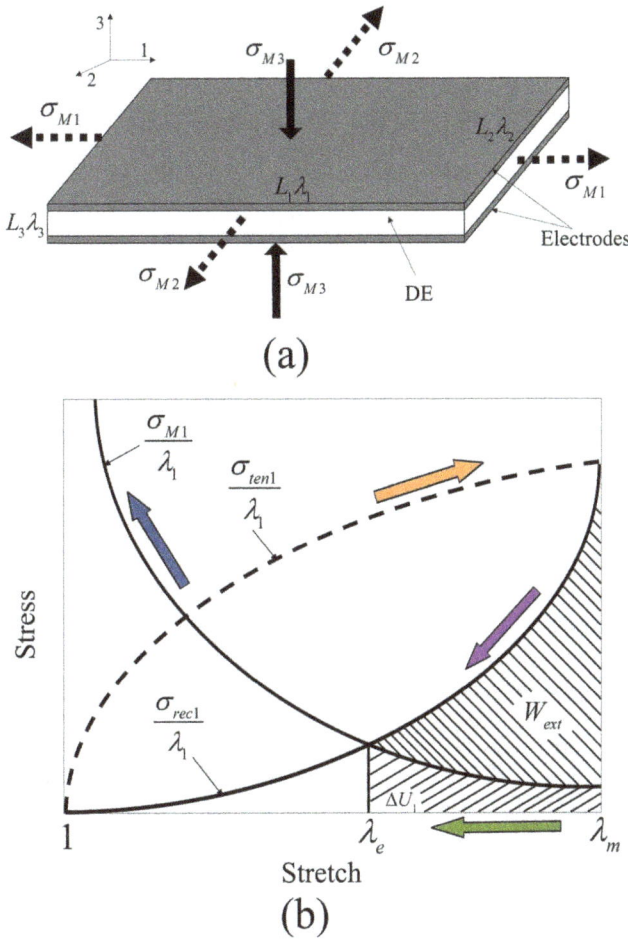

Figure 3. (a) Maxwell stress in direction 3 can be equivalent to directions 1 and 2, that is $\sigma_{M1} = \sigma_{M2} = -\sigma_{M3}$. (b) Schematic of the tension stress σ_{ten1} and recovery stress σ_{rec1} and Maxwell stress σ_{M1} in a tension-recovery process (direction 1).

where λ_1 denotes the stretch in direction 1 and λ_2 denotes the stretch in direction 2; Q is the charge accumulated on the surface of DE; $A_0 = L_1 L_2$ is the initial area of the DE, $\mu = 5.2 \times 10^4$ Pa [28] is the shear modulus; $\varepsilon_0 = 8.85 \times 10^{-12} F/m$ and $\varepsilon_r = 4.2$ [28] denote vacuum permittivity and the relative permittivity of DE, respectively. The Maxwell stress in direction 3 can be expressed as

$$\sigma_{M3} = \frac{Q^2}{\varepsilon_r \varepsilon_0 A_0^2 \sqrt{1 + (Q/(A_0 \sqrt{\mu \varepsilon_r \varepsilon_0}))^2} \lambda_1} \tag{2}$$

As the experiments is processed at a steady speed condition, during the recovery of DE, the relationship of the stresses in direction 1 is

$$\sigma_{rec1} = \sigma_{M1} + \sigma_{ext} \tag{3}$$

where σ_{rec1} denotes the true elastic stress of DE during recovery and σ_{ext} denotes the external stress which is used to keep the speed steady.

Figure 3(b) plots a schematic diagram showing the elastic stress and Maxwell stress in direction 1 during the tension-recovery process. The elastic stress of DE is determined by the stretch and corresponding stretch rate. Thus, the elastic stress in the recovery process is independent of the charge on the DE surface as long as the stretch and stretch rate are given. The total mechanical energy is equal to the work done by the tension stress. The generated electrical energy in direction 1, ΔU_1, is the work done by the Maxwell stress in this direction.

During the recovery process, if the total stress becomes negative (elastic stress is smaller than Maxwell stress), a DE is in a condition of loss of tension and may arise wrinkle instability [10,27]. So the stress equilibrium point, λ_e, is selected as the position to harvest electrical energy [10,27], as shown in Figure 3(b). After the charge is harvested, Maxwell stress will decay to 0 for a sufficient time since the presence of dissipation [16]. The DE membrane can be restored to the initial configuration, and a second cycle of energy harvesting is guaranteed. In other words, the device can achieve recycling of energy harvesting. In recovery process, part of the reduced elastic energy is transformed to electrical energy, and the rest of the elastic energy is used to do work on the external environment, W_{ext}, as shown in Figure 3(b). After obtaining the equilibrium stretch λ_e, the net generated electrical energy can be calculated. Since the equivalent Maxwell stresses exist in both directions 1 and 2, σ_{M1} does negative work to increase electrical energy while σ_{M2} does positive work to decrease electrical energy. The differential equations of the work density for σ_{M1} and σ_{M2} are expressed as

$$dU_1 = \frac{1}{V}\sigma_{M1}\lambda_2 L_{20}\lambda_3 L_{30}d(\lambda_1 L_{10}) \tag{4}$$

$$dU_2 = \frac{1}{V}\sigma_{M2}\lambda_1 L_{10}\lambda_3 L_{30}d(\lambda_2 L_{20}) \tag{5}$$

where $V = L_1 L_2 L_3$, is the volume of the DE membrane. Combined with Equation (1), Equations (4) and (5) can be simplified as

$$dU_1 = \frac{\sigma_{M1}}{\lambda_1}d\lambda_1 \tag{6}$$

$$dU_2 = -\frac{\sqrt[4]{1 + (Q/(A_0\sqrt{\mu\varepsilon_r\varepsilon_0}))^2}}{2}\frac{\sigma_{M2}}{\lambda_1}d\lambda_1 \tag{7}$$

Basically, the net generated electrical energy is defined as the difference between output electrical energy and input electrical energy [12] and should be equal to the sum of the work done by σ_{M1} and σ_{M2}. The density of generated electrical energy can be obtained as

$$\Delta U = \left(1 - \frac{\sqrt[4]{1 + (Q/(A_0\sqrt{\mu\varepsilon_r\varepsilon_0}))^2}}{2}\right)\int_{\lambda_e}^{\lambda_m}\frac{\sigma_{M1}}{\lambda_1}d\lambda_1 \tag{8}$$

During the recovery, the external mechanical load is needed to keep the uniform speed; thus, the elastic stress of DE does mechanical work to the external environment. The work density can be written as

$$W_{ext} = \int_{\lambda_e}^{\lambda_m} \frac{\sigma_{rec1} - \sigma_{M1}}{\lambda_1} d\lambda_1 \qquad (9)$$

The density of total mechanical energy during the tension process can be calculated by the tension stress as

$$W_{total} = \int_{1}^{\lambda_m} \frac{\sigma_{ten1}}{\lambda_1} d\lambda_1 \qquad (10)$$

Using Equation (8), the electrical energy density can be obtained. Furthermore, the energy conversion efficiency during the energy harvesting process can be given as

$$\eta = \frac{\Delta U}{W_{total} - W_{ext}} \qquad (11)$$

According to the developed equations and the uniaxial tensile experiments, the effect of viscoelasticity on the generated electrical energy density and the energy conversion efficiency can be obtained. Based on the developed equations, we can maximize the energy conversion efficiency by tuning the stretch rate and the surface charge, which is shown in the following.

4. Results and discussions

Combining with both the experimental data and above analysis of energy harvesting process, in the following, we discuss the effect of viscoelasticity on the energy harvesting performance of VHB-based DE. Figure 4(a) plots the three-dimensional images of the generated electrical energy density versus the stretch rate and surface charge. When the surface charge is kept constant, the generated electrical energy density increases slightly with the increasing stretch rate. Especially when the surface charge is $Q = 4 \times 10^{-6}$ C, the extent of the enlargement of electrical energy density reaches the maximum. For a constant stretch rate, the electrical energy density increases first and then decreases subsequently with the increasing surface charge. The reason is that large surface charge adds the dissipation of electrical energy in direction 2, resulting in the reduction of the total generated electrical energy, as shown by Equations (7) and (8). Figure 4(b) plots the three-dimensional image of energy conversion efficiency versus the stretch rate and surface charge. When the surface charge is given a specific value, energy conversion efficiency decreases gradually with the increasing stretch rate, although the electrical energy density increases slightly as the stretch rate increases (Figure 4(a)). The reason is large stretch rate increases the input mechanical energy significantly, reducing the energy conversion efficiency. Under a constant stretch rate, when the surface charge increases, the energy conversion efficiency increases first and a subsequent decrease followed, which is consistent with the trend of the generated electrical energy density. The electrical energy density and corresponding energy conversion efficiency in our results is lower than the work reported before [11,12]. The reason is that the uniaxial tension mode is utilized in this article and the equivalent Maxwell stress in direction 2 dissipates

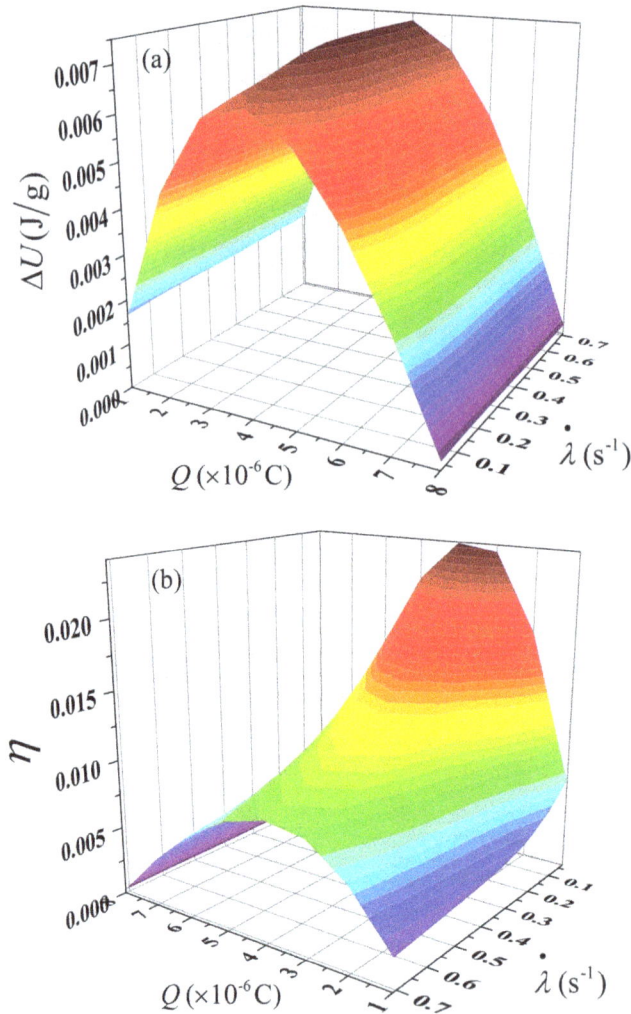

Figure 4.   The three-dimensional images of the generated electrical energy density (a) and the energy conversion efficiency (b) versus the stretch rate and surface charge.

the total electrical energy, causing the reduction of the generated electrical energy density and corresponding energy conversion efficiency.

As seen in Figure 4, under the same stretch rate, the energy conversion efficiency can be tuned by the surface charge. When the stretch rate is kept at a constant, with the increase of surface charge, both the generated electrical energy density and the energy conversion efficiency initially increase and subsequently decrease. Specifically, when the surface charge is $Q = 4 \times 10^{-6}$ C, both the generated electrical energy density and the energy conversion efficiency reach the maximum. That is, by tuning the value of surface charge and stretch rate, a higher electrical energy density and energy conversion efficiency can be achieved, which is equivalent to the suppression of the effect of viscoelasticity.

Furthermore, the maximum surface charge in our simulation is $Q = 8 \times 10^{-6}$ C, as seen in Figure 4. If the surface is above $Q = 8 \times 10^{-6}$ C, there will be no electrical energy generated because the equivalent Maxwell stress in direction 2 dissipates the electrical energy completely. As shown in Equation (8), when the surface charge exceeds a critical level, ΔU will be a negative number. The negative work done by the equivalent Maxwell stress in direction 2 causes the lower generated electrical energy density and energy conversion efficiency. Thus, in order to add the values of the two quantities, the equibiaxial tension mode should be applied in the further research and applications.

5. Conclusions

In summary, combined with the uniaxial tension-recovery experimental results, the impact of viscoelasticity is incorporated into the study of VHB-based DE energy harvesting. By applying a method of calculating the work done by the elastic stress and the Maxwell stress respectively, the generated electrical energy density and corresponding energy conversion efficiency are quantitatively obtained. According to the numerical results, in the proposed range of stretch rate, the maximum electrical energy density of the uniaxial stretched DE energy generator is approximately below 0.01 J/g, and the energy conversion efficiency is no more than 3%, which are far less than both the theoretical values in ideal case and the experimental results reported before. Such a comparison demonstrates that the viscoelasticity and the tension mode play an important role in determination of the energy harvesting performance. This article provides a potential theoretical prediction which may contribute to the analysis, design and application of DE energy generators. Reasonable stretch rate and tension mode can therefore reduce the viscoelastic dissipation and achieve high electrical energy output and energy conversion efficiency.

Disclosure statement
No potential conflict of interest was reported by the authors.

Funding
This work was supported by the National Natural Science Foundation of China [grant number 11321062], [grant number 51375376].

References
[1] R. Pelrine, R. Kornbluh, Q. Pei, and J. Joseph, *High-speed electrically actuated elastomers with strain greater than 100%*, Science 287 (2000), pp. 836–839. doi:10.1126/science.287.5454.836

[2] P. Brochu and Q. Pei, *Advances in dielectric elastomers for actuators and artificial muscles*, Macromol. Rapid Commun. 31 (2010), pp. 10–36. doi:10.1002/marc.v31:1

[3] F. Carpi, S. Bauer, and D. Rossi, *Stretching dielectric elastomer performance*, Science 330 (2010), pp. 1759–1761. doi:10.1126/science.1194773

[4] J. Zhang, H. Chen, L. Tang, B. Li, J. Sheng, and L. Liu, *Modelling of spring roll actuators based on viscoelastic dielectric elastomers*, Appl. Phys. A 119 (2015), pp. 825–835. doi:10.1007/s00339-015-9034-2

[5] J. Zhang and H. Chen, *Electromechanical performance of a viscoelastic dielectric elastomer balloon*, Int. J. Smart Nano Mater. 5 (2014), pp. 76–85. doi:10.1080/19475411.2014.893930

[6] B.M. O'Brien, T.G. McKay, T.A. Gisby, and I.A. Anderson, *Rotating turkeys and self-commutating artificial muscle motors*, Appl. Phys. Lett. 100 (2012), p. 074108. doi:10.1063/1.3685708

[7] Q. Pei, M. Rosenthal, S. Stanford, H. Prahlad, and R. Pelrine, *Multiple-degrees-of-freedom electroelastomer roll actuators*, Smart Mater. Struct. 13 (2004), pp. N86–N92. doi:10.1088/0964-1726/13/5/N03

[8] C. Keplinger, J.-Y. Sun, C.C. Foo, P. Rothemund, G.M. Whitesides, and Z. Suo, *Stretchable, transparent, ionic conductors*, Science 341 (2013), pp. 984–987. doi:10.1126/science.1240228

[9] F. Carpi, G. Frediani, S. Turco, and D. De Rossi, *Bioinspired tunable lens with muscle-like electroactive elastomers*, Adv. Funct. Mater. 21 (2011), pp. 4152–4158. doi:10.1002/adfm.v21.21

[10] S.J.A. Koh, X. Zhao, and Z. Suo, *Maximal energy that can be converted by a dielectric elastomer generator*, Appl. Phys. Lett. 94 (2009), p. 262902. doi:10.1063/1.3167773

[11] R. Kaltseis, C. Keplinger, R. Baumgartner, M. Kaltenbrunner, T. Li, P. Mächler, R. Schwödiauer, Z. Suo, and S. Bauer, *Method for measuring energy generation and efficiency of dielectric elastomer generators*, Appl. Phys. Lett. 99 (2011), p. 162904. doi:10.1063/1.3653239

[12] J. Huang, S. Shian, Z. Suo, and D.R. Clarke, *Maximizing the energy density of dielectric elastomer generators using equi-biaxial loading*, Adv. Funct. Mater. 23 (2013), pp. 5056–5061. doi:10.1002/adfm.v23.40

[13] R. Pelrine, R. Kornbluh, J. Eckerle, P. Jeuck, S. Oh, Q. Pei, and S. Stanford, *Dielectric elastomers: Generator mode fundamentals and applications*, Proc. SPIE. 4329 (2001), pp. 148–156.

[14] J. Zhang, Y. Wang, D. McCoul, Q. Pei, and H. Chen, *Viscoelastic creep elimination in dielectric elastomer actuation by preprogrammed voltage*, Appl. Phys. Lett. 105 (2014), p. 212904. doi:10.1063/1.4903059

[15] C.C. Foo, S.J.A. Koh, C. Keplinger, R. Kaltseis, S. Bauer, and Z. Suo, *Performance of dissipative dielectric elastomer generators*, J. Appl. Phys. 111 (2012), p. 094107. doi:10.1063/1.4714557

[16] J. Zhang, H. Chen, and B. Li, *A method of tuning viscoelastic creep in charge-controlled dielectric elastomer actuation*, EPL 108 (2014), p. 57002. doi:10.1209/0295-5075/108/57002

[17] J. Zhang, H. Chen, B. Li, D. McCoul, and Q. Pei, *Coupled nonlinear oscillation and stability evolution of viscoelastic dielectric elastomers*, Soft Matter (2015). doi:10.1039/c5sm01436k

[18] Y. Wang, H. Xue, H. Chen, and J. Qiang, *A dynamic visco-hyperelastic model of dielectric elastomers and their energy dissipation characteristics*, Appl. Phys. A 112 (2013), pp. 339–347. doi:10.1007/s00339-013-7740-1

[19] J. Zhang, L. Tang, B. Li, Y. Wang, and H. Chen, *Modeling of the dynamic characteristic of viscoelastic dielectric elastomer actuators subject to different conditions of mechanical load*, J. Appl. Phys. 117 (2015), p. 084902. doi:10.1063/1.4913384

[20] T. Li, S. Qu, and W. Yang, *Energy harvesting of dielectric elastomer generators concerning inhomogeneous fields and viscoelastic deformation*, J. Appl. Phys. 112 (2012), p. 034119. doi:10.1063/1.4745049

[21] J.-S. Plante and S. Dubowsky, *Large-scale failure modes of dielectric elastomer actuators*, Int. J. Solids Struct. 43 (2006), pp. 7727–7751. doi:10.1016/j.ijsolstr.2006.03.026

[22] M. Pharr, J.-Y. Sun, and Z. Suo, *Rupture of a highly stretchable acrylic dielectric elastomer*, J. Appl. Phys. 111 (2012), p. 104114. doi:10.1063/1.4721777

[23] J. Zhang, H. Chen, J. Sheng, L. Liu, Y. Wang, and S. Jia, *Constitutive relation of viscoelastic dielectric elastomer*, Theor. Appl. Mech. Lett. 3 (2013), p. 054011. doi:10.1063/2.1305411

[24] T. Mckay, B. O'Brien, E. Calius, and I. Anderson, *Self-priming dielectric elastomer generators*, Smart Mater. Struct. 19 (2010), p. 055025. doi:10.1088/0964-1726/19/5/055025

[25] S.J.A. Koh, C. Keplinger, T. Li, S. Bauer, and Z. Suo, *Dielectric elastomer generators: How much energy can be converted?* IEEE/ASME Trans. Mechatron. 16 (2011), pp. 33–41. doi:10.1109/TMECH.2010.2089635

[26] P. Brochu, H. Stoyanov, X. Niu, and Q. Pei, *Energy conversion efficiency of dielectric elastomer energy harvesters under pure shear strain conditions*, Proc. SPIE. 8340 (2012), p. 83401W.

[27] Y. Wang, J.-X. Zhou, X.-H. Wu, B. Li, and L. Zhang, *Energy diagrams of dielectric elastomer generators under different types of deformation*, Chin. Phys. Lett. 30 (2013), p. 066103. doi:10.1088/0256-307X/30/6/066103

[28] R. Kaltseis, C. Keplinger, S.J.A. Koh, R. Baumgartner, Y.F. Goh, W.H. Ng, A. Kogler, A. Tröls, C.C. Foo, Z. Suo, and S. Bauer, *Natural rubber for sustainable high-power electrical energy generation*, RSC Adv. 4 (2014), pp. 27905–27913. doi:10.1039/C4RA03090G

14

Silicone rubbers for dielectric elastomers with improved dielectric and mechanical properties as a result of substituting silica with titanium dioxide

Liyun Yu and Anne Ladegaard Skov*

The Danish Polymer Centre, Department of Chemical and Biochemical Engineering, Technical University of Denmark, 2800 Kgs. Lyngby, Denmark

One prominent method of modifying the properties of dielectric elastomers (DEs) is by adding suitable metal oxide fillers. However, almost all commercially available silicone elastomers are already heavily filled with silica to reinforce the otherwise rather weak silicone network and the resulting metal oxide filled elastomer may contain too much filler. We therefore explore the replacement of silica with titanium dioxide to ensure a relatively low concentration of filler. Liquid silicone rubber (LSR) has relatively low viscosity, which is favorable for loading inorganic fillers. In the present study, four commercial LSRs with varying loadings of silica and one benchmark room-temperature vulcanizable rubber (RTV) were investigated. The resulting elastomers were evaluated with respect to their dielectric permittivity, tear and tensile strengths, electrical breakdown, thermal stability and dynamic viscosity. Filled silicone elastomers with high loadings of nano-sized titanium dioxide (TiO_2) particles were also studied. The best overall performing formulation had 35 wt.% TiO_2 nanoparticles in the POWERSIL® XLR LSR, where the excellent ensemble of relative dielectric permittivity of 4.9 at 0.1 Hz, breakdown strength of 160 V μm^{-1}, tear strength of 5.3 MPa, elongation at break of 190%, a Young's modulus of 0.85 MPa and a 10% strain response (simple tension) in a 50 V μm^{-1} electric field was obtained.

Keywords: silicone rubber; titanium dioxide; dielectric permittivity; mechanical properties; electrical breakdown

1. Introduction

Polydimethylsiloxanes (PDMSs) and other silicone-based elastomers are widely used in dielectric elastomer (DE) formulation due to their favorable electro-mechanical properties.[1] Dielectric elastomers which consist of an elastomer film with deposited electrodes on both sides have lately gained increased interest as materials for actuators, generators, and sensors. Several promising materials have been developed based on silicone elastomers, and approaches are currently limited to two categories of elastomers, namely 1) composites with semiconductive metal oxides or screened conductive particles and 2) molecular modified elastomers.[2,3,4]. We here explore the most common approach of composites with high permittivity metal oxides as this method by far results in the cheapest elastomers due to the relatively low cost of e.g. TiO_2 and the ease of processing.

*Corresponding author. Email: al@kt.dtu.dk

There are two types of silylation-cure based commercial silicone elastomer formulations which are based on curing temperatures, namely 1) high-temperature vulcanizable (HTV) and 2) room-temperature vulcanisable (RTV) rubbers. HTV rubbers can be further classified as 1) liquid silicone rubbers (LSRs) and 2) millable silicone rubbers, depending on the degree of polymerization.[5]

LSRs are a family of tough and versatile silicones, enabling fast-cured, high-precision injection molding for high-performance parts such as transducer devices.[6] They can be molded into everything from o-rings to intricate geometries with thin walls, and they remain elastic without tearing.[7] Since they are cured by addition reaction, the formulations enable a high degree of control over the network topology – and thus on the mechanical properties – of the final material.[8] LSRs may be clear or translucent and can be manufactured to have unique properties such as solvent resistance, increased thermal stability or low outgassing.[9] Furthermore, they can incorporate additives and unique fillers, such as pigments and active pharmaceutical ingredients, with minimal changes to the key characteristics of the cured elastomer, i.e. toughness and elasticity.[10]

In most cases LSRs use a platinum catalyzed addition cure system and require a curing temperature above 80°C.[11] They typically comprise ~ 75 wt.% linear silicone polymers and ~ 23 wt.% fumed silica, with the remainder consisting of a combination of curing additives such as hydride crosslinkers and inhibitors.[12] The majority of commercially available LSRs are formulated as two-part systems: 1:1 mix ratio silicones with viscosities ranging from 50,000 to >1,000,000 cps.[13] Part A contains a platinum catalyst and Part B contains a crosslinker and an inhibitor. Fumed silica improves the ultimate mechanical properties of the silicone and has the ability to interact non-covalently with the polymer, thereby allowing stress relief when a shear is applied in an uncured or a cured state. Moreover, fumed silica also increases the tear growth resistance strongly. Semi-volatile inhibitors control the reaction rate, in order to provide an acceptable liquid elastomer pot life until heat is applied.[14] The combination of medium-high viscosity polymers and silica gives uncured LSRs a creamy consistency similar to petroleum jelly.[15] Processing LSRs successfully depends on using equipment designed to mix and pump thick materials without introducing air.[16] For thin films this requirement becomes even more important.

The dielectric and mechanical properties of LSR elastomers are dependent on many factors due to the molecular 'architecture' formed by the polymers in the elastomer and their interaction with additives.[17] Polymer viscosity, filler type and filler quantity will affect the tensile strength, maximum elongation and toughness of the cured silicone. From a transducer point of view LSRs should possess high dielectric permittivity, little dielectric loss, high tear strength, large elongation at the break, a low Young's modulus (in the case of an actuator), little viscous dissipation and the films should be perfectly homogeneous. [18] Crosslink density can be optimized by adjusting the ratio of vinyl and hydride groups in the LSR formulation, which offers the advantage of optimizing hardness from very soft (Shore hardness 40 '00') to hard (Shore hardness 90 'A') with minimal changes in the uncured viscosity. The filled soft materials with high permittivity and low dielectric loss could also be beneficial to the electromechanical coupling studies, such as microwave applications.[19] It was shown in our previous study that titanium dioxide (TiO_2) nanoparticles increase both the dielectric permittivity and breakdown strength of LSR formulations.[20] The rutile TiO_2 exhibits a higher dielectric constant (ε_r ~110) than the anatase TiO_2 (ε_r ~45), which therefore makes it a good candidate for achieving higher nanocomposite elastomer permittivity.[21] Several other studies on similar systems have been conducted but without varying the silica content as the commercial elastomers do not usually provide such handle. Amongst this large ensemble of work the pioneering work of

Carpi and Rossi [21] on titanium dioxide in silicone for dielectric elastomer application as well as more recent work of Wang et al [22] on more complex titanates can be mentioned. In the present study Wacker Chemie has provided commercial elastomers with varying silica content. XLR 630 is the commercially available elastomer and MJK 5/13 and 4/13 contain reduced amounts of silica. Thereby it is possible to replace silica with titanate without varying significantly on crosslinking density, type of silica, catalyst etc.

Four commercial LSRs, with different viscosities, and one RTV were chosen in the present work, in which the dielectric and mechanical properties of pure materials and hydrophobic rutile TiO_2 composite formulations are reported. The elastomers were evaluated based on two figures of merit, namely with respect to actuation $F_{om}(DEA)$ [23] and to generation $F_{om}(DEG)$.[24] The benchmark normalization material used in this instance was the pure RTV ELASTOSIL® RT 625, from Wacker Chemie, such that a value of 1 or higher corresponded to improved actuation/generation properties.[25]

2.   Experimental section

2.1.   Materials

Four liquid silicone rubbers (LSRs) were supplied by Wacker Chemie AG, Germany. The LSRs were LR3043/30 (denoted LR3043) and 3 types of XLR (extra liquid rubber), namely XLR 630 (denoted XLR) and its two derivates MJK 5/13 and MJK 4/13 with respectively 25 wt.% and 50 wt.% less silica compared to the XLR. All other constituents in the elastomer formulations remained identical. The mixing ratios of parts A and B were 1:1.

Additionally, one room-temperature vulcanizable rubber (RTV) RT 625 was supplied by Wacker Chemie AG, Germany. The mixing ratio of parts A and B was 9:1.

The dynamic viscosity η (Pa·s) with shear rate 10 s^{-1} at 23°C of these four LSRs and one RTV mixtures are 430 (LR 3043), 12 (XLR), 6.5 (MJK 5/13), 2.9 (MJK 4/13), and 12 (RT 625), respectively.

The two utilized hydrophobic TiO_2 nanofillers were:

SACHTLEBEN® R420: Rutile TiO_2 was supplied by Sachtleben Chemie GmbH, Germany. Primary particle size was 250–290 nm.

AEROXIDE® T805: Anatase/rutile TiO_2 was supplied by Evonik Industries AG, Germany. Primary particle size was 25 nm. We denote both types of filler as nanofillers. The size of R420 is, however, on the upper limit for being regarded as a nanofiller.[26]

Solvent OS-20 (an ozone-safe volatile methylsiloxane (VMS) fluid) was purchased from Dow Corning, USA. OS-20 was not added to the three pure LSR (XLR, MJK 5/13 and 4/13) or RT 625 formulations, since their viscosities allow for coating without solvent. The mixing ratio of LR3043 premix A and B and OS-20 was 5:5:7 by mass in the following, while OS-20 addition in the XLR–TiO_2 formulations was 50 wt.% of the TiO_2 loading.

R420 TiO_2 (250–290 nm) fillers were added into RT625, LR3043 and XLR, respectively. The increased viscosity of the mixtures constitutes a problem when small nanoscale fillers such as T805 TiO_2 (25 nm) are applied. Therefore this type of filler was solely added to the LR with the lowest viscosity, namely MJK 4/13.

2.2.   Sample preparation

Nanofillers were mixed into the silicone premix A using a Speedmixer (DAC 150FVZ, Hauschild Co., Germany) for 2 minutes at 3000 rpm. Premix B and the OS-20 solvent were then mixed in to the first mixture for another 2 minutes at 3000 rpm with the

Speedmixer. The films were coated with a film applicator (3540 bird, Elcometer, Germany) on a glass substrate and cured in an oven at 110°C for 8 minutes. The thickness of the film was approximately 40–80 μm.

2.3. Characterization

2.3.1. Thermal analyses

The thermal stability of the films was evaluated by thermal gravimetric analysis (TGA Q500, TA Instruments, USA), and measurements were carried out under a nitrogen atmosphere at a heating rate of $10°C$ min^{-1} from room temperature to 900°C.

2.3.2. Tear test

The tear strength of the elastomers was measured using a material tester (Zwick/Roell Zmart.pro, Zwick GmbH & Co. KG, Germany). The sample of 110 mm length and 10 mm width with a 0.5 mm nick at its middle edge was placed between two clamps and initially separated by a distance of 33 ± 2 mm. The test specimen was elongated at 500 ± 50 mm min^{-1} (i.e. strain rate of 25%/s) until failure along the nick. Each sample was subjected to four tear measurements and then averaged.

2.3.3. Tensile stress–strain

The tensile stress–strain of the elastomers was measured using a material tester (Zwick/ Roell Zmart.pro, Zwick GmbH & Co. KG, Germany). The sample of 115 mm length and 6 mm width was placed between two clamps and initially separated by a distance of 33 ± 2 mm. The test specimen was elongated uniaxially at 500 ± 50 mm min^{-1} with respect to length and forced throughout the test to a level of accuracy within ±2% until sample failure at the middle part. Each composition was subjected to four tensile measurements which were then averaged.

2.3.4. Breakdown strength test

Electrical breakdown tests were performed on an in-house-built device based on international standards (IEC 60,243–1 (1998) and IEC 60,243–2 (2001)) with DC voltage output power. Film thicknesses were measured through the microscopy of cross-sectional cuts, and the distance between the spherical electrodes was set accordingly with a micrometer stage and gauge. An indent of less than 5% of sample thickness was added, to ensure that the spheres were in contact with the sample. The polymer film was slid between the two spherical electrodes (radius of 20 mm), and the breakdown was measured at the point of contact with a stepwise increasing voltage applied (50–100 V $step^{-1}$) at a rate of 0.5–1 steps s^{-1}. Each sample was subjected to 12 breakdown measurements, and an average of these values was given as the breakdown strength of the sample.

2.3.5. Viscosity measurements

The dynamic viscosities of mixtures were tested by the flow sweep of an ARES-G2 rheometer (TA Instruments, USA) with a shear rate ranging from 10^{-3} to 10 s^{-1} at 23°C in an ambient atmosphere using a parallel-plate geometry of 25 mm in diameter.

2.3.6. Dielectric characterization

Dielectric relaxation spectroscopy (DRS) was performed on a Novocontrol Alpha-A high-performance frequency analyzer (Novocontrol Technologies GmbH & Co. KG, Germany) operating in the frequency range 10^{-1}–10^{6} Hz at 23°C. The sample diameters tested were 25 mm, while thickness was approximately 0.5–1.0 mm.

2.3.7. Actuation investigation

The actuation performance of films of thickness 60–80 μm was measured by the actuation set-up, which has been described in previous work.[20] Both surfaces of films were metallized with silver 100 nm-thick electrodes (width 30 mm × length 90 mm). Conductive electrode tapes were attached to make a connection with the high voltage amplifier. The top side of the film was fixed with a clamp on a stage. While the bottom side of the film was hung with a certain weight to get 10% pre-stretch. The voltage was applied to the films in steps of 500 V, starting from 0 V up to 9 kV. The film length under different voltage was recorded by a camera. The thickness strain S_z is calculated by the decrease of film length (under different voltage) divided by the electrode length (90 mm). Tensile stress–strain, breakdown strength and actuation measurements are shown in Figure 1.

2.3.8. Morphology observation

The morphology of the fillers and film surfaces was examined by a scanning electron microscopy (SEM) (FEI Inspect S, USA). All samples were coated with gold under vacuum before testing.

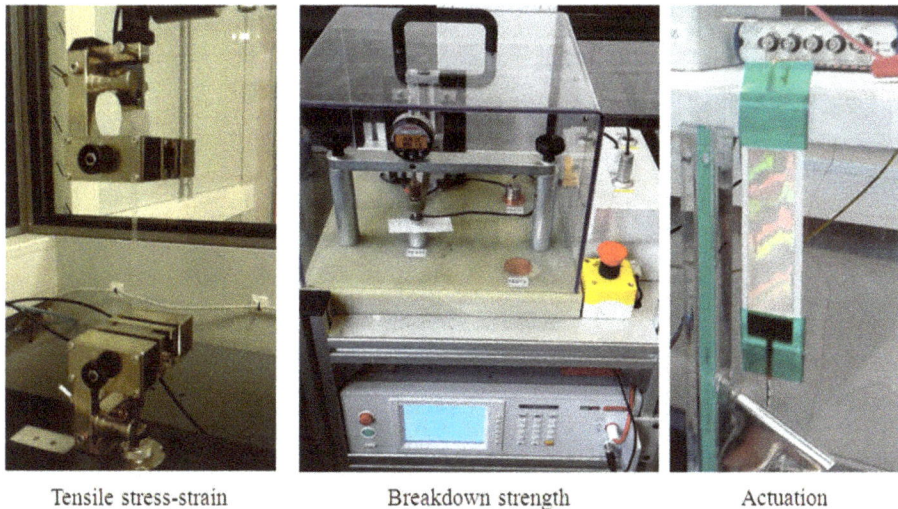

| Tensile stress-strain | Breakdown strength | Actuation |

Figure 1.   Tensile stress–strain, breakdown strength and actuation measurements.

3. Results and discussion

In the following, the properties of the LSRs and the benchmark RTV are discussed, followed by a discussion on LSR composites.

3.1. *LSR and RTV formulations with no additional fillers*

In the following sections four different commercially available LSRs, with varying amounts of silica fillers, are investigated and discussed, with a focus on properties relevant to DEs.

3.1.1. *Thermal analysis*

Thermogravimetric analysis (under nitrogen atmosphere) of the silicone films enables the determination of filler content.[27] In purely commercial silicone formulations, the majority of residue at 900°C corresponds to the initial silica addition.[6] Figure 2 presents the degradation curves under N_2 of the four LSRs (LR3043, XLR, MJK 5/13 and MJK 4/13) and one RTV (RT625) film. All formulations behave similarly, with no significant degradation taking place before 250°C, followed by slow degradation up to around 500°C, where weight losses of 5–8% are detected. LR3043 produces the most residue (72 wt.%), while XLR and RT625 present 37 wt.% and 27 wt.% residues, respectively, and MJK 5/13 and MJK 4/13 contain the least amount of inorganic residues at 21 wt.% and 18 wt.%. This agrees well with the viscosities from section 2.1.

A summary of the data obtained from TGA is shown in Table 1. Key values are defined here as temperatures at 2% weight loss ($T_{d2\%}$) and the temperature (T_{max}) at which the fastest derivative weight ($\%°C^{-1}$) was observed, as well as the final inorganic residue at 900°C.[28] The temperature of $T_{d2\%}$ is especially important with respect to production scenarios. One serious drawback with a low $T_{d2\%}$ would be bubble formation during production or the elastomer curing weakening the produced materials or causing

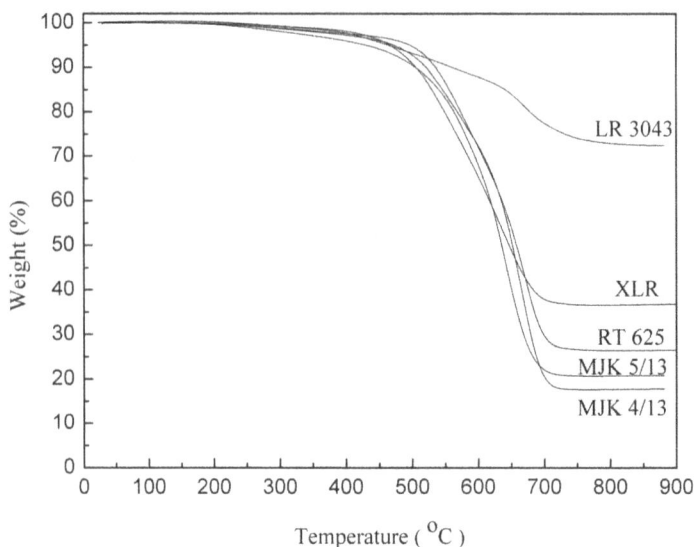

Figure 2.   Thermal degradation analysis of the LSR and RTV films.

Table 1.   Thermal properties of the LSR and RTV films.

Silicone	Formulation	$T_{d2\%}(°C)$	$T_{max}(°C)$	Inorganic residue (wt.%)
LSR	LR 3043	406	683	72
	XLR	389	671	37
	MJK 5/13	349	643	21
	MJK 4/13	303	639	18
RTV	RT 625	365	665	27

premature electrical breakdowns.[27] The overall thermal stability of films increases in line with an increase in silica content – as expected.

3.1.2. Breakdown strength

Electrical breakdown might be caused by several factors, such as intrinsic breakdown, thermal breakdown, electromechanical breakdown and partial discharge breakdown.[29] In thin dielectric films, when the heat generated within the materials cannot be dissipated sufficiently, a critical voltage will be reached when the applied voltage is increased, termed 'the maximum thermal voltage,' i.e. the voltage before the thermal runway occurs.[30] Breakdown strength reveals – besides true material properties – the homogeneity of samples as well as dielectric losses present in the materials. Imperfections may lead to premature treeing [31] and local heating, due to variations in thermal conductivity.[30]

The relationship between breakdown strength and silica content in the investigated silicone rubbers is displayed in Figure 3. The breakdown strength of the LSR films is generally much higher than that of the reference RTV. Moreover, breakdown strength increases with increased filler content for the three LSRs. However, for even higher filler

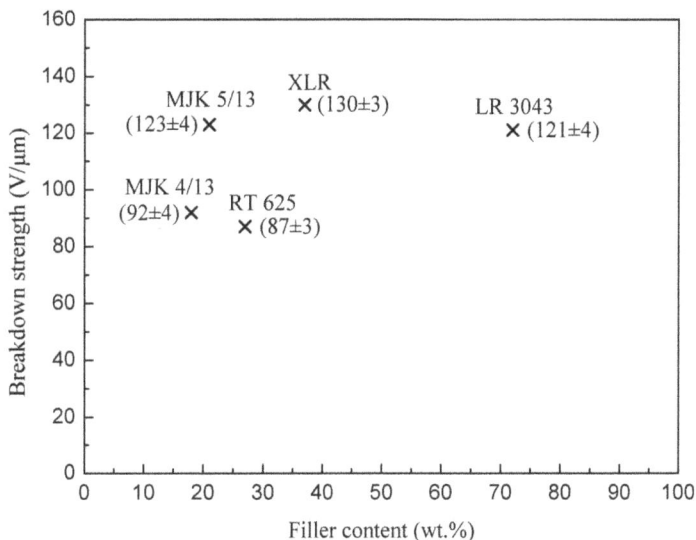

Figure 3.   Breakdown strength versus filler content as determined from the TGA of LSR and RTV films. All films are of similar thickness (60–80 μm).

Table 2.   Weibull analysis parameters for LSR and RTV samples.

Silicone	Formulation	β	η	r^2
LSR	ELASTOSIL® LR 3043 A/B	35	123	0.90
	POWERSIL® XLR A/B	46	132	0.92
	MJK® 135 5/13 A/B	33	125	0.96
	MJK® 135 4/13 A/B	25	94	0.92
RTV	ELASTOSIL® RT 625 A/B	32	88	0.87

loading, such as LR3043 (72 wt.% as determined from TGA), the films are probably less homogeneous due to the high viscosity of the reaction mixture. Thus the electrical breakdown strength reaches a maximum at a certain loading where homogeneity is lost partly and therefore a decrease in the electrical breakdown strength is observed.

In order to get more insight into the electrical stability as well as the homogeneity of the samples as Weibull analysis was performed on the breakdown strength data. The data is fitted to the Weibull cumulative distribution:

$$\mathbf{P(E)} = 1 - \mathbf{exp}\left(\frac{E}{\eta}\right)^{\beta}, \tag{1}$$

where P is the cumulative distribution at a given electrical field E, β refers to shape parameter, and η refers to scale parameter.[32,33]

The fitted data is shown in Table 2. For the elastomers the XLR is by far the best performing elastomer with respect to electrical reliability. Both shape and scale parameters of the XLR are the highest. The scale parameter is closely related to the determined breakdown strengths. The shape parameter is somewhat more complicatedly connected to the standard deviations but from the Weibull analysis important information can be obtained. The steeper the Weibull plot, the higher βand the narrower probability distribution of breakdown, thus a more homogeneous material. As the silica content is dropped 25% (MJK 5/13) (compared to the commercial elastomer) both parameters drop significantly. For the MJK 4/13 where the silica content is lowered 50% compared to XLR is clear that both Weibull parameters are even further decreased. It is therefore clear that it is not trivial to decrease the silica content while maintaining product performance.

The RT625 elastomer shows relatively good homogeneity (large β) but lower η. The latter is most likely due to the decreased Young's modulus of this type of formulation. The LR3043 behaves comparable to the XLR with 25% reduction of silica (MJK 5/13). Data for all samples agree very well with the Weibull distribution with no indication of bimodality in the distributions.

3.1.3. Mechanical properties

Tear strengths of MJK 5/13 and MJK 4/13 elastomers are 64% and 49%, respectively, of the XLR elastomer, according to product information from the supplier. The XLR exhibits good tear characteristics for thin film production, but poorer properties cannot be accepted for the production of micro-structured thin films. Therefore, these two LSRs were discarded for mechanical testing and the other three silicone formulations (LR3043, XLR, and RT625) were chosen in the following studies to compare mechanical performances.

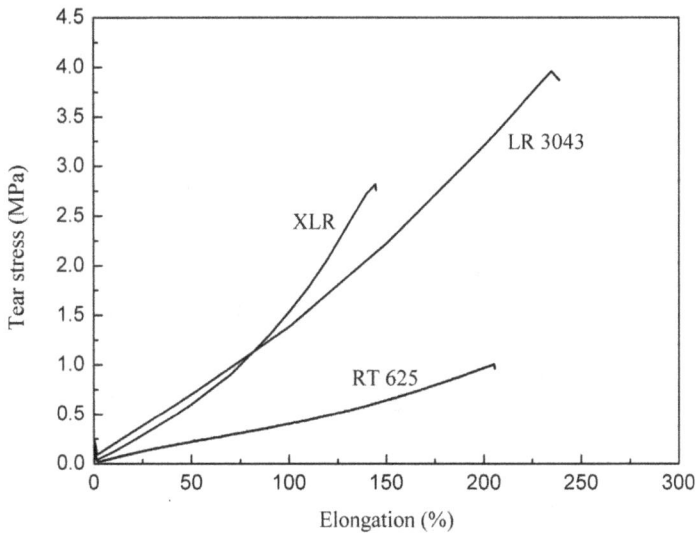

Figure 4.   Tear strength of the LSR and RTV films. The results are summarized in Table 3.

Tear strength and elongation at the break point of various silicone films are displayed in Figure 4. LSR elastomer tear strengths are much higher than RT625 and increase in line with any increase in silica content, while elongation at the break point of XLR is slightly lower than that seen for RT625 and LR3043. Both tear strength and elongation at the break point in the LR3043 tear experiment are the highest amongst the investigated elastomers, which is in agreement with the good inter-particle interaction of silica particles with the silicone matrix.[34]

The tensile and tear results are summarized in Table 3. High inorganic silica loading results in high tensile strength, high tear strength and a high Young's modulus. Moreover, the breakdown results shown in Figure 3 suggest that the resistance of the elastomers to dielectric breakdown is enhanced by an increase in the Young's modulus (as seen in Table 3), which has been reported in previous work on polyethylenes [24] and LSR elastomers.[20]

3.2. LSR–TiO$_2$ formulations

LSR materials enable a fast curing process, and the resulting elastomers possess favorable dielectric strengths and mechanical properties – as seen from studies in this work and in previous work.[20] XLR with the overall best properties for mixing in particles is chosen

Table 3.   The mechanical properties of LSR and RTV films.

Formulation	Sample thickness (μm)	Young's modulus (MPa)	Tensile strength (MPa)	Tear strength (MPa)	Elongation at break in tear experiment (%)	Permittivity (ε_r) @ 0.1 Hz
LR 3043	73 ± 2	0.80 ± 0.02	8.8 ± 0.2	3.96 ± 0.09	235 ± 6	2.70 ± 0.01
XLR	80 ± 2	0.76 ± 0.02	7.0 ± 0.2	2.82 ± 0.07	145 ± 4	3.06 ± 0.01
RT 625	74 ± 2	0.56 ± 0.01	6.5 ± 0.2	1.01 ± 0.02	206 ± 5	2.97 ± 0.01

as a matrix to investigate the TiO_2 filled elastomers, in order to further improve the properties of the elastomers with respect to energy densities. High-permittivity TiO_2 particles are blended into the LSRs to improve the dielectric constants of the composites. While previous results indicate that low TiO_2 content cannot increase permittivity significantly [20], too high TiO_2 loading results in aggregates, and thus other properties deteriorate sharply due to poor film formation properties.[35]

3.2.1. Viscosity and thermal stability

The dynamic viscosity of pre-mixtures plays an important role in the processing of thin films in relation to coating ability and resulting film thickness. In addition, the viscosity of mixtures is an important factor when fillers aim at being homogeneously distributed in the matrix.

The dynamic viscosity profiles of pure XLR and XLR–TiO_2 mixtures at the shear rate of 10^{-3}–10 s^{-1} are shown in Figure 5. OS-20 solvent is needed in XLR–TiO_2 mixtures to allow for the coating of thin films, and the mixing ratio of TiO_2 and OS-20 is 2:1 by mass in all XLR–TiO_2 mixtures. A large viscosity increase appears at low shear rates in line with increased filler addition (30–40 wt.%) at low shear rates. However, the difference decreases in line with increasing shear rate, and curves of 30–35 wt.% especially are very closed to pure XLR after 0.1 s^{-1}. Finally, all of the viscosity and shear stress curves coincide at 10 s^{-1}, which is considered the shear rate of a conventional coating process.[36] The results indicate that the filler content of 30–40 wt.% chosen in this work does not cause evident limitations in processing.

3.2.2. Dielectric permittivities

Dielectric permittivity variations, as a function of frequency for both pure XLR and XLR–TiO_2 elastomers, are shown in Figure 6. The results show a clear increase in the relative permittivity of the composite films in line with increased TiO_2 loading within the measured frequency range. It is well-known that the TiO_2 filler has much higher permittivity ($\varepsilon_r \sim 110$) compared to that of pure silicone ($\varepsilon_r \sim 2.8$).[37] The influence of the TiO_2 filler's inherent high permittivity on the overall permittivities of the composites increases

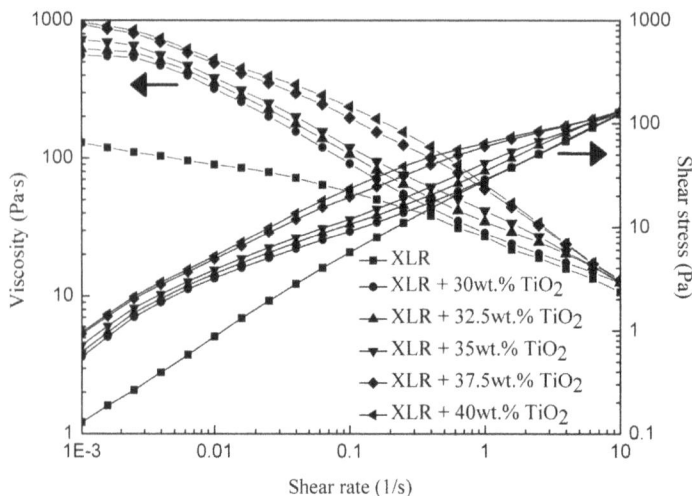

Figure 5.   Dynamic viscosity of the XLR–TiO_2 mixtures at 23°C.

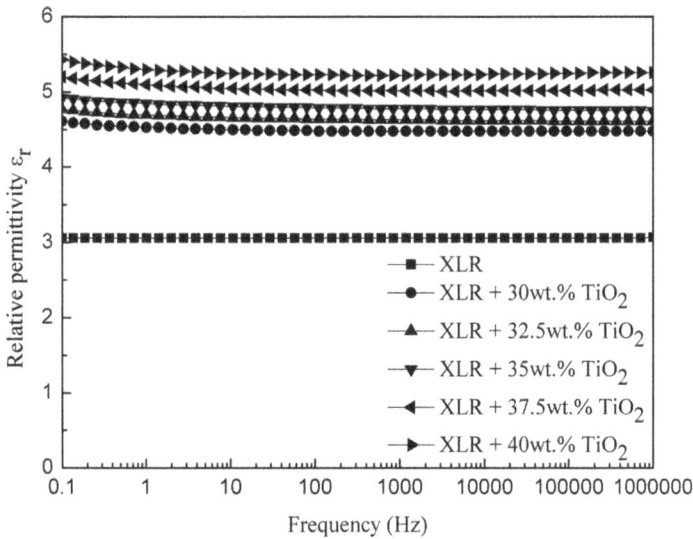

Figure 6. Frequency-dependent relative permittivity spectra of the pure XLR and XLR–TiO$_2$ elastomers at 23°C.

as loading concentration increases. On the other hand, the relative permittivity of the filled elastomer is found to be governed by the polarization associated with silicone and particles as well, as it is strongly influenced by interfacial polarization at the interface between PDMS and the TiO$_2$ nanoparticles.[38] The presence of more interfacial polarization at the interface between PDMS and TiO$_2$ results in high permittivity at low frequencies, as TiO$_2$ loading concentration increases.

Variations in dielectric losses, expressed as tan delta, in both pure XLR and XLR–TiO$_2$ films, are presented in Figure 7. It can be observed that the tanδ value of pure XLR

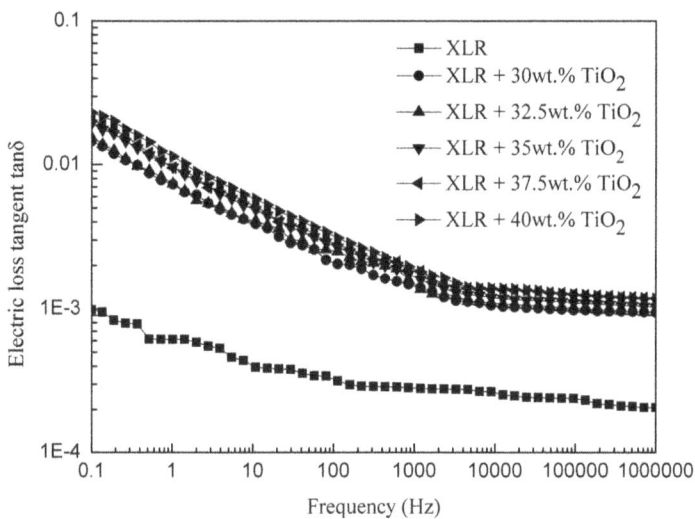

Figure 7. Frequency-dependent dielectrical loss tangent spectra of the pure XLR and XLR–TiO$_2$ elastomers at 23°C. The observed relaxations are Maxwell relaxations.

is almost frequency-independent, while the $\tan\delta$ values of composite films increase in line with decreasing frequency. Therefore, the presence of TiO_2 fillers has a stronger influence over the electric loss tangent of composites at lower frequency ranges (< 100 Hz), whereas losses are negligible (i.e. comparable to pure silicone) above 100 Hz. The level of the dielectric losses is, however, still acceptable.

3.2.3. Mechanical properties

It is well-known that polymer molecular weight (chain length), particle loading and specific particle characteristics (specific surface area and morphology) affect the mechanical performances of composites.[39] Active particles interact with the polymer matrix through the adsorption of polymer molecules on their surface (immobilization) and with each other. However, the deterioration of the mechanical strength of composites by fillers was also observed, which is mainly due to agglomerates. Most often, reinforcing particles are mechanically blended into PDMS polymers prior to composite curing, thereby resulting in loss of control over the degree of particle dispersion that leads to large agglomerates.[40]

Tear strength curves for the pure XLR and XLR–TiO_2 composite films are shown in Figure 8, while Figure 9 displays the relation between tear strength and elongation on breaking, respectively, of the films and TiO_2 content in the mixtures. Both tear strength and elongation at the break point increase in line with increased filler concentration for loadings of TiO_2 < 35 wt.%. Once the particle content increases to more than 35 wt.%, neither stress nor elongation at breaking decreases. This can be attributed to two factors: softening effects due to the interference of cross-linking (major effect for TiO_2 concentrations lower than 35 wt.%) and hardening effects due to the intrinsic high elastic modulus of TiO_2 (major effect for TiO_2 concentrations higher than 35 wt.%). At 35 wt.% of filler content, the two effects balance; therefore, the composite film exhibits excellent tear strength and elasticity. In Figure 10 SEM images of R420 TiO_2 filler, pure XLR film and XLR film with 35 wt.% TiO_2 (R420) are shown. It is clear that with the composite

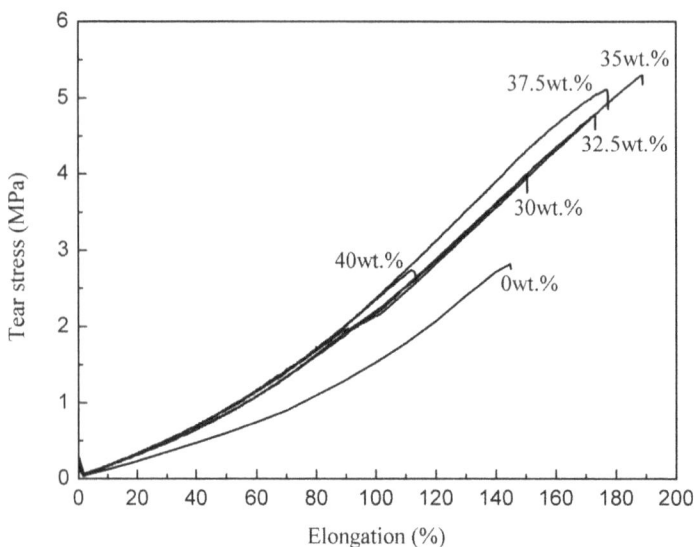

Figure 8.   Tear strength of the pure XLR and XLR–TiO_2 elastomers.

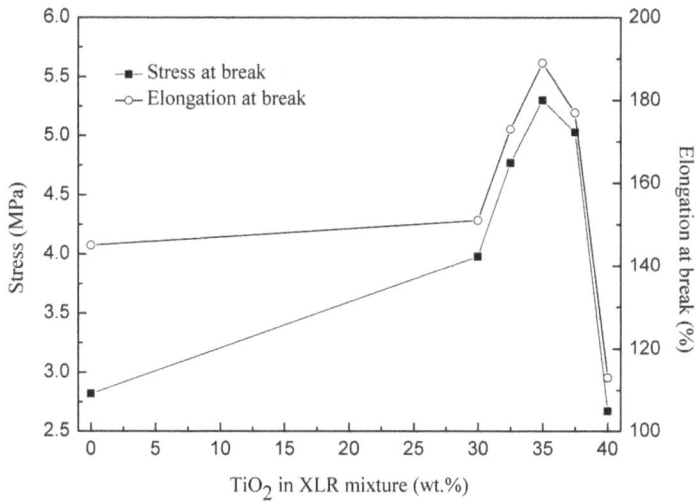

Figure 9.   Stress and elongation at breaking in the tear experiment on the pure XLR and XLR–TiO$_2$ elastomers.

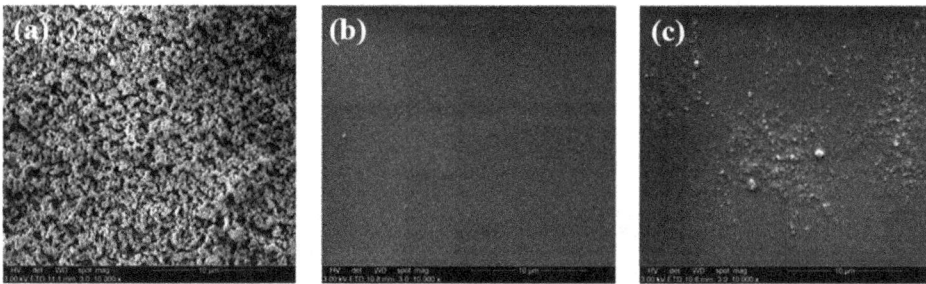

Figure 10.   SEM images of R420 TiO$_2$ filler (a), pure XLR film (b) and XLR+35 wt.% R420 TiO$_2$ film (c).

XLR there are some non-homogeneities but the size of these remain rather low with the example given here showing one of the more non-homogeneous areas.

The mechanical and breakdown performances of the pure XLR and XLR–TiO$_2$ films are listed in Table 4. The Young's modulus stated here is the ratio of stress to strain measured at 5% strain. It is clear that at 35 wt.% TiO$_2$ concentration, the composite system shows the highest strength and maximal elasticity, as well as outstanding break-down strength and a favorable Young's modulus. Therefore, 35 wt.% TiO$_2$ content is also chosen to be added into both high viscous LSR LR3043 and RTV RT625 formulations, in order to compare the effect of the elastomer matrix.

3.2.4.   Relation between dielectric and mechanical properties

Figure 11 shows relations between TiO$_2$ content in the XLR system, the Young's modulus and the breakdown strength of the film. It is evident that the addition of TiO$_2$ fillers enhances the dielectric breakdown strength of the investigated XLR composites, partly due to well-dispersed nanofillers leading to larger field enhancement. The variation in data is partly due

Table 4. Mechanical properties and breakdown strength of the XLR–TiO$_2$ films. The optimum of a given property is marked in gray.

Formulation	Sample thickness (μm)	Young's modulus (MPa)	Tear strength (MPa)	Elongation at break (%)	Breakdown strength (V/μm)
XLR	80 ± 2	0.76 ± 0.02	2.82 ± 0.07	145 ± 4	130 ± 3
XLR + 30 wt.% TiO$_2$	65 ± 1	0.80 ± 0.02	3.98 ± 0.08	151 ± 4	144 ± 4
XLR + 32.5 wt.% TiO$_2$	45 ± 1	0.82 ± 0.02	4.77 ± 0.1	173 ± 4	146 ± 3
XLR + 35 wt.% TiO$_2$	62 ± 1	0.85 ± 0.02	5.30 ± 0.09	189 ± 5	158 ± 4
XLR + 37.5 wt.% TiO$_2$	57 ± 1	0.88 ± 0.02	5.03 ± 0.1	177 ± 4	160 ± 5
XLR + 40 wt.% TiO$_2$	53 ± 1	0.90 ± 0.03	2.67 ± 0.1	113 ± 3	163 ± 5

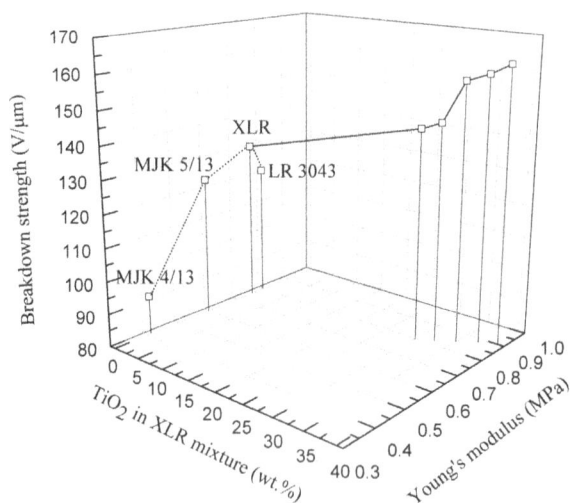

Figure 11. 3D plot of breakdown strength as a function of the Young's modulus and filler amount for the XLR–TiO$_2$ (R420) films and the pure LSRs.

to the fact that not all films have identical thicknesses due to very different coating behavior. A recent study showed the opposite phenomenon that the concentration of charge carriers was strongly increased upon addition of fillers.[41] However, as the commercial silicone elasto-mers are complex formulations, one formulation may differ significantly from another due to addition of various stabilizers and additives. It can also be seen in Figure 11 that the Young's modulus of the film plays a signification role in breakdown strength. The introduction of additional inorganic fillers enhances the rigidity of the composites and consequently increases the Young's modulus of the film. The results from Figure 11 indicate that the resistance of the elastomers to dielectric breakdown is indeed enhanced by the increase in Young's modulus (Y) for the given composites. The data that the electrical breakdown strength increases progressively with the increase in Young's modulus is also reported by Kollosche and Kofod as well as a previous study of ours.[20,42]

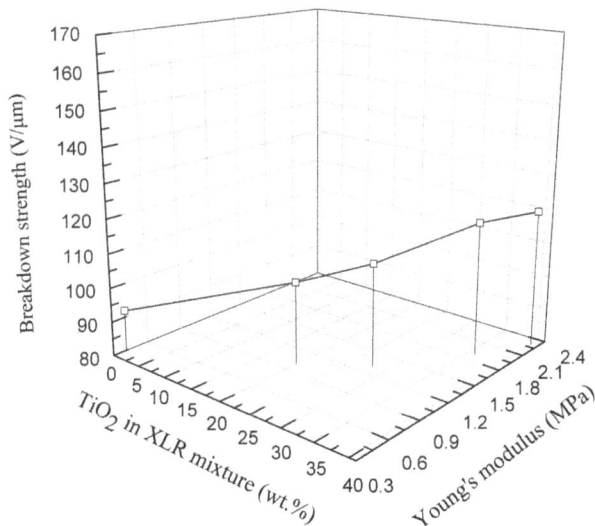

Figure 12.   3D plot of breakdown strength as a function of the Young's modulus and filler amount for the MJK 4/13-TiO$_2$ (T805) films.

Films with high Y have more compactly structured chains which can trap charges. This is most likely due to lower mobility of trapped charges, which then require a higher electric field in order to introduce electron avalanches.[20] Also electromechanical instability may be reduced upon increasing the Young's modulus. However, it is not possible to increase Y unlimitedly for a given material, since the formulations start to become inhomogeneous, with the result that the opposite effect (deterioration) is encountered.[24,43]

MJK 4/13 formulation is furthermore sought, strengthened by strongly reinforcing TiO$_2$ particles (T805) [44], to evaluate whether or not the silica content in a traditionally formulated LSR can be replaced with TiO$_2$, in order to obtain simultaneously acceptable mechanical properties and improved dielectric performances. Figure 12 shows the 3D plot for MJK 4/13-TiO$_2$ (T805) films. The relation between TiO$_2$ content and breakdown (BD) and Young's modulus (Y) is similar to that of the XLR–TiO$_2$ (R420) formulations. A much smaller type of filler is applied here (T805 TiO$_2$ (25 nm) compared to R420 TiO$_2$ (250–290 nm)), since MJK 4/13 allows for small and strongly reinforcing particles. The more specific surface area of smaller T805 TiO$_2$ causes greater van der Waals forces between the fillers, resulting in the higher viscosity of the mixtures and larger modulus of the films. From Figure 12 it is clear that the Young's moduli of MJK 4/13-TiO$_2$ films are 1.0–2.3 MPa, with 20–40 wt.% TiO$_2$ (T805) loading, which is unfavorable for actuators, while the Young's modulus of XLR film with 40 wt.% TiO$_2$ (R420) is 0.9 MPa (as seen in Figure 11). Additionally, more agglomerates are formed in the highly viscous mixture, which might negatively affect other properties such as lifetime. Also other effects may play a role such as the photo-catalytic properties of TiO$_2$ that can destroy the polymer chains [45] as well as the lower OH-group surface density of TiO$_2$ compared with SiO$_2$.[46] It is, however, clear no matter the effect that the replacement of silica with TiO$_2$ is not straight forward when the requirements to the elastomer remain so fierce.

The Weibull analysis of the data (Table 5) clearly reveals that the shape parameter is increased steadily upon increased TiO$_2$ loading. This is in good agreement with the increase in the Young's modulus. However, it seems that there is a local minimum in the scale parameter. The data at 35 wt.% loading furthermore – as the only sample – show

Table 5.   Weibull analysis parameters for XLR–TiO$_2$ and RTV–TiO$_2$ samples.

Formulation	β	η	r^2
XLR	46	132	0.92
XLR + 30 wt.% TiO$_2$	37	146	0.97
XLR + 32.5 wt.% TiO$_2$	50	148	0.96
XLR + 35 wt.% TiO$_2$	48	160	0.89
XLR + 37.5 wt.% TiO$_2$	36	162	0.97
XLR + 40 wt.% TiO$_2$	35	165	0.97
LR3043 + 35 wt.% TiO$_2$	33	116	0.93
RT625 + 35 wt.% TiO$_2$	16	91	0.92

a tendency toward fitting an S-shape (bimodal breakdown distribution) and this may be an experimental artifact due to insufficient mixing.

In Table 5 the data for the 3 types of elastomers with 35 wt.% TiO$_2$ is shown. From that it can easily be observed that XLR by far accommodates 35 wt.% TiO$_2$ the best. This results show that obtaining good performance of the composite relies on finding the best performing elastomer in the pristine state.

3.3.   *LSR–TiO$_2$ and RTV–TiO$_2$ formulations*

LSR and RTV composites, as well as pure RT625 elastomers, are evaluated based on two figures of merit, namely with respect to actuation F$_{om}$(DEA) and to generation F$_{om}$(DEG). The theory and derivation of F$_{om}$, to assess elastomer performance as a dielectric elastomer actuator (DEA), have been performed by Sommer-Larsen and Larsen [23] and based on this F$_{om}$ for a dielectric elastomer generator (DEG) presented by McKay et al. [24]

$$F_{om}(DEA) = \frac{3\varepsilon_r\varepsilon_0 E_{breakdown}^2}{Y}, \tag{2}$$

$$F_{om}(DEG) = \frac{\varepsilon_r\varepsilon_0 E_{breakdown}^2}{2\varphi}, \tag{3}$$

where $E_{breakdown}$ (V μm^{-1}) is the electrical field at which electrical breakdown occurs, ε_r the relative dielectric permittivity, ε_0 the permittivity of free space (8.85×10^{-12} F m^{-1}) and φ the strain energy function of the elastomer, which is assumed to be equal for each silicone formulation.

The figure of merit provides important information on the performance quality of the dielectric elastomer material, since the figure takes into account not only actuation (break-down strength) but also maximum voltage. Transducers are usually operated at significantly lower voltages than their breakdown strengths. F$_{om}$ depends upon the dielectric constant, dielectric breakdown strength and the elastic modulus of the elastomer material, which highlights the most important properties for dielectric materials.[25] Maximum elongation, in theory, should also be considered, but due to the high maximum elongations of the investigated elastomers in this study, electrical breakdown strength is the limiting factor.

Table 6 lists the mechanical and dielectric properties of the LSR and RTV composite films, as well as the figures of merit. The figures of merit of the TiO$_2$-filled formulations all exceed that of the unfilled RTV RT625 formulation.

Table 6. Dielectric and mechanical performances and figures of merit (F_{om}) of the LSR–TiO_2 and RTV–TiO_2 formulations. Optimal properties are marked in gray.

Formulation	Sample thickness (μm)	Young's modulus (MPa)	Tear strength (MPa)	Elongation at break (%)	Breakdown strength (V/μm)	Permittivity (ε_r) @ 0.1 Hz	F_{om}(DEA)@ 0.1 Hz	F_{om}(DEG)@ 0.1 Hz
XLR + 35 wt.% TiO_2	62 ± 1	0.85 ± 0.02	5.30 ± 0.09	189 ± 5	158 ± 4	4.92 ± 0.02	3.60 ± 0.04	5.46 ± 0.05
LR3043 + 35 wt.% TiO_2	64 ± 1	1.30 ± 0.03	2.90 ± 0.07	200 ± 5	114 ± 4	4.69 ± 0.02	1.17 ± 0.01	2.71 ± 0.03
RT625 + 35 wt.% TiO_2	68 ± 1	0.80 ± 0.02	2.25 ± 0.06	180 ± 5	87.9 ± 6	4.03 ± 0.02	0.97 ± 0.01	1.39 ± 0.02
RT625 [a]	74 ± 1	0.56 ± 0.01	1.01 ± 0.03	206 ± 4	87.0 ± 3	2.97 ± 0.01	1 ± 0.01 *	1 ± 0.01 *

Note: [a] Benchmark material, which is used for normalization, such that a value of 1 for both F_{om}(DEA) @ 0.1 Hz and F_{om}(DEG) @ 0.1 Hz of RT625 elastomer.

However, it can be seen from Table 6 that both the breakdown strength and the relative permittivity of LR3043-TiO$_2$ are lower than the XLR–TiO$_2$ formulation. One possibility for such a phenomenon is that the reduction of TiO$_2$ dipolar group mobility within the high viscous LR3043 reduces polarization within the composite. When a large amount of filler is loaded into LSR, more immobile nanolayers form and the mobility of the chain decreases continuously, resulting in a reduction in the composite's permittivity. [41] On the other hand, more particles in LR3043 act to suppress surface erosion caused by partial discharges, which are also possible sources of breakdown.[47]

Furthermore, it is clear that LSR composite formulations have much better dielectric and mechanical performances than the RTV composite. Thus, both the F$_{om}$(DEA) and F$_{om}$(DEG) of LSR films are greater than RT625-TiO$_2$, and XLR–TiO$_2$ is certainly the best candidate due to the significantly higher F$_{om}$ than its counterparts.

3.4. *Electromechanical properties*

In any given electrical field (E), the mechanical response (strain, p) of the elastomer can be expressed by [10]:

$$P = \frac{\varepsilon_0 \varepsilon_r E^2}{Y} = \frac{\varepsilon_0 \varepsilon_r}{Y}\left(\frac{U}{d}\right)^2, \tag{4}$$

where ε_0 is the permittivity of free space (8.85×10^{-12} F m^{-1}), ε_r the relative dielectric permittivity, Y the Young's modulus (Pa), U the voltage (V), and d (m) the original thickness of the elastomer.

Pure XLR (film thickness = 78 μm, Y = 0.76 MPa, ε_r = 3.07@0.1 Hz) and XLR +35 wt.% TiO$_2$ (film thickness = 75 μm, Y = 0.85 MPa and ε_r = 4.92@0.1 Hz) films were tested for electromechanical strain response. The electric field, as a function of the strain, is illustrated in Figure 13, which highlights, as expected, that the strain increases faster for

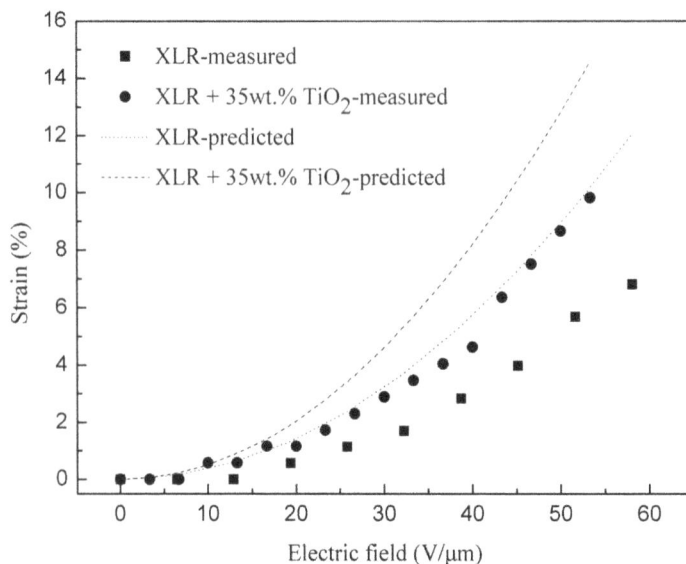

Figure 13.   Actuation curves for pure XLR and XLR+35 wt.% TiO$_2$ films at 23°C.

XLR films with 35 wt.%TiO$_2$ in line with increasing electric field compared to the pure XLR film, e.g. XLR-35 wt.%TiO$_2$ with 9.8% strain @ 53 V μm^{-1} and pure XLR with 6.8% strain @ 58 V μm^{-1}.

The predicted strain is calculated from the measured thickness, Young's modulus and relative permittivity using Equation (4). Figure 13 shows that the actual measured strains are not consistent with the predicted measurements, one likely reason for which is that dielectric loss dissipates as heat rather than being stored in the form of elastic energy.[20] The samples tested for actuation are furthermore laminated by two single films, and air voids at the interfaces may be introduced upon stretching, thereby leading to the slightly lower relative permittivity of the non-ideal laminate. However, it is obvious that actuation is greatly enhanced by the addition of 35 wt.% titanium dioxide particles (almost a factor of two).

4. Conclusions

Liquid silicone rubbers (LSRs) exhibit better thermal, dielectric and mechanical properties than room-temperature vulcanizable (RTV) rubbers when applied in DE transduction. The low intrinsic viscosity of XLR LSR formulation enables high TiO$_2$ loading of up to 35 wt.%, with minimal destruction of mechanical properties and the best elastomers were obtained with this elastomer as basis. However, too high TiO$_2$ loading (>35 wt.%) results in aggregates with the current processing equipment, and thus other properties of the elastomers deteriorate accordingly. The resulting 35 wt.% loading composite exhibits the highest figures of merit for both actuation F$_{om}$(DEA) and generation F$_{om}$(DEG) in all studied formulations. The XLR-35 wt.%TiO$_2$ elastomer possesses superior overall performance, with a maximum tear strength of 5.3 MPa and a relative permittivity of 4.92 @ 0.1 Hz, dielectric breakdown of 158 V μm^{-1} and a favorable Young's modulus of 0.85 MPa, as well as a 10% strain response in a 50 V μm^{-1} electric field.

The study furthermore showed that replacing silica with TiO$_2$ in commercial elastomers was not straight-forward as the resulting elastomer suffered from several shortcomings with the major problem being that the tear strength was largely reduced already in the original non-filled elastomers. Also it was found that in order to obtain the most reliable elastomer with respect to electrical properties, the elastomer without permittivity-enhancing fillers should be very reliable as the increased Young's modulus from the filler loading cannot compensate for a poor base material. This may be due to the photocatalytic properties of TiO$_2$ that can destroy the polymer chains as well as the lower OH-group surface density of TiO$_2$ compared with SiO$_2$.

Disclosure statement

No potential conflict of interest was reported by the authors.

Funding

The authors gratefully acknowledge the financial support of InnovationsFonden.

References

[1] A.L. Larsen, K.K. Hansen, P. Sommer-Larsen, O. Hassager, A. Bach, S. Ndoni, and M. Jørgensen, *Elastic properties of nonstoichiometric reacted PDMS networks*, Macromolecules 36 (2003), pp. 10063–10070. doi:10.1021/ma034355p

[2] F.B. Madsen, A.E. Daugaard, S. Hvilsted, M.Y. Benslimane, and A.L. Skov, *Dipolar cross-linkers for PDMS networks with enhanced dielectric permittivity and low dielectric loss*, Smart Mater. Struct. 22 (2013), pp. 104002. doi:10.1088/0964-1726/22/10/104002

[3] F.B. Madsen, L. Yu, A.E. Daugaard, S. Hvilsted, M.Y. Benslimane, and A.L. Skov, *A new soft dielectric silicone elastomer matrix with high mechanical integrity and low losses*, RSC Adv. 5 (2015), pp. 10254–10259. doi:10.1039/C4RA13511C

[4] B. Kussmaul, H. Krueger, S. Risse, and G. Kofod, *Synergistic improvement of actuation properties with compatibilized high permittivity filler*, Adv. Funct. Mater. 22 (2012), pp. 3958–3962. doi:10.1002/adfm.201200320

[5] S. Vudayagiri, L. Yu, and A.L. Skov, *Techniques for hot embossing microstructures on liquid silicone rubbers with fillers*, J. Elastomers Plast. (2014). doi:10.1177/0095244314526743

[6] E. Delebecq and F. Ganachaud, *Looking over liquid silicone rubbers: (1) Network topology vs chemical formulations*, ACS Appl. Mater. Interfaces. 4 (2012), pp. 3340–3352. doi:10.1021/am300502r

[7] C. Hopmann, C. Behmenburg, U. Recht, and K. Zeuner, *Injection molding of superhydrophobic liquid silicone rubber surfaces*, Silicon 6 (2014), pp. 35–43. doi:10.1007/s12633-013-9164-0

[8] E. Delebecq, N. Hermeline, A. Flers, and F. Ganachaud, *Looking over liquid silicone rubbers: (2) Mechanical properties vs network topology*, ACS Appl. Mater. Interfaces. 4 (2012), pp. 3353–3363. doi:10.1021/am300503j

[9] W.L. Wu and J.Y. Cai, *Study on short basalt fiber reinforced silicone rubber composites*, Advanced Mater. Res. 800 (2013), pp. 383–386. doi:10.4028/www.scientific.net/AMR.800

[10] A.L. Skov, A.G. Bejenariu, J. Bøgelund, M. Benslimane, and A.D. Egede, *Influence of micro- and nanofillers on electro-mechanical performance of silicone EAPs*, SPIE Proceedings 8340, SPIE, San Diego, CA, 2012, pp. 83400M-1–83400M-10.

[11] S.S. Hassouneh, A.E. Daugaard, and A.L. Skov, *Design of elastomer structure to facilitate incorporation of expanded graphite in silicones without compromising electromechanical integrity*, Macromol. Mater. Eng. 300 (2015), pp. 542–550. doi:10.1002/mame.201400383

[12] G. Rajesh, P.K. Maji, M. Bhattacharya, A. Choudhury, N. Roy, A. Saxena, and A.K. Bhowmick, *Liquid silicone rubber vulcanizates: Network structure – property relationship and cure kinetics*, Polym. Polym. Compos. 18 (2010), pp. 477–487.

[13] E. Haberstroh, W. Michaeli, and E. Henze, *Simulation of the filling and curing phase in injection molding of liquid silicone rubber (LSR)*, J. Reinf. Plast. Compos. 21 (2002), pp. 461–471. doi:10.1177/0731684402021005476

[14] Z. Chen, D. Li, A.J. Shi, Y. Li, and S.X. Wu, *Properties characteristics of silicone inhibitors*, Advanced Mater. Res. 2527 (2013), pp. 43–46.

[15] L.M. Lopez, A.B. Cosgrove, J.P. Hernandez-Ortiz, and T.A. Osswald, *Modeling the vulcanization reaction of silicone rubber*, Polymer Eng. Sci. 47 (2007), pp. 675–683. doi:10.1002/(ISSN)1548-2634

[16] M. Benslimane, H.E. Kiil, and M.J. Tryson, *Electromechanical properties of novel large strain PolyPower film and laminate components for DEAP actuator and sensor applications.* SPIE Proceedings 7642, SPIE, San Diego, CA, 2010, pp. 764231-1–764231-11.

[17] F.B. Madsen, I. Dimitrov, A.E. Daugaard, S. Hvilsted, and A.L. Skov, *Novel cross-linkers for PDMS networks for controlled and well distributed grafting of functionalities by click chemistry*, Polym Chem. 4 (2013), pp. 1700–1707. doi:10.1039/c2py20966g

[18] L. Yu and A.L. Skov, *Monolithic growth of partly cured polydimethylsiloxane thin film layers*, Polymer J. 46 (2014), pp. 123–129. doi:10.1038/pj.2013.72

[19] C. Brosseau and P. Talbot, *Instrumentation for microwave frequency-domain spectroscopy of filled polymers under uniaxial tension*, Meas. Sci. Technol. 16 (2005), pp. 1823–1832. doi:10.1088/0957-0233/16/9/015

[20] S. Vudayagiri, S. Zakaria, L. Yu, S.S. Hassouneh, M. Benslimane, and A.L. Skov, High breakdown-strength composites from liquid silicone rubbers, Smart Mater. Struct. 23 (2014), pp. 105017. doi:10.1088/0964-1726/23/10/105017

[21] F. Carpi and D.D. Rossi, *Dielectrics and Electrical Insulation*, IEEE Trans. Dielectr. Electr.l Insul. 12 (2005), pp. 835–843. doi:10.1109/TDEI.2005.1511110

[22] G.L. Wang, Y.Y. Zhang, L. Duan, K.H. Ding, Z.F. Wang, and M. Zhang, *Property reinforcement of silicone dielectric elastomers filled with self-prepared calcium copper titanate particles*, J. Appl. Polym. Sci. (2015). doi:10.1002/APP.42613

[23] P. Sommer-Larsen and A.L. Larsen, *Materials for dielectric elastomer actuators*, SPIE Proc. 5385 (2004), pp. 68–77.

[24] T.G. McKay, E. Calius, and I.A. Anderson, *Dielectric constant of 3M VHB: A parameter in dispute*. SPIE Proceedings 7287, SPIE, San Diego, CA, 2009, pp. 72870P-1–72870P-10.

[25] L. Yu, S. Vudayagiri, S.B. Zakaria, and M.Y. Benslimane, *Filled liquid silicone rubbers: Possibilities and challenges*. SPIE Proceedings 9056, SPIE, San Diego, CA, 2014, pp. 90560S-1–90560S-9.

[26] M. Hosokawa, K. Nogi, M. Naito, and T. Yokoyama, *Nanoparticle Technology Handbook*, M. Hosokawa, ed., Elsevier, Amsterdam, 2007. pp. 5.

[27] L. Yu, L.B. Gonzalez, S. Hvilsted, and A.L. Skov, *Soft silicone based interpenetrating networks as materials for actuators*. SPIE Proceedings 9056, SPIE, San Diego, CA, 2014, pp. 90560C-1–90560C-9.

[28] L. González, A.L. Skov, and S. Hvilsted, *Ionic networks derived from the protonation of dendritic amines with carboxylic acid end-functionalized PEGs*, Polym Chem. 51 (2013), pp. 1359–1371. doi:10.1002/pola.26503

[29] L.A. Dissado and J.C. Fothergill, *Electrical Degradation and Breakdown in Polymers*, Peter Peregrinus Publisher, London, UK, 1992, pp. 63–65.

[30] S.B. Zakaria, P.H.F. Morshuis, M.Y. Benslimane, V.G. Krist, and A.L. Skov, *The electrical breakdown of thin dielectric elastomers: Thermal effects*. SPIE Proceedings 9056, SPIE, San Diego, CA, 2014, pp. 90562V-1–90562V-12.

[31] M.H. Ahmad, H. Ahmad, N. Bashir, Y.Z. Arief, Z. Abdul-Malek, R. Kurnianto, and F. Yusof, *A new statistical approach for analysis of tree inception voltage of silicone rubber and epoxy resin under AC ramp voltage*, Int. J. Electrical Eng. Inform. 4 (2012), pp. 27–39. doi:10.15676/ijeei

[32] S.B. Zakaria, P.H.F. Morshuis, M.Y. Benslimane, L. Yu, and A.L. Skov, *The electrical breakdown strength of pre-stretched elastomers, with and without sample volume conservation*, Smart Mater. Struct. 24 (2015), pp. 055009. doi:10.1088/0964-1726/24/5/055009

[33] R. Kochetov, I.A. Tsekmes, and P.H.F. Morshuis, *Electrical conductivity, dielectric response and space charge dynamics of an electroactive polymer with and without nanofiller reinforcement*, Smart Mater. Struct. 24 (2015), pp. 075019. doi:10.1088/0964-1726/24/7/075019

[34] A. Camenzind, T. Schweizer, M. Sztucki, and S.E. Pratsinis, *Structure & strength of silica-PDMS nanocomposites*, Polymer 51 (2010), pp. 1796–1804. doi:10.1016/j.polymer.2010.02.030

[35] V.P. Silva, M.P. Paschoalino, M.C. Gonçalves, M.I. Felisberti, W.F. Jardim, and I.V.P. Yoshida, Silicone rubbers filled with TiO_2: Characterization and photocatalytic activity, Mater. Chem. Phys. 113 (2009), pp. 395–400. doi:10.1016/j.matchemphys.2008.07.104

[36] U. Eriksson, G. Engstrom, and M. Rigdahl, *Viscosity of some clay-based coating colors at high shear rates*, Rheol. Acta 29 (1990), pp. 352–359. doi:10.1007/BF01339890

[37] B. Hudec, K. Husekova, E. Dobrocka, T. Lalinsky, J. Aarik, and K. Frohlich, *High-permittivity metal-insulator-metal capacitors with TiO_2 rutile dielectric and RuO_2 bottom electrode*, Mater. Sci. Eng. 8 (2010), pp. 012024–012027.

[38] G.Z. Liu, C. Wang, C.C. Wang, J. Qiu, M. He, J. Xing, K.J. Jin, H.B. Lu, and G.Z. Yang, *Effects of interfacial polarisation on the dielectric properties of $BiFeO_3$ thin film capacitors*, Appl. Phys. Lett. 92 (2008), pp. 122903-1–122903-3. doi:10.1063/1.2900989

[39] H. Liu, L. Zhang, D. Yang, Y. Yu, L. Yao, and M. Tian, *Mechanical, dielectric and actuated strain of silicone elastomer filled with various types of TiO_2*, Soft Mater. 11 (2013), pp. 363–370. doi:10.1080/1539445X.2012.661821

[40] D. Tan, Y. Cao, E. Tuncer, and P. Irwin, *Nanofiller dispersion in polymer dielectrics*, Mater. Sci. Appl. 4 (2013), pp. 6–15. doi:10.4236/msa.2013.44A002

[41] Q. Wang and G. Chen, *Effect of nanofillers on the dielectric properties of epoxy nanocomposites*, Adv. Mater. Res. 1 (2012), pp. 93–107. doi:10.12989/amr.2012.1.1.093

[42] M. Kollosche and G. Kofod, *Electrical failure in blends of chemically identical, soft thermoplastic elastomers with different elastic stiffness*, Appl. Phys. Lett. 96 (2010), pp. 071904-1. doi:10.1063/1.3319513

[43] T. He, X. Zhao, and Z. Suo, *Dielectric elastomer membranes undergoing inhomogeneous deformation*, J. Appl. Phys. 106 (2009), pp. 083522. doi:10.1063/1.3253322

[44] J.L. Yang, Z. Zhang, A.K. Schlarb, and K. Friedrich, On the characterization of tensile creep resistance of polyamide 66 nanocomposites. Part I. Experimental results and general discussions, Polymer 47 (2006), pp. 2791–2801. doi:10.1016/j.polymer.2006.02.065

[45] L.H. Lin, H.J. Liu, J.J. Hwang, K.M. Chen, and J.C. Chao, *Photocatalytic effects and surface morphologies of modified silicone-TiO₂ polymer composites*, Mater. Chem. Phys. 127 (2011), pp. 248–252. doi:10.1016/j.matchemphys.2011.01.069

[46] P. Paoprasert, S. Kandala, D.P. Sweat, R. Ruther, and P. Gopalan, *Versatile grafting chemistry for creation of stable molecular layers on oxides*, J. Mater. Chem. 22 (2012), pp. 1046–1053. doi:10.1039/C1JM13293H

[47] Z. Li, K. Okamoto, Y. Ohki, and T. Tanaka, *Effects of nano-filler addition on partial discharge resistance and dielectric breakdown strength of micro-Al₂O₃/epoxy composite*, IEEE Trans. Dielectr. Electr.l Insul. 17 (2010), pp. 653–661. doi:10.1109/TDEI.2010.5492235

Permissions

All chapters in this book were first published in IJSNM, by Taylor & Francis Online; hereby published with permission under the Creative Commons Attribution License or equivalent. Every chapter published in this book has been scrutinized by our experts. Their significance has been extensively debated. The topics covered herein carry significant findings which will fuel the growth of the discipline. They may even be implemented as practical applications or may be referred to as a beginning point for another development.

The contributors of this book come from diverse backgrounds, making this book a truly international effort. This book will bring forth new frontiers with its revolutionizing research information and detailed analysis of the nascent developments around the world.

We would like to thank all the contributing authors for lending their expertise to make the book truly unique. They have played a crucial role in the development of this book. Without their invaluable contributions this book wouldn't have been possible. They have made vital efforts to compile up to date information on the varied aspects of this subject to make this book a valuable addition to the collection of many professionals and students.

This book was conceptualized with the vision of imparting up-to-date information and advanced data in this field. To ensure the same, a matchless editorial board was set up. Every individual on the board went through rigorous rounds of assessment to prove their worth. After which they invested a large part of their time researching and compiling the most relevant data for our readers.

The editorial board has been involved in producing this book since its inception. They have spent rigorous hours researching and exploring the diverse topics which have resulted in the successful publishing of this book. They have passed on their knowledge of decades through this book. To expedite this challenging task, the publisher supported the team at every step. A small team of assistant editors was also appointed to further simplify the editing procedure and attain best results for the readers.

Apart from the editorial board, the designing team has also invested a significant amount of their time in understanding the subject and creating the most relevant covers. They scrutinized every image to scout for the most suitable representation of the subject and create an appropriate cover for the book.

The publishing team has been an ardent support to the editorial, designing and production team. Their endless efforts to recruit the best for this project, has resulted in the accomplishment of this book. They are a veteran in the field of academics and their pool of knowledge is as vast as their experience in printing. Their expertise and guidance has proved useful at every step. Their uncompromising quality standards have made this book an exceptional effort. Their encouragement from time to time has been an inspiration for everyone.

The publisher and the editorial board hope that this book will prove to be a valuable piece of knowledge for researchers, students, practitioners and scholars across the globe.

List of Contributors

Rajesh K. Agrawalla, Rima Paul, Amit K. Chakraborty and Apurba Krishna Mitra
Physics, NIT Durgapur, Durgapur 713209, India

Pratap K. Sahoo
Physics, NISER, Bhubaneswar, India

Elena F. Sheka
General Physics Department, Peoples' Friendship University of Russia, Moscow 117198, Russian Federation

Natalia N. Rozhkova
Institute of Geology, Karelian Research Centre RAS, Petrozavodsk, Russian Federation

Y.C. Su and C.T. Sun
School of Aeronautics and Astronautics, Purdue University, West Lafayette, IN 47907, USA

Chaudhary Ravi Prakash Patel, Prashant Tripathi, O.N. Srivastava and T.P. Yadav
Department of Physics (Centre of Advanced Studies), Nanoscience Centre, Banaras Hindu University, Varanasi 220005, India

R. Kempegowda and P. Malingappa
Department of Chemistry, Bangalore University, Central College Campus, Bangalore 560001, India

D. Antony
Raman Research Institute, C.V. Raman Avenue, Bangalore 560080, India

Pei Li, Xuebing Chen, Xiaoming Zhou and Gengkai Hu
Key Laboratory of Dynamics and Control of Flight Vehicle, Ministry of Education, and School of Aerospace Engineering, Beijing Institute of Technology, 100081, Beijing, China

Ping Xiang
Systems Engineering Research Institute, 100036, Beijing, China

Viljar Palmre
Department of Mechanical Engineering, University of Nevada-Reno, Reno, NV, USA; Department of Mechanical Engineering, University of Nevada-Las Vegas, Las Vegas, NV, USA

Seong Jun Kim and David Pugal
Department of Mechanical Engineering, University of Nevada-Reno, Reno, NV, USA;

Kwang Kim
Department of Mechanical Engineering, University of Nevada-Las Vegas, Las Vegas, NV, USA

H.E. Misak and S. Mall
Department of Aeronautics and Astronautics, Air Force Institute of Technology, 2950 Hobson Way, Wright-Patterson AFB, OH 45433-7765, USA

R. Asmatulu and E. Jurak
Department of Mechanical Engineering, Wichita State University, 1845 Fairmount, Wichita, KS 67260-0133, USA

M. O' Malley
Air Force Research Laboratory, Materials and Manufacturing Directorate, 2941 Hobson Way, Wright-Patterson AFB, OH 45433-7750, USA

H. Zhu and F. Semperlotti
Department of Aerospace and Mechanical Engineering, University of Notre Dame, Notre Dame, Indiana 46556, USA

Ankit Srivastava
Department of Mechanical, Materials, and Aerospace Engineering, Illinois Institute of Technology, Chicago, IL 60616, USA

T. Islam
Department of Electrical Engineering, Faculty of Engineering & Technology, Jamia Millia Islamia (A Central University), Maulana Mohammed Ali Jauhar Marg, New Delhi 110025, India

Upendra Mittal
Department of Electrical Engineering, Faculty of Engineering & Technology, Jamia Millia Islamia (A Central University), Maulana Mohammed Ali Jauhar Marg, New Delhi 110025, India
Solid State Physics Laboratory, Lucknow Road, Timarpur, Delhi 110054, India

A.T. Nimal and M.U. Sharma
Solid State Physics Laboratory, Lucknow Road, Timarpur, Delhi 110054, India

Bismark Mensah, Han Gil Kim, Jong-Hwan Lee and Changwoon Nah
BK21 Haptic Polymer Composite Research Team, Department of Polymer-Nano Science and Technology, Chonbuk National University, Jeonju 561-756, Republic of Korea

Sivaram Arepalli
Nano Science Consultants LLC, Hampton, VA 23666-6701, USA

Junshi Zhang
State Key Laboratory for Strength and Vibration of Mechanical Structures, Xi'an Jiaotong University, Xi'an 710049, China
School of Aerospace, Xi'an Jiaotong University, Xi'an 710049, China

Yongquan Wang, Hualing Chen and Bo Li
State Key Laboratory for Strength and Vibration of Mechanical Structures, Xi'an Jiaotong University, Xi'an 710049, China
School of Mechanical Engineering, Xi'an Jiaotong University, Xi'an 710049, China

Liyun Yu and Anne Ladegaard Skov
The Danish Polymer Centre, Department of Chemical and Biochemical Engineering, Technical University of Denmark, 2800 Kgs. Lyngby, Denmark

Index

www.ingramcontent.com/pod-product-compliance
Lightning Source LLC
Chambersburg PA
CBHW080242230326
41458CB00096B/2892